Andy Warner's
Oddball Histories

PESTS and PETS

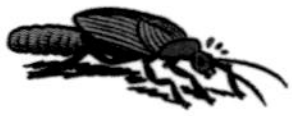

About This Book

This book was edited by Andrea Colvin and designed by Neil Swaab
The production was supervised by Bernadette Flinn, and the
production editor was Lindsay Walter-Greaney.
The text was set in fonts created by Andy Warner from his handwriting.

Colorist: Luke Healy

Little, Brown and Company
Hachette Book Group
1290 Avenue of the Americas, New York, NY 10104
Visit us at LBYR.com

First Edition: September 2021

Little, Brown and Company is a division of Hachette Book Group, Inc.
The Little, Brown name and logo are trademarks of Hachette Book Group, Inc.

The publisher is not responsible for websites (or their content) that are not owned by the publisher.

Library of Congress Cataloging-in-Publication Data
Names: Warner, Andy, author, illustrator.
Title: Pests and pets / Andy Warner.
Description: First edition. | New York, NY : Little, Brown and Company, 2021. | Series: Andy Warner's oddball histories | Summary: "Find out more than you ever thought possible about creatures both cute and weird, large and small, while discovering new stories about human history from the perspective of our animal companions"—Provided by publisher.

Identifiers: LCCN 2021004738 | ISBN 9780316498234 (hardcover) | ISBN 9780316463386 (trade paperback) | ISBN 9780316498241 (ebook) | ISBN 9780316220736 (ebook other)

Subjects: LCSH: Animals—Juvenile literature. | Animals—Comic books, strips, etc. | Pets—Juvenile literature. | Livestock—Juvenile literature. | Pests—Juvenile literature.

Classification: LCC QL49 .W375 2021 | DDC 590—dc23
LC record available at https://lccn.loc.gov/2021004738

ISBNs: 978-0-316-49823-4 (hardcover), 978-0-316-46338-6 (paperback), 978-0-316-49824-1 (ebook), 978-0-316-22305-8 (ebook), 978-0-316-15229-7 (ebook)

Printed in China

1010

Hardcover: 10 9 8 7 6 5 4 3 2 1
Paperback: 10 9 8 7 6 5 4 3 2 1

For my parents, Bob and Isabel;
my siblings, Toby and Olivia;
and all our animals.

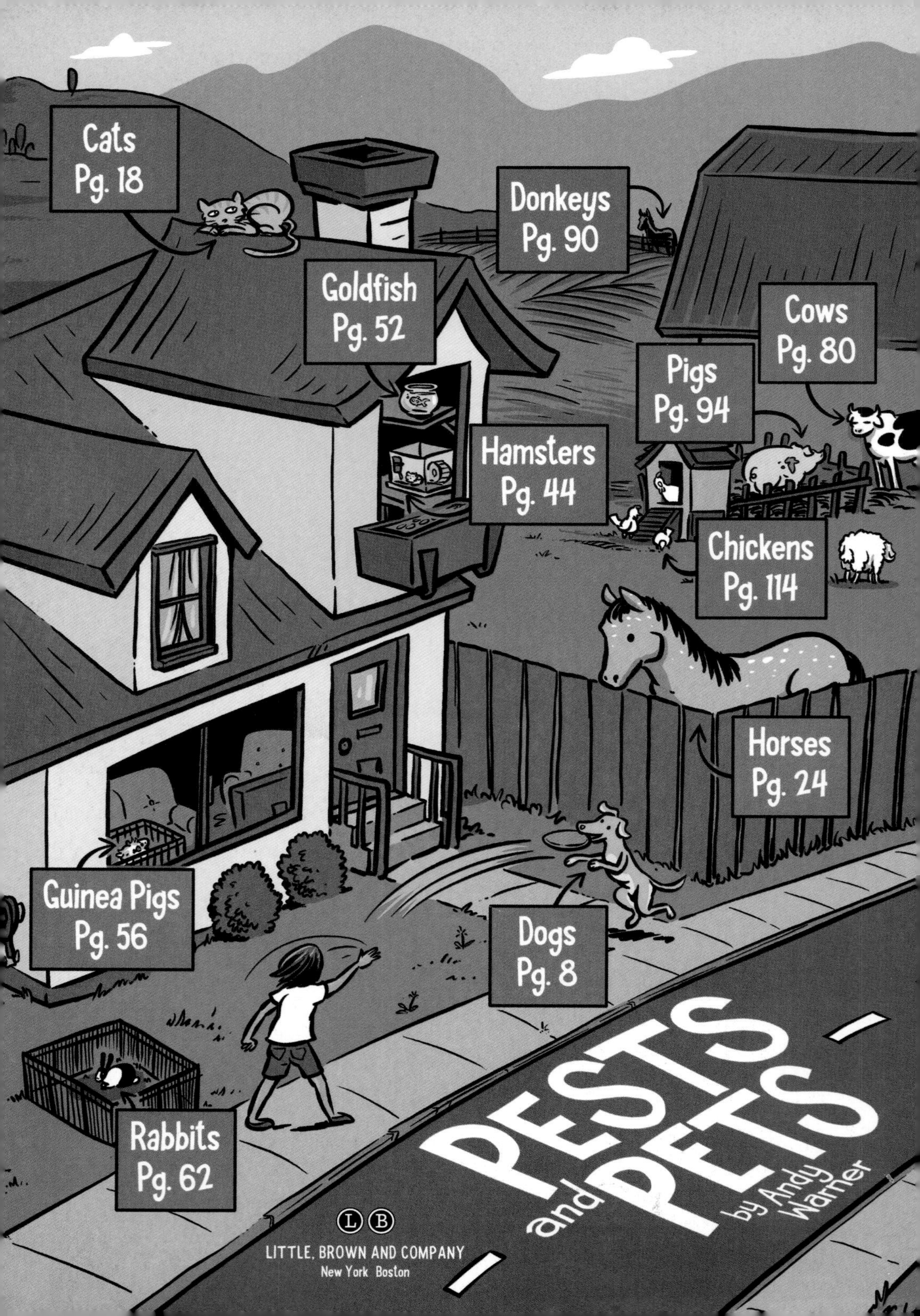

Cats
Pg. 18
Donkeys
Pg. 90
Goldfish
Pg. 52
Cows
Pg. 80
Pigs
Pg. 94
Hamsters
Pg. 44
Chickens
Pg. 114
Horses
Pg. 24
Guinea Pigs
Pg. 56
Dogs
Pg. 8
Rabbits
Pg. 62
PESTS and PETS
by Andy Warner

LITTLE, BROWN AND COMPANY
New York Boston

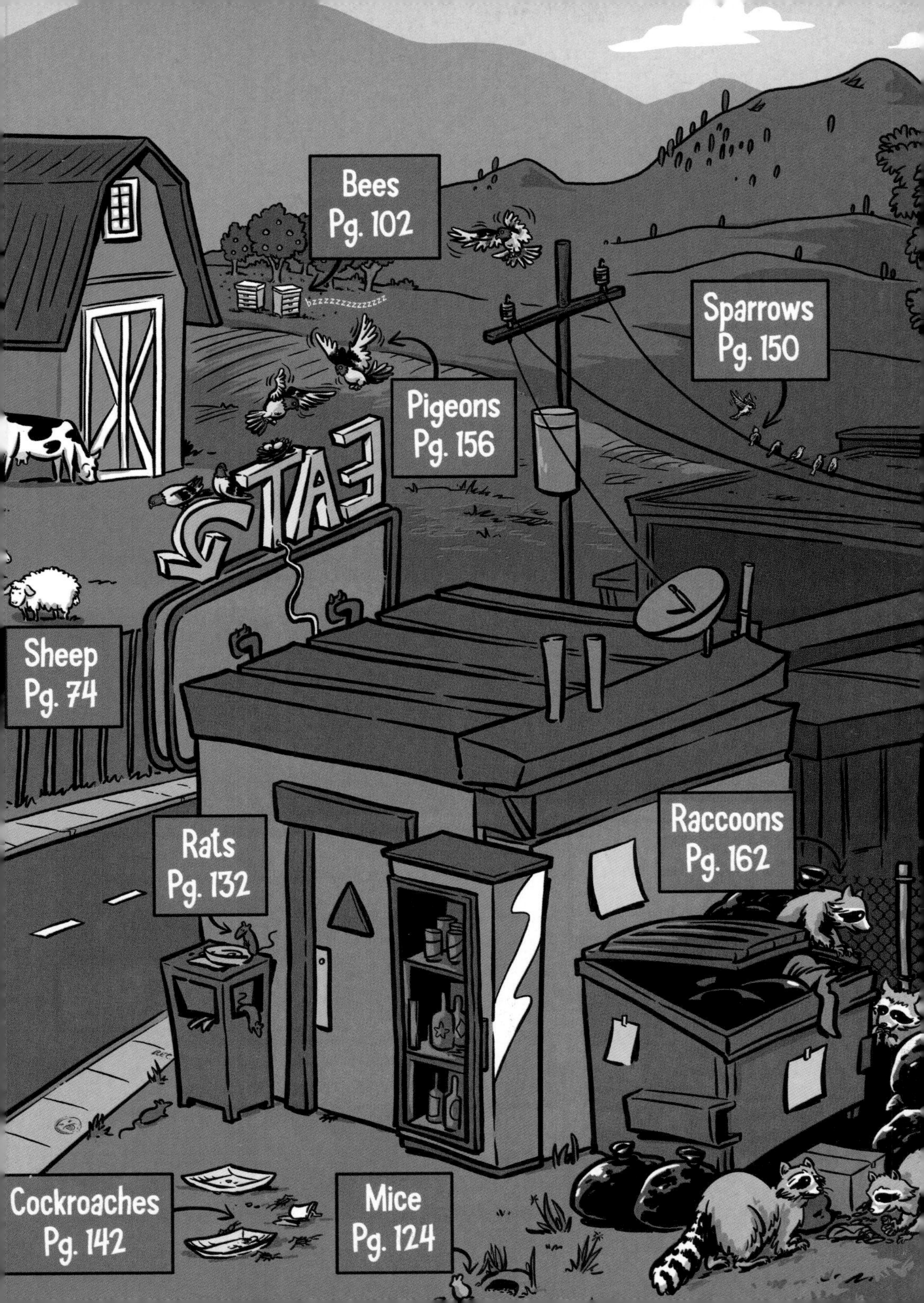

Bees
Pg. 102
Sparrows
Pg. 150
Pigeons
Pg. 156
Sheep
Pg. 74
Rats
Pg. 132
Raccoons
Pg. 162
Cockroaches
Pg. 142
Mice
Pg. 124

INTRODUCTION

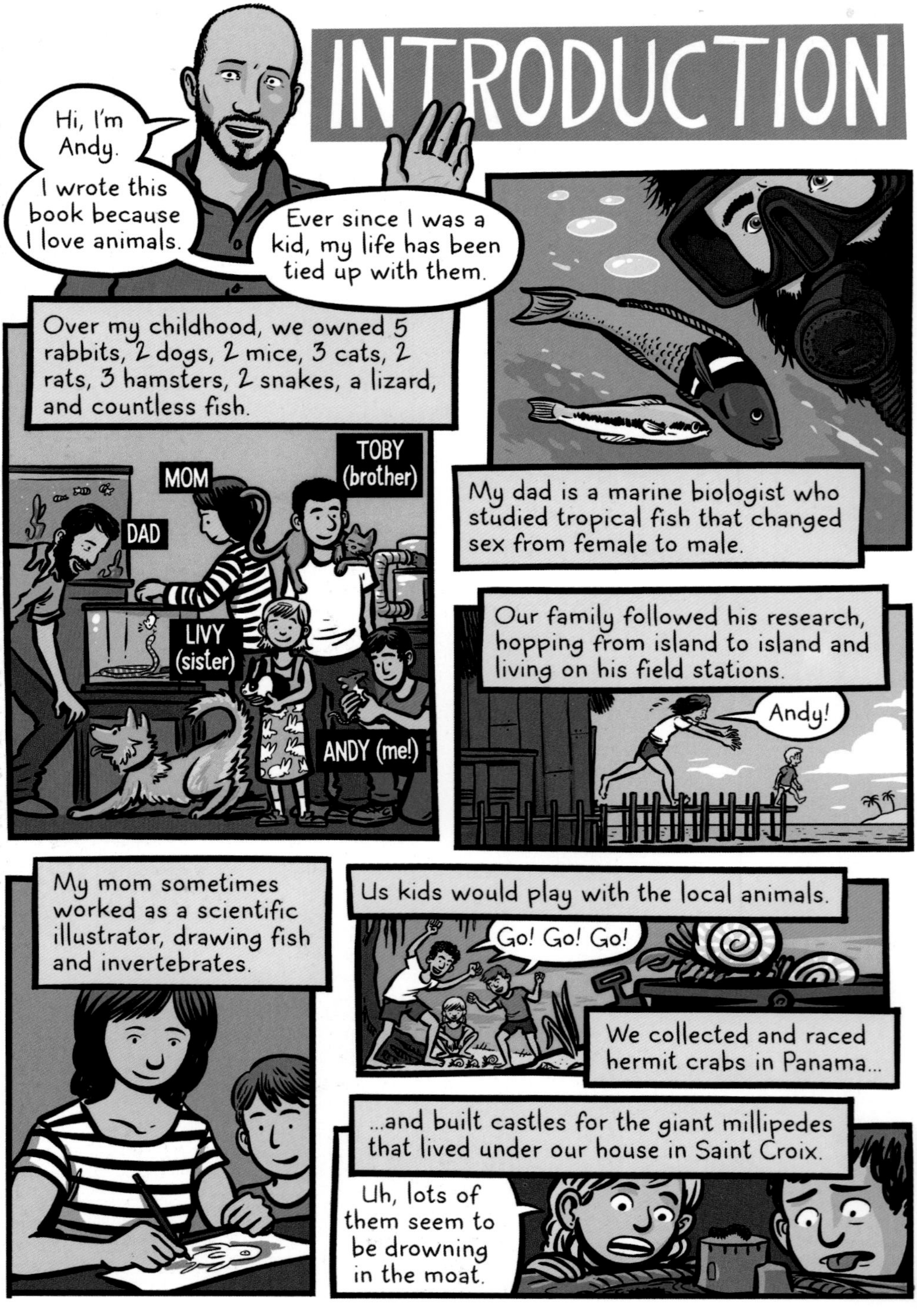
Hi, I'm Andy.
I wrote this book because I love animals.
Ever since I was a kid, my life has been tied up with them.
Over my childhood, we owned 5 rabbits, 2 dogs, 2 mice, 3 cats, 2 rats, 3 hamsters, 2 snakes, a lizard, and countless fish.
MOM
TOBY (brother)
DAD
LIVY (sister)
ANDY (me!)
My dad is a marine biologist who studied tropical fish that changed sex from female to male.
Our family followed his research, hopping from island to island and living on his field stations.
Andy!
My mom sometimes worked as a scientific illustrator, drawing fish and invertebrates.
Us kids would play with the local animals.
Go! Go! Go!
We collected and raced hermit crabs in Panama...
...and built castles for the giant millipedes that lived under our house in Saint Croix.
Uh, lots of them seem to be drowning in the moat.

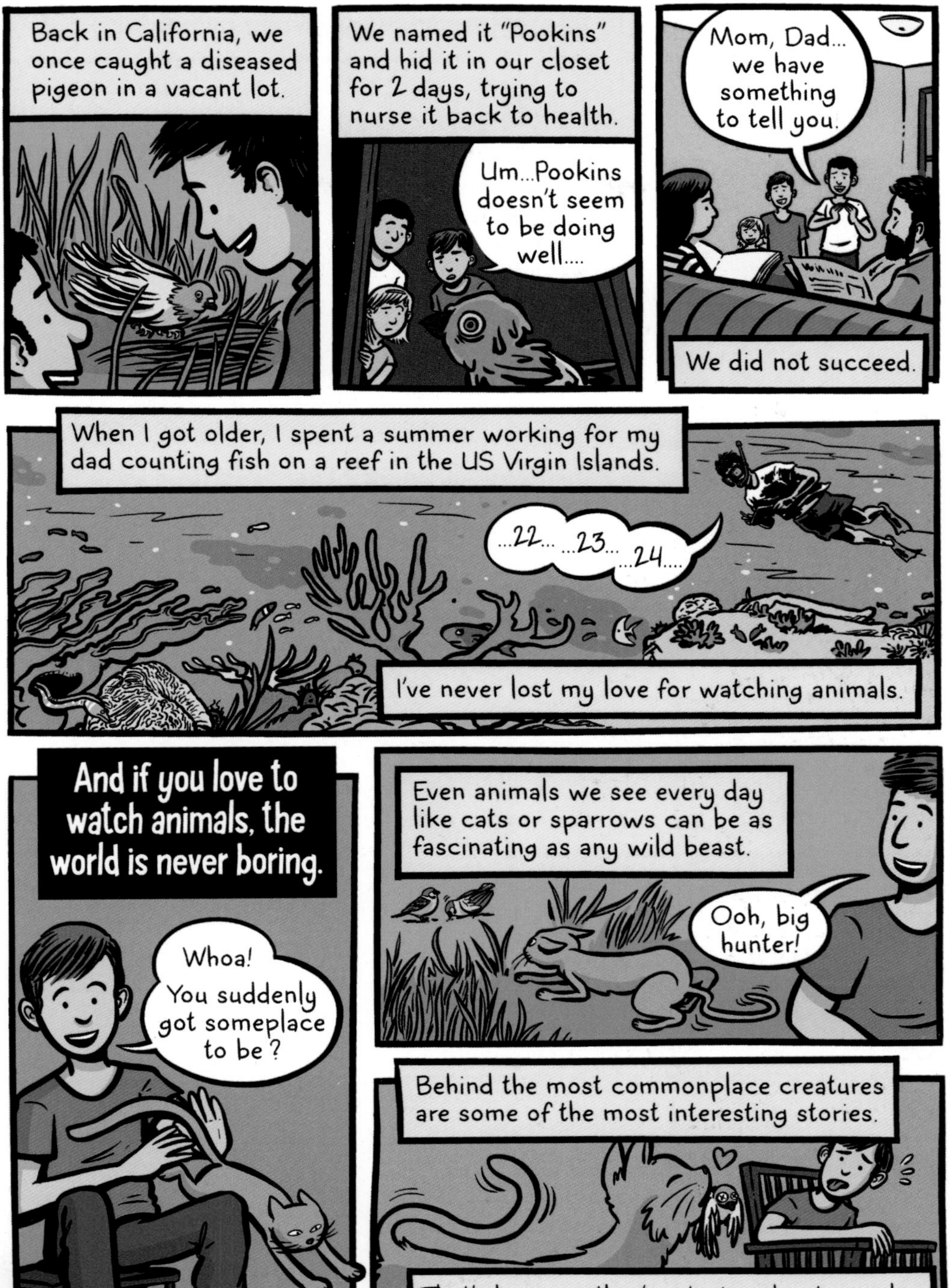
Back in California, we once caught a diseased pigeon in a vacant lot.
We named it "Pookins" and hid it in our closet for 2 days, trying to nurse it back to health.
Um...Pookins doesn't seem to be doing well....
Mom, Dad... we have something to tell you.
We did not succeed.
When I got older, I spent a summer working for my dad counting fish on a reef in the US Virgin Islands.
...22... ...23... ...24....
I've never lost my love for watching animals.
And if you love to watch animals, the world is never boring.
Whoa! You suddenly got someplace to be?
Even animals we see every day like cats or sparrows can be as fascinating as any wild beast.
Ooh, big hunter!
Behind the most commonplace creatures are some of the most interesting stories.
That's because they're stories about people.

People are pretty interesting animals in their own right.

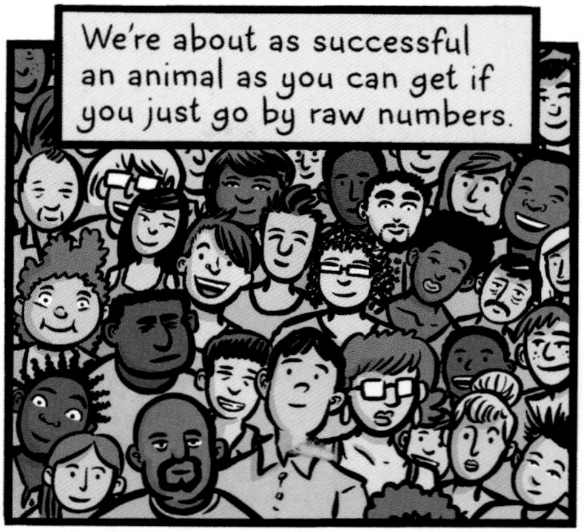
We're about as successful an animal as you can get if you just go by raw numbers.

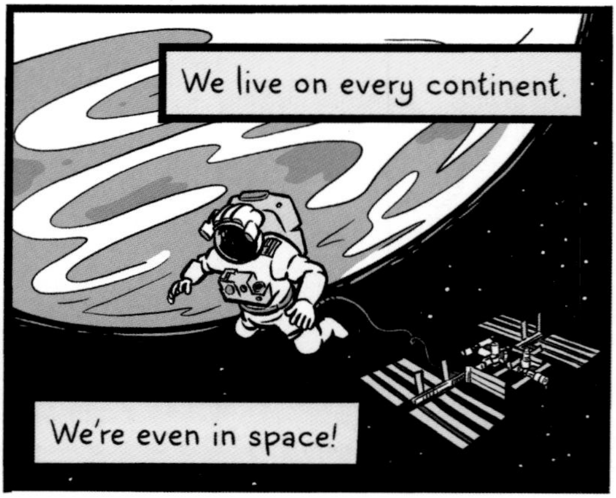
We live on every continent.
We're even in space!

And everywhere we've gone, we've brought other creatures with us.

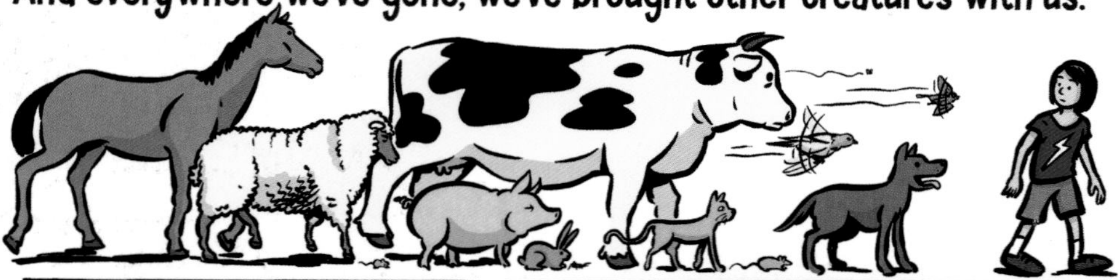

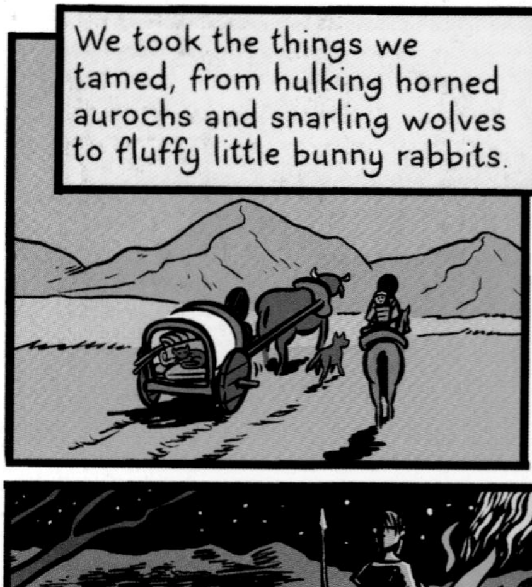
We took the things we tamed, from hulking horned aurochs and snarling wolves to fluffy little bunny rabbits.

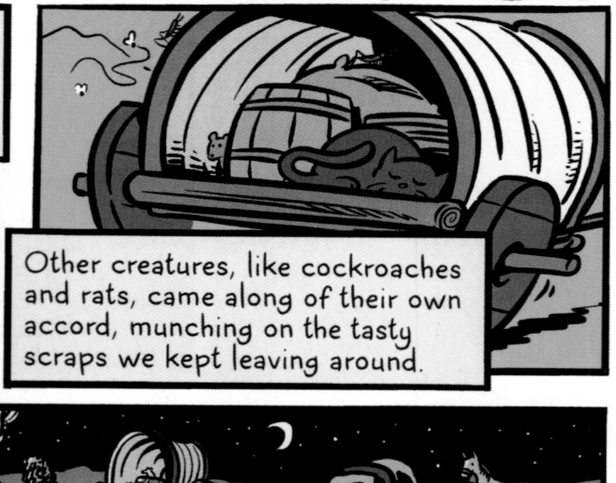
Other creatures, like cockroaches and rats, came along of their own accord, munching on the tasty scraps we kept leaving around.

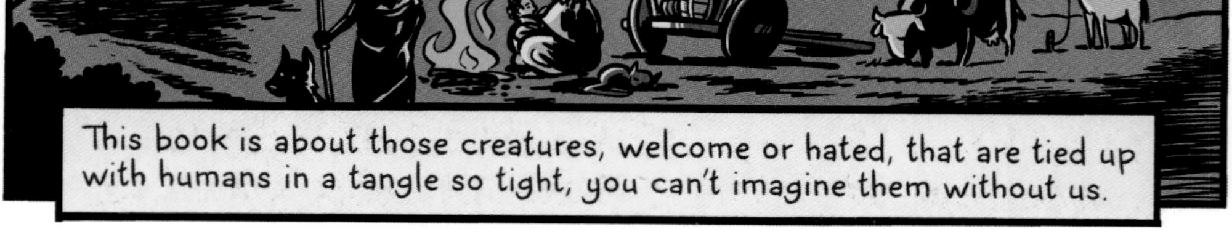
This book is about those creatures, welcome or hated, that are tied up with humans in a tangle so tight, you can't imagine them without us.

The creatures in this book are
organized into 3 categories:

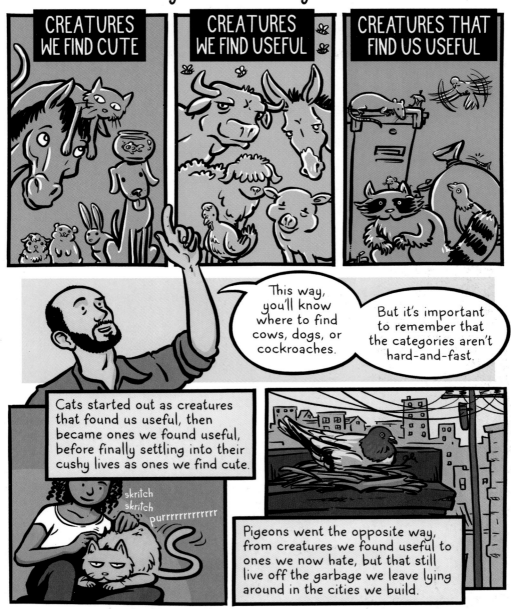

CREATURES WE FIND CUTE

CREATURES WE FIND USEFUL

CREATURES THAT FIND US USEFUL

This way, you'll know where to find cows, dogs, or cockroaches.

But it's important to remember that the categories aren't hard-and-fast.

Cats started out as creatures that found us useful, then became ones we found useful, before finally settling into their cushy lives as ones we find cute.

skritch
skritch
purrrrrrrrrrrrr

Pigeons went the opposite way, from creatures we found useful to ones we now hate, but that still live off the garbage we leave lying around in the cities we build.

What unites them, though, is that we take them for granted.
But read on, and you'll see that their stories hide our own history.

CREATURES WE FIND CUTE

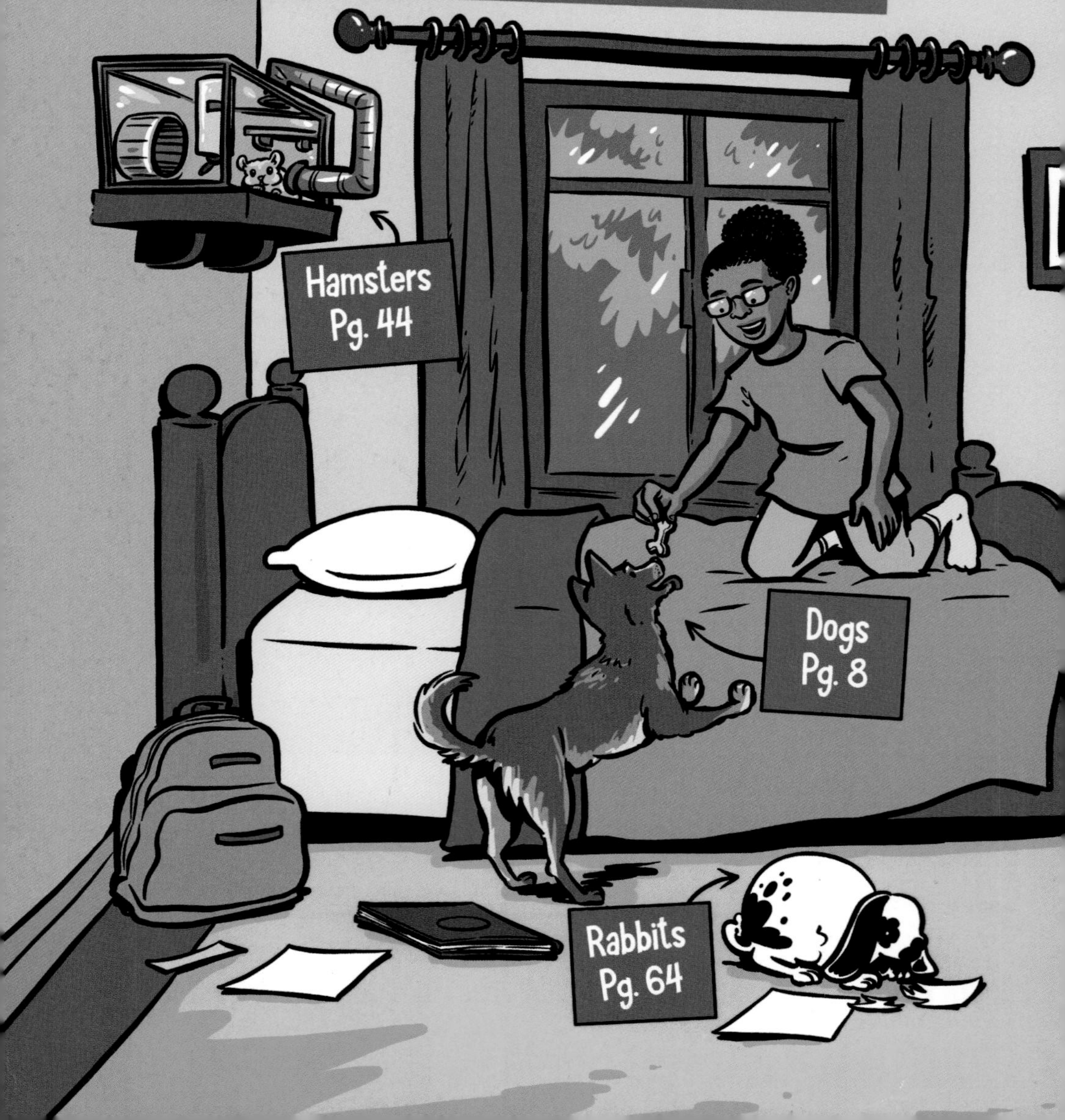
Hamsters
Pg. 44
Dogs
Pg. 8
Rabbits
Pg. 64

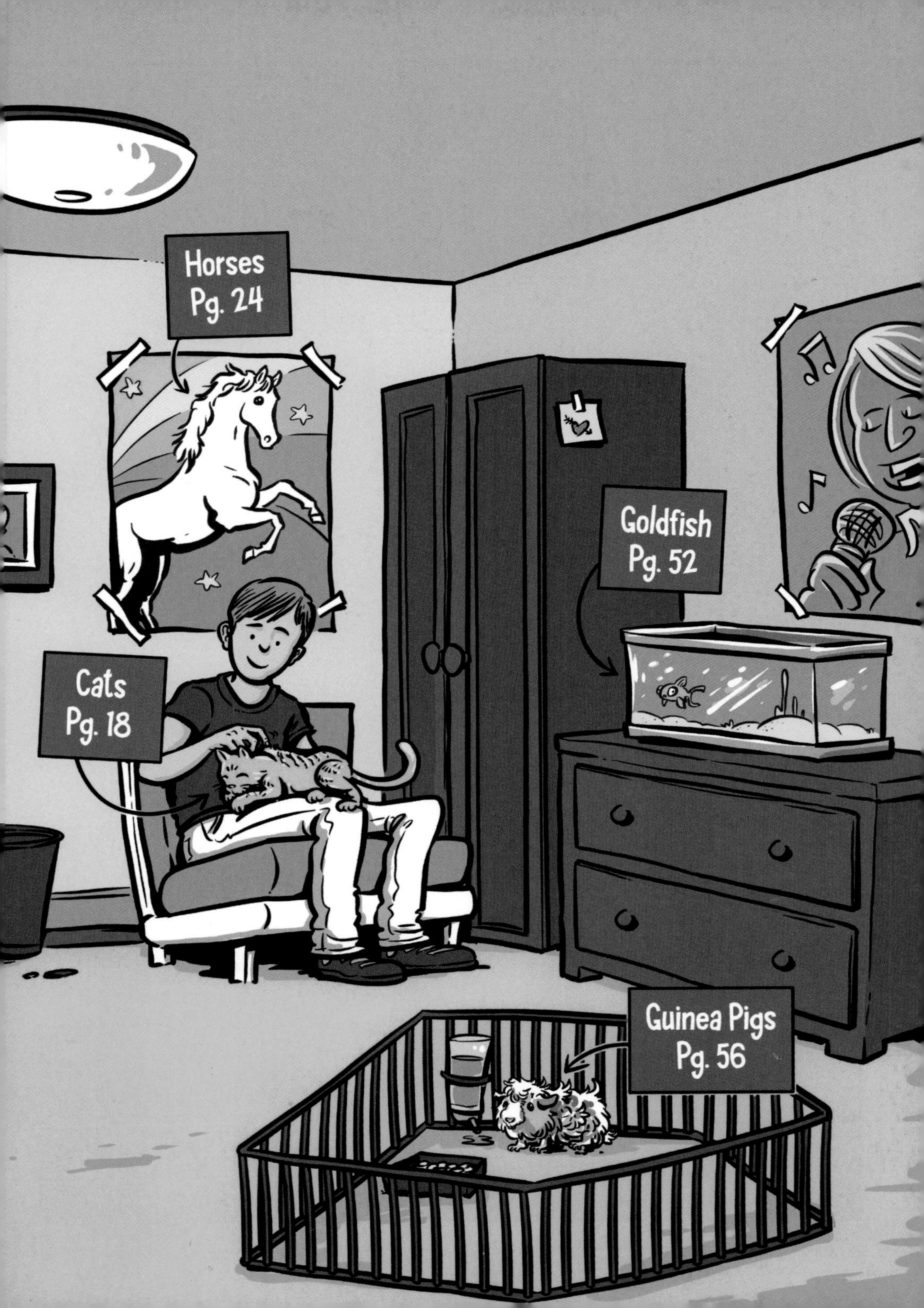
Horses
Pg. 24
Goldfish
Pg. 52
Cats
Pg. 18
Guinea Pigs
Pg. 56

DOGS

Canis lupus familiaris

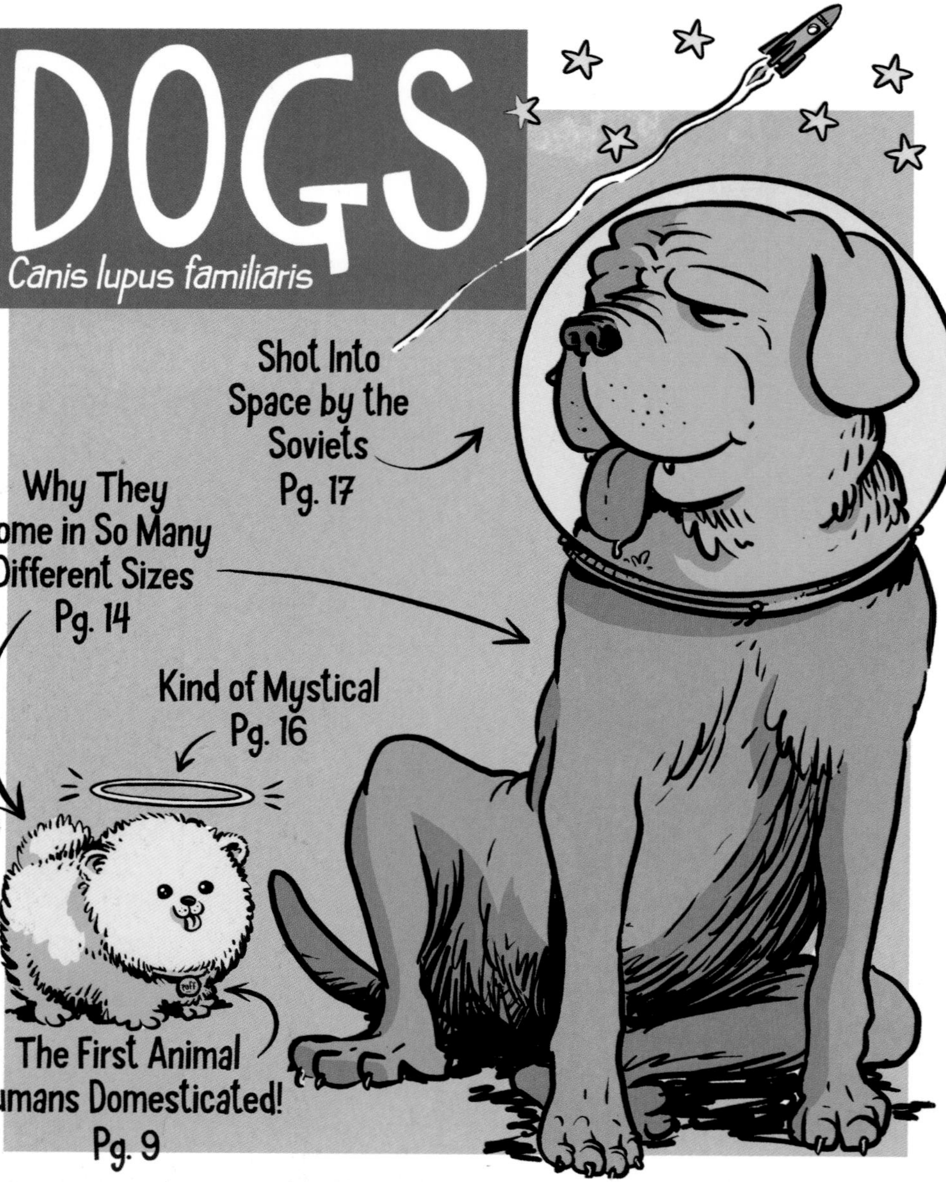
Shot Into Space by the Soviets Pg. 17
Why They Come in So Many Different Sizes Pg. 14
Kind of Mystical Pg. 16
ruff
The First Animal Humans Domesticated! Pg. 9

Size: 2.5–42 in (at the shoulder)

Weight: 0.5–345 lbs

Global Population: ~900 million (includes feral dogs)

Diet: Dog food, treats, trash, other dogs' poop

(Likely) **Wild Origin:**

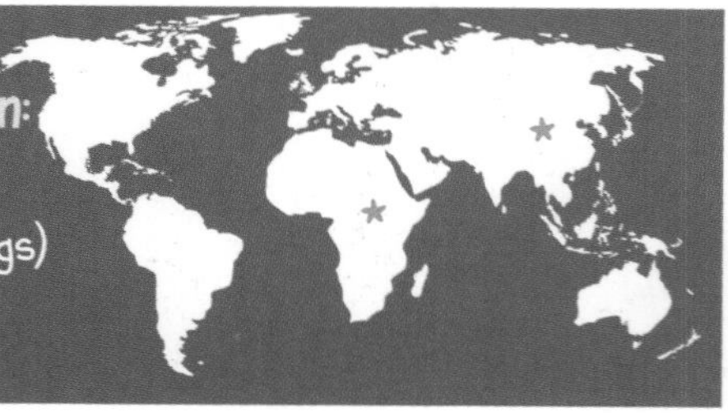

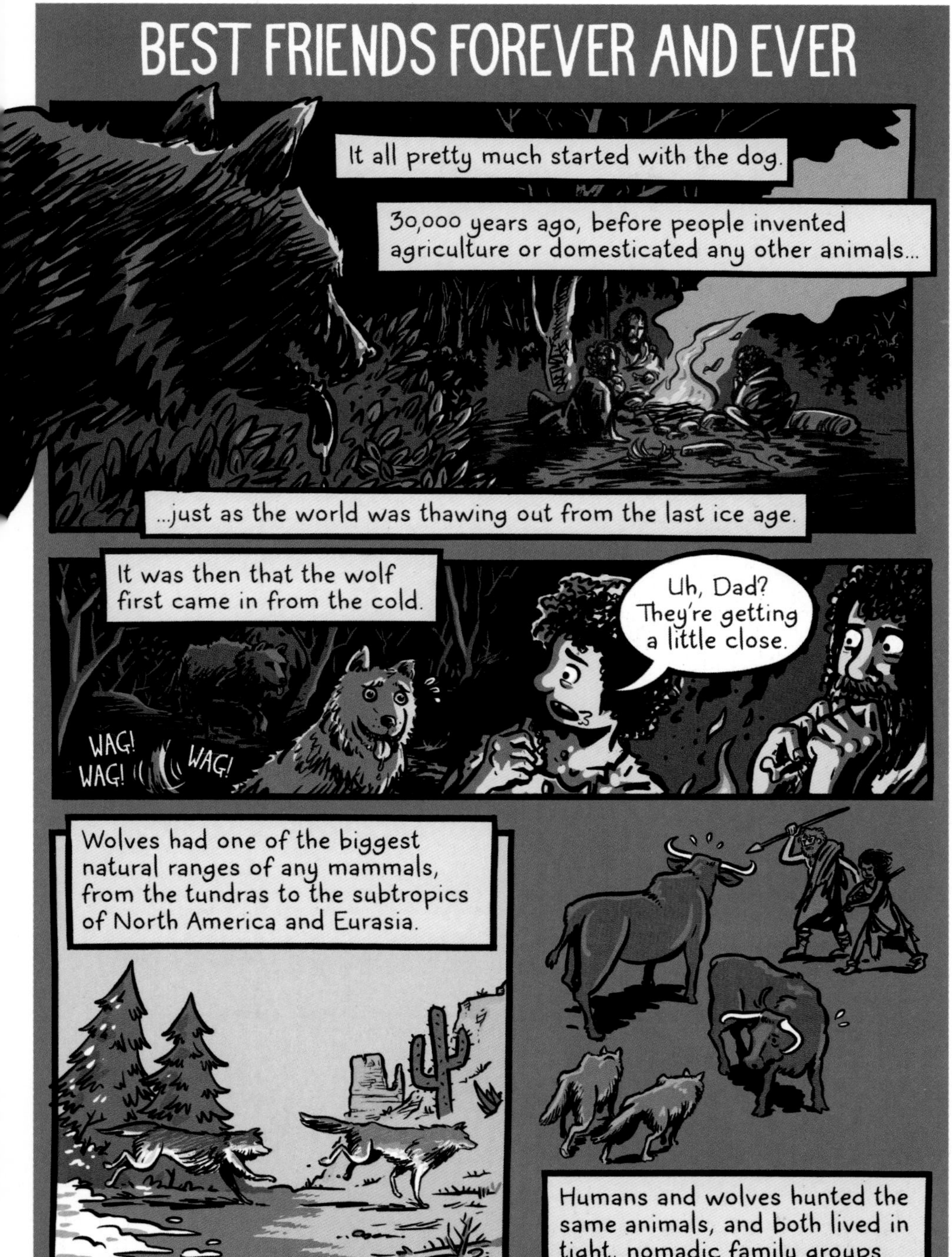
BEST FRIENDS FOREVER AND EVER
It all pretty much started with the dog.
30,000 years ago, before people invented agriculture or domesticated any other animals...
...just as the world was thawing out from the last ice age.
It was then that the wolf first came in from the cold.
WAG! WAG! WAG!
Uh, Dad? They're getting a little close.
Wolves had one of the biggest natural ranges of any mammals, from the tundras to the subtropics of North America and Eurasia.
Humans and wolves hunted the same animals, and both lived in tight, nomadic family groups.

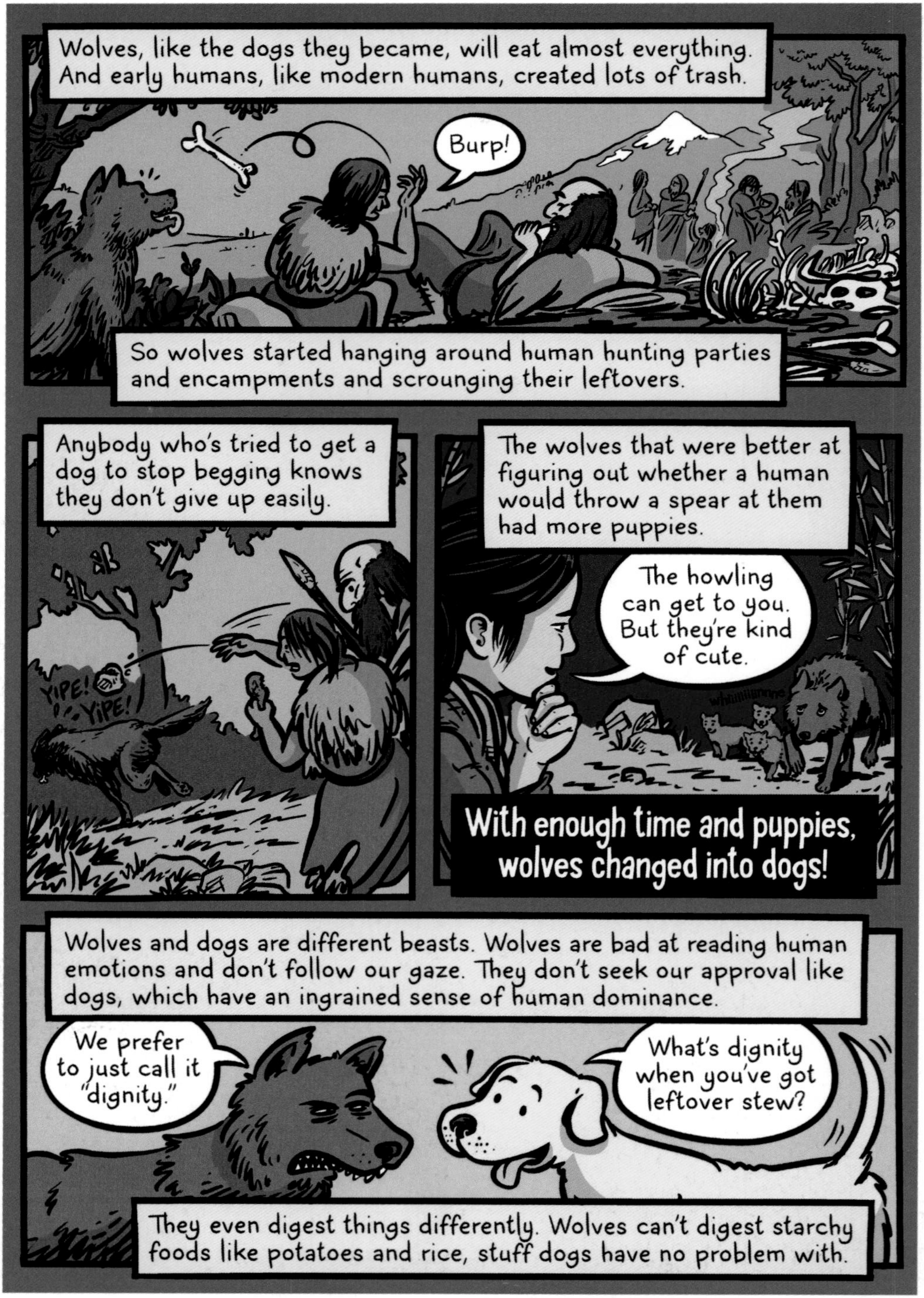
Wolves, like the dogs they became, will eat almost everything. And early humans, like modern humans, created lots of trash.
Burp!
So wolves started hanging around human hunting parties and encampments and scrounging their leftovers.
Anybody who's tried to get a dog to stop begging knows they don't give up easily.
YIPE! YIPE!
The wolves that were better at figuring out whether a human would throw a spear at them had more puppies.
The howling can get to you. But they're kind of cute.
With enough time and puppies, wolves changed into dogs!
Wolves and dogs are different beasts. Wolves are bad at reading human emotions and don't follow our gaze. They don't seek our approval like dogs, which have an ingrained sense of human dominance.
We prefer to just call it "dignity."
What's dignity when you've got leftover stew?
They even digest things differently. Wolves can't digest starchy foods like potatoes and rice, stuff dogs have no problem with.

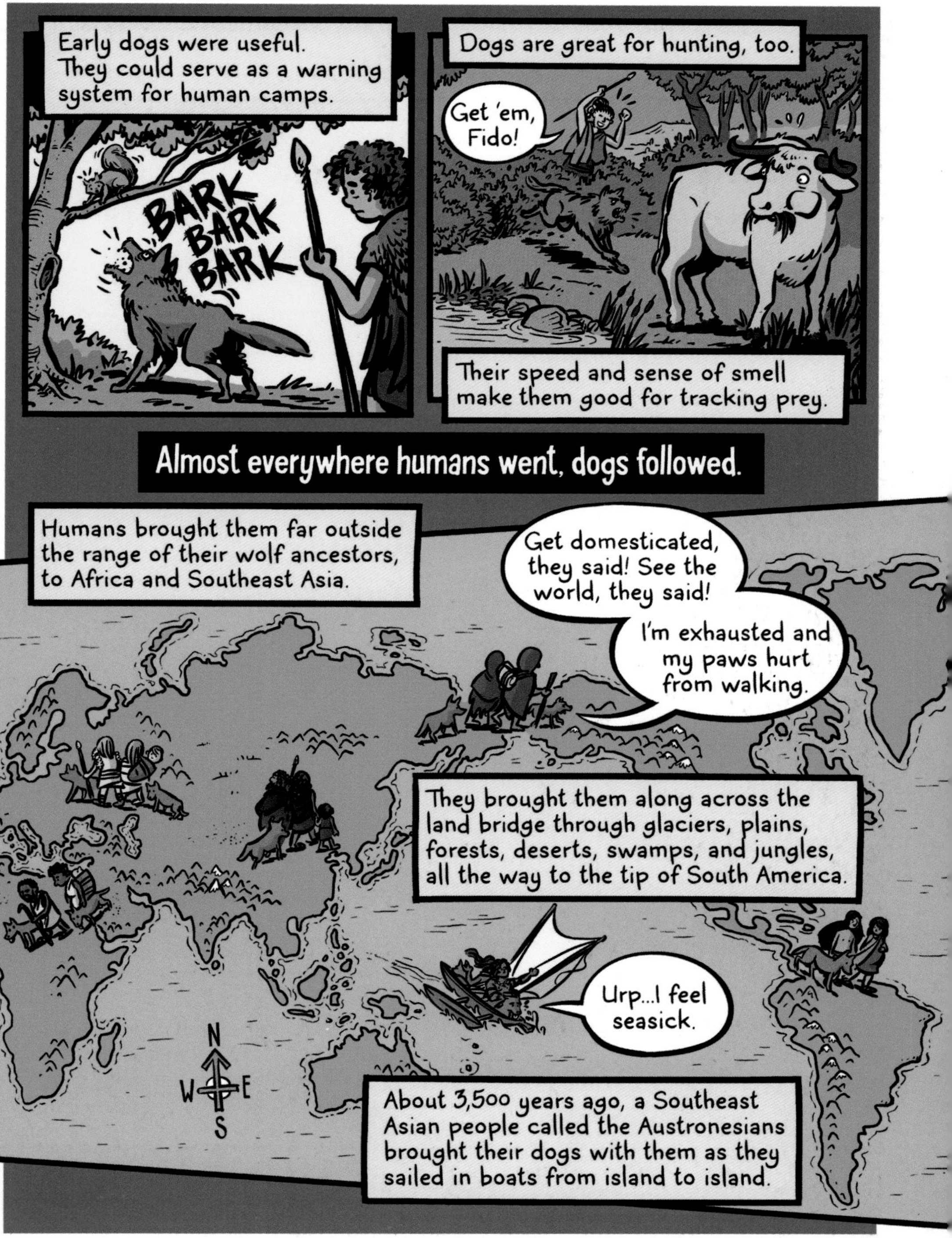
Early dogs were useful. They could serve as a warning system for human camps.
BARK BARK BARK
Dogs are great for hunting, too.
Get 'em, Fido!
Their speed and sense of smell make them good for tracking prey.
Almost everywhere humans went, dogs followed.
Humans brought them far outside the range of their wolf ancestors, to Africa and Southeast Asia.
Get domesticated, they said! See the world, they said!
I'm exhausted and my paws hurt from walking.
They brought them along across the land bridge through glaciers, plains, forests, deserts, swamps, and jungles, all the way to the tip of South America.
Urp...I feel seasick.
N
W
E
S
About 3,500 years ago, a Southeast Asian people called the Austronesians brought their dogs with them as they sailed in boats from island to island.

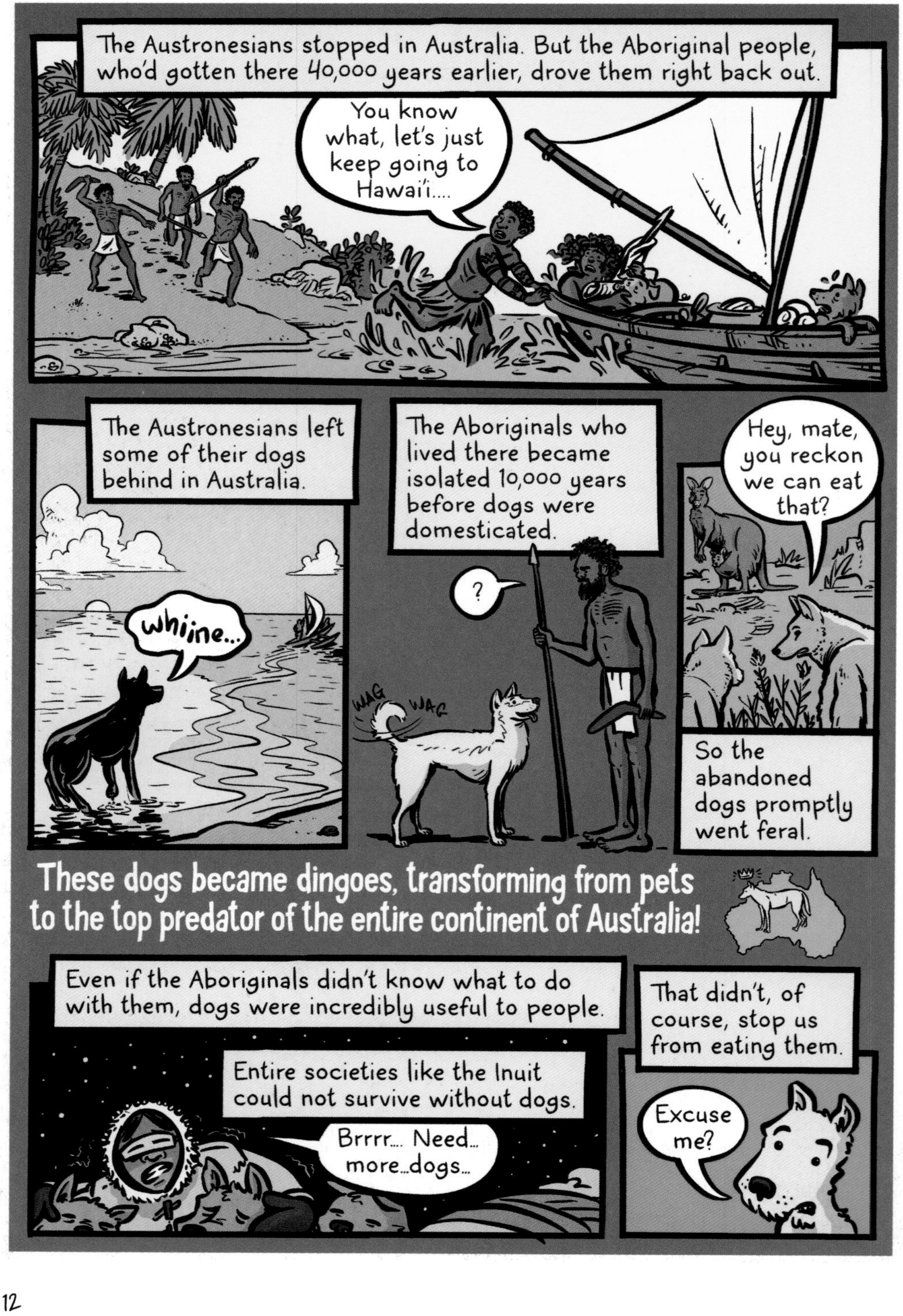
The Austronesians stopped in Australia. But the Aboriginal people, who'd gotten there 40,000 years earlier, drove them right back out.
You know what, let's just keep going to Hawai'i....
The Austronesians left some of their dogs behind in Australia.
whiine...
The Aboriginals who lived there became isolated 10,000 years before dogs were domesticated.
?
WAG WAG
Hey, mate, you reckon we can eat that?
So the abandoned dogs promptly went feral.
These dogs became dingoes, transforming from pets to the top predator of the entire continent of Australia!
Even if the Aboriginals didn't know what to do with them, dogs were incredibly useful to people.
Entire societies like the Inuit could not survive without dogs.
Brrrr... Need... more...dogs...
That didn't, of course, stop us from eating them.
Excuse me?

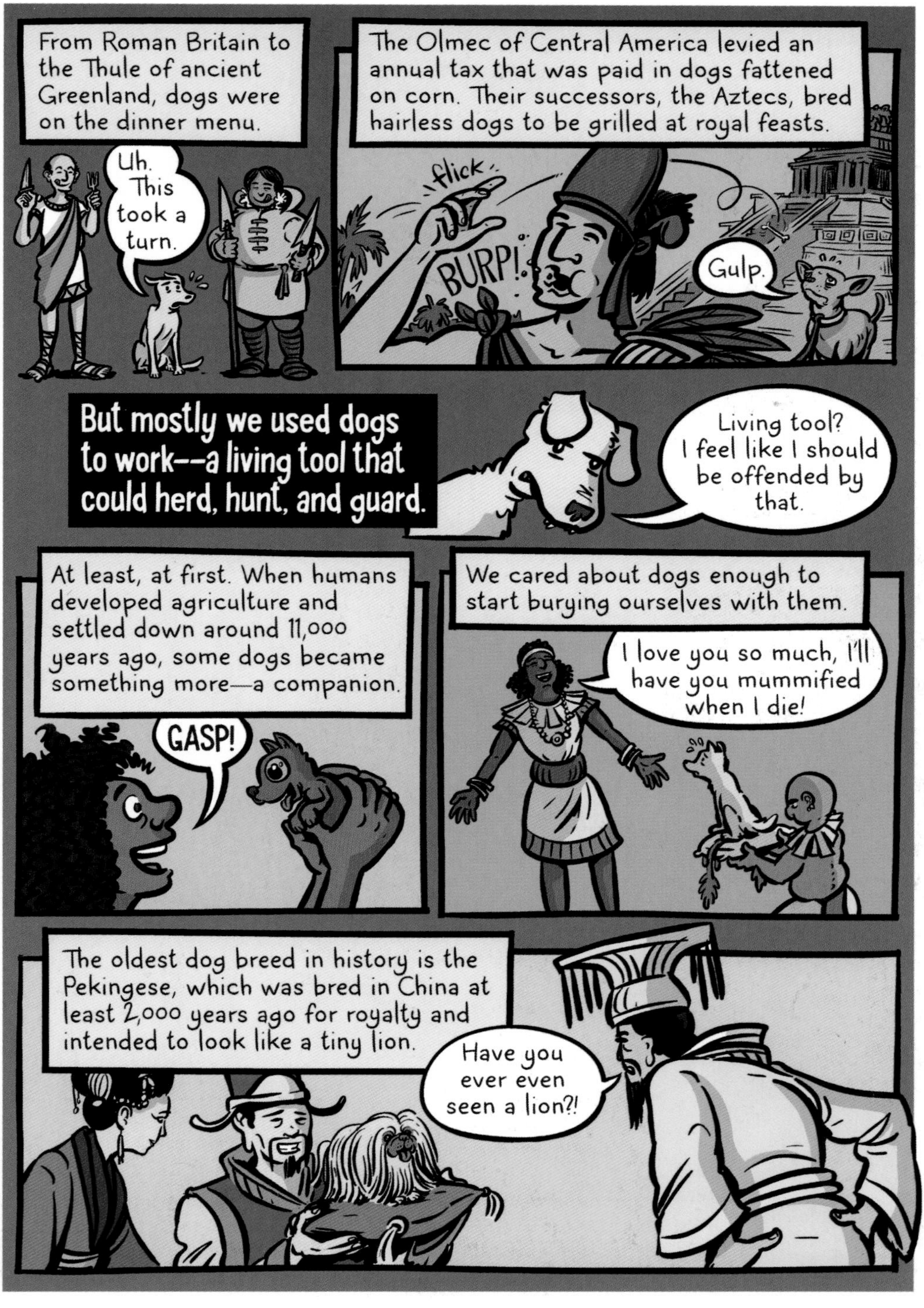
From Roman Britain to the Thule of ancient Greenland, dogs were on the dinner menu.
Uh. This took a turn.
The Olmec of Central America levied an annual tax that was paid in dogs fattened on corn. Their successors, the Aztecs, bred hairless dogs to be grilled at royal feasts.
flick
BURP!
Gulp.
But mostly we used dogs to work--a living tool that could herd, hunt, and guard.
Living tool? I feel like I should be offended by that.
At least, at first. When humans developed agriculture and settled down around 11,000 years ago, some dogs became something more—a companion.
GASP!
We cared about dogs enough to start burying ourselves with them.
I love you so much, I'll have you mummified when I die!
The oldest dog breed in history is the Pekingese, which was bred in China at least 2,000 years ago for royalty and intended to look like a tiny lion.
Have you ever even seen a lion?!

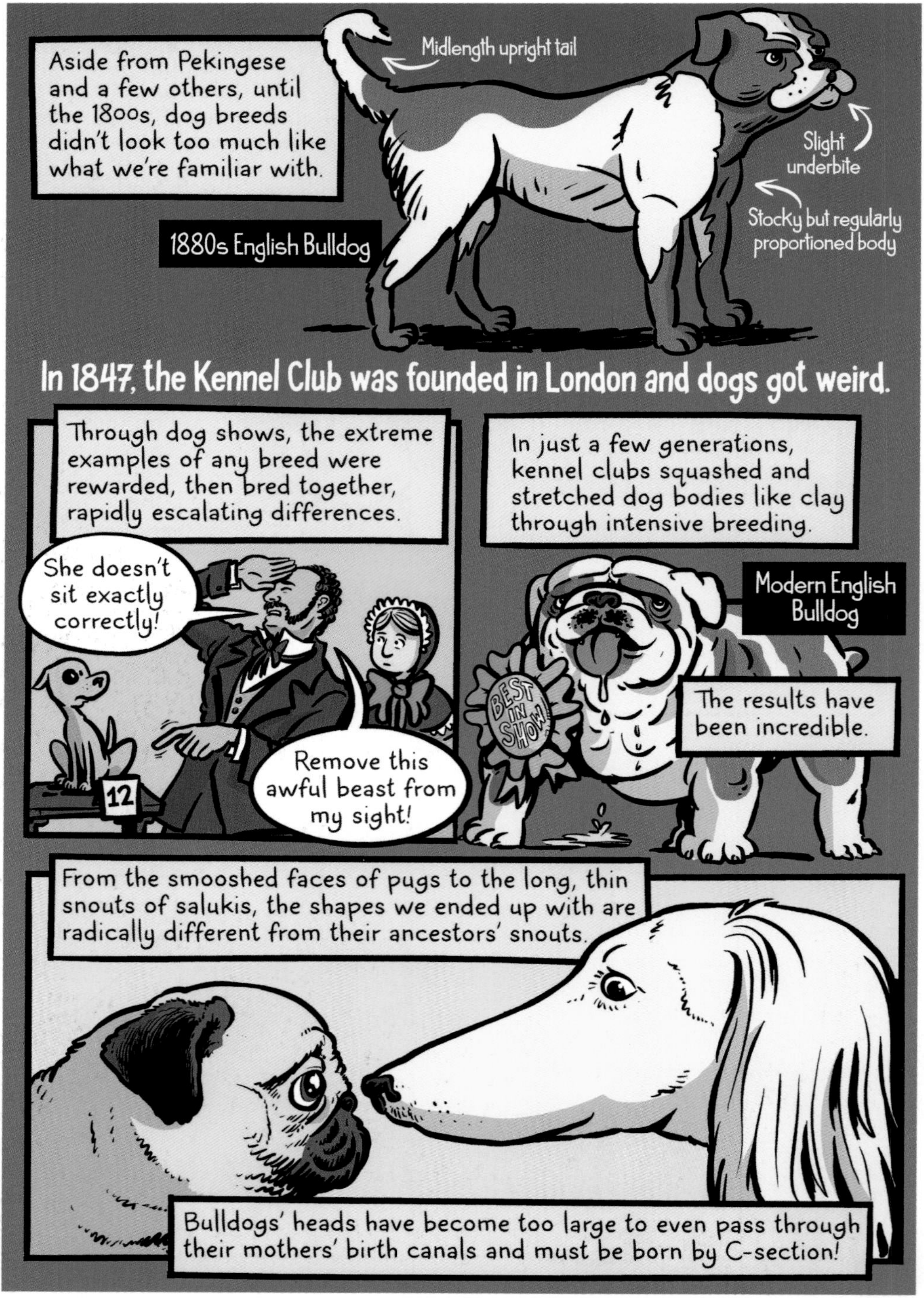

Aside from Pekingese and a few others, until the 1800s, dog breeds didn't look too much like what we're familiar with.
Midlength upright tail
Slight underbite
Stocky but regularly proportioned body
1880s English Bulldog
In 1847, the Kennel Club was founded in London and dogs got weird.
Through dog shows, the extreme examples of any breed were rewarded, then bred together, rapidly escalating differences.
She doesn't sit exactly correctly!
Remove this awful beast from my sight!
12
In just a few generations, kennel clubs squashed and stretched dog bodies like clay through intensive breeding.
Modern English Bulldog
BEST IN SHOW
The results have been incredible.
From the smooshed faces of pugs to the long, thin snouts of salukis, the shapes we ended up with are radically different from their ancestors' snouts.
Bulldogs' heads have become too large to even pass through their mothers' birth canals and must be born by C-section!

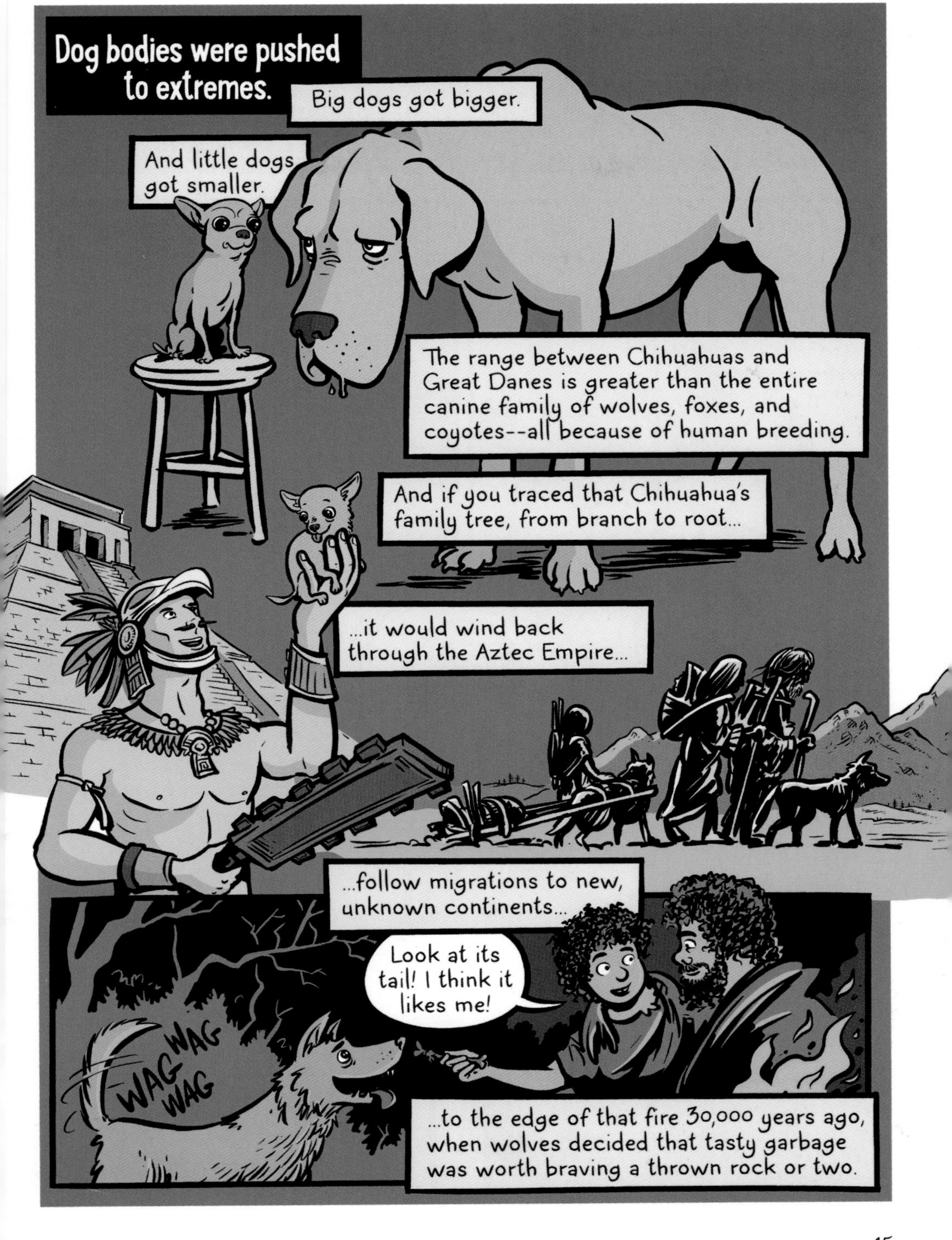
Dog bodies were pushed to extremes.
Big dogs got bigger.
And little dogs got smaller.
The range between Chihuahuas and Great Danes is greater than the entire canine family of wolves, foxes, and coyotes--all because of human breeding.
And if you traced that Chihuahua's family tree, from branch to root...
...it would wind back through the Aztec Empire...
...follow migrations to new, unknown continents...
Look at its tail! I think it likes me!
WAG WAG WAG WAG
...to the edge of that fire 30,000 years ago, when wolves decided that tasty garbage was worth braving a thrown rock or two.

MYSTICAL MUTTS

With 30,000 years together, we've had a lot of time to come up with some pretty fun beliefs about dogs. Here are some great ones:

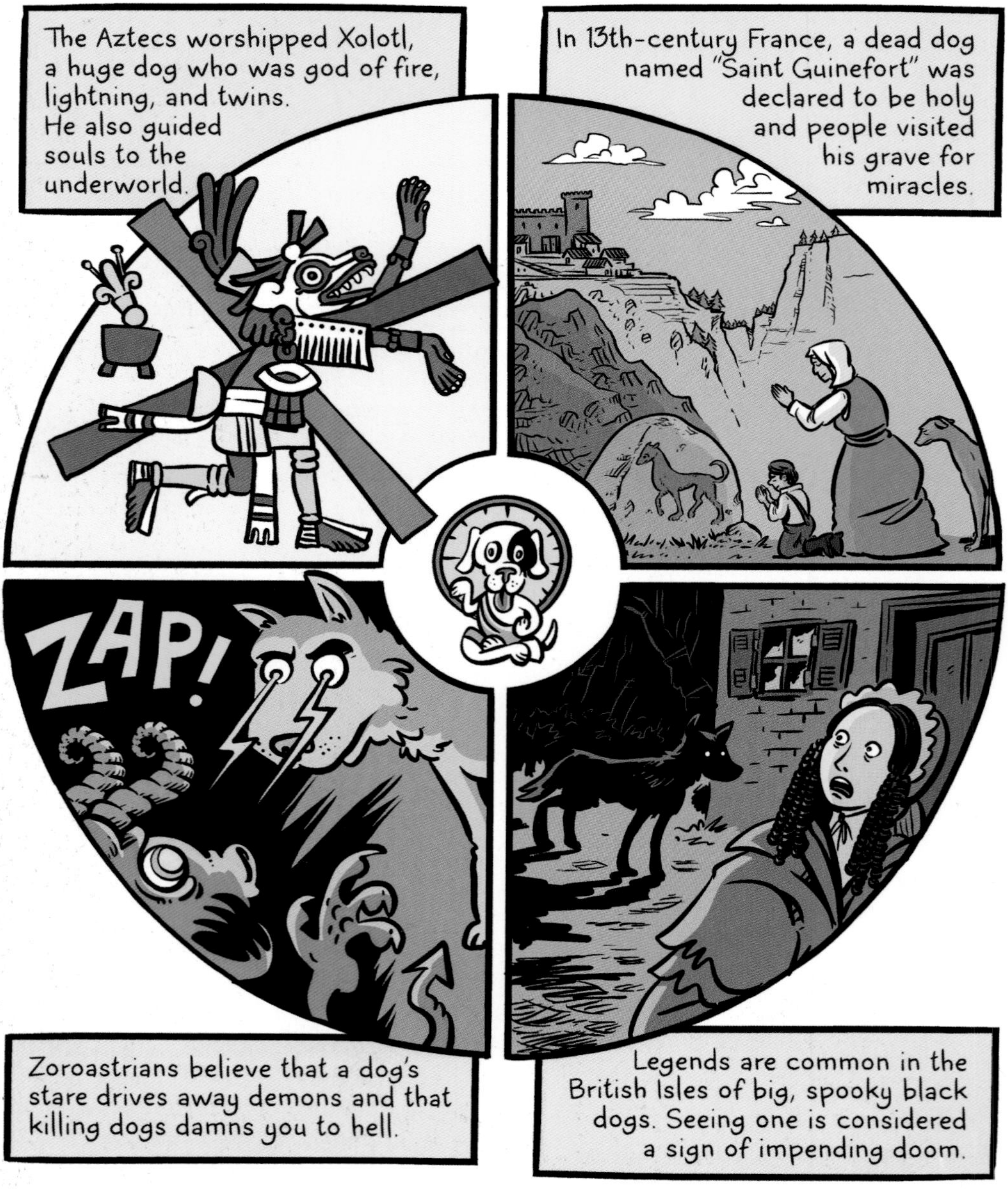
The Aztecs worshipped Xolotl, a huge dog who was god of fire, lightning, and twins. He also guided souls to the underworld.
In 13th-century France, a dead dog named "Saint Guinefort" was declared to be holy and people visited his grave for miracles.
ZAP!
Zoroastrians believe that a dog's stare drives away demons and that killing dogs damns you to hell.
Legends are common in the British Isles of big, spooky black dogs. Seeing one is considered a sign of impending doom.

THE SPACE DOGS OF RUSSIA

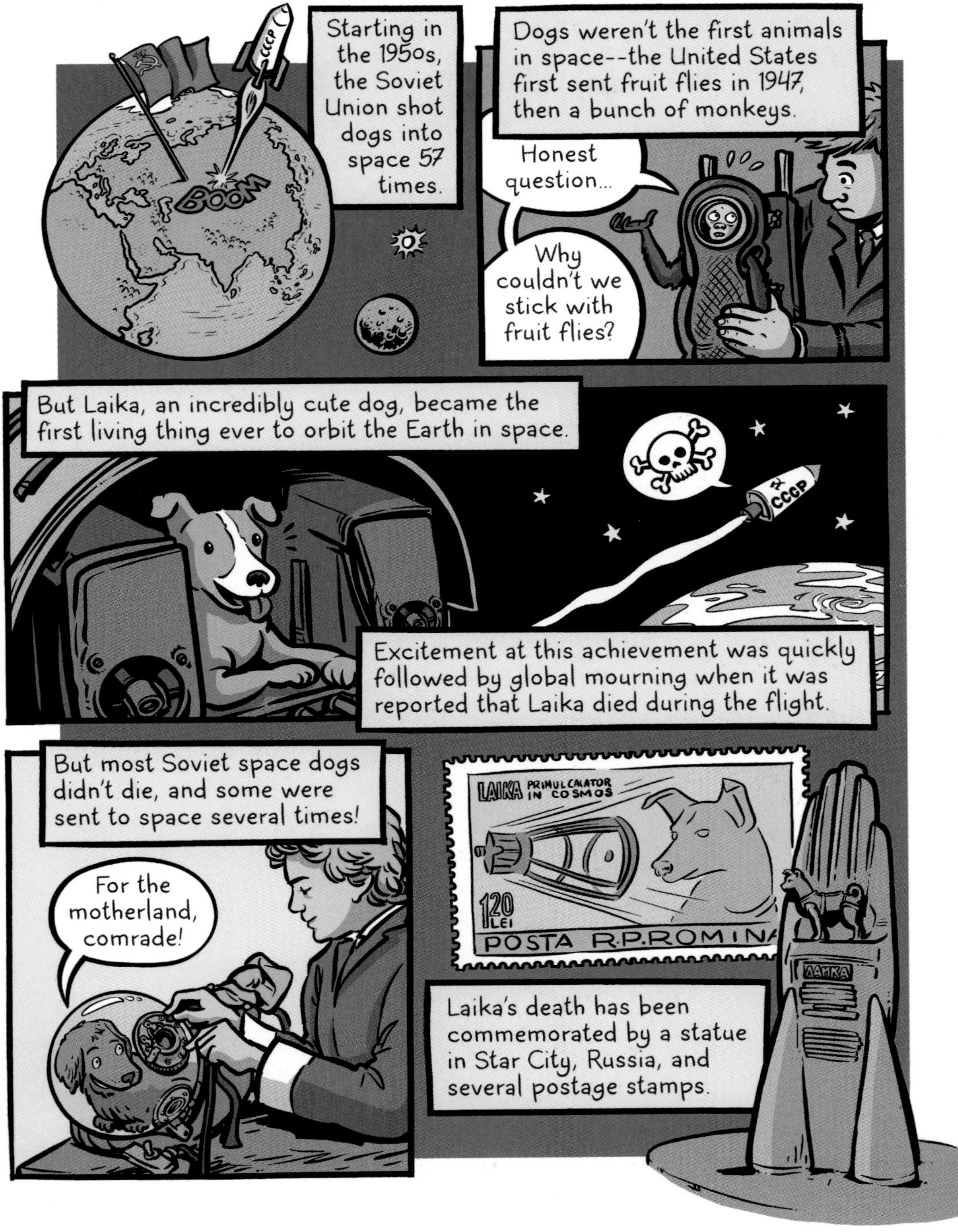

Starting in the 1950s, the Soviet Union shot dogs into space 57 times.
CCCP
BOOM
Dogs weren't the first animals in space--the United States first sent fruit flies in 1947, then a bunch of monkeys.
Honest question...
Why couldn't we stick with fruit flies?
But Laika, an incredibly cute dog, became the first living thing ever to orbit the Earth in space.
CCCP
Excitement at this achievement was quickly followed by global mourning when it was reported that Laika died during the flight.
But most Soviet space dogs didn't die, and some were sent to space several times!
For the motherland, comrade!
LAIKA PRIMUL CALATOR IN COSMOS
1.20 LEI
POSTA R.P.ROMINA
ЛАЙКА
Laika's death has been commemorated by a statue in Star City, Russia, and several postage stamps.

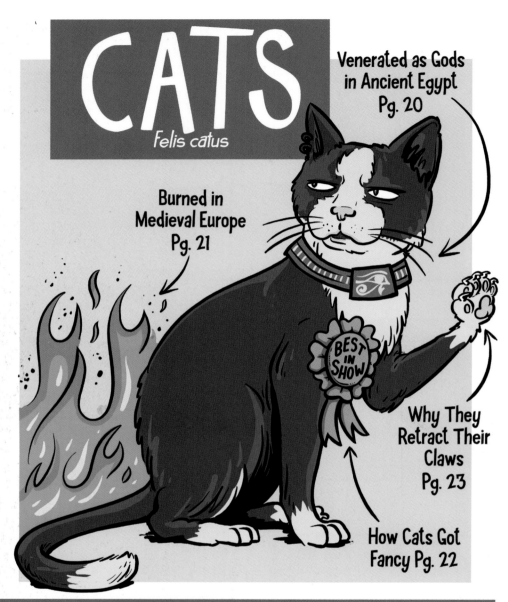

CATS

Felis catus

Venerated as Gods in Ancient Egypt Pg. 20

Burned in Medieval Europe Pg. 21

BEST IN SHOW

Why They Retract Their Claws Pg. 23

How Cats Got Fancy Pg. 22

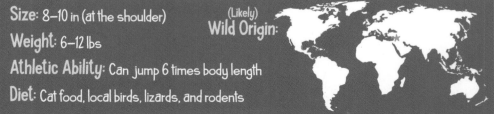

Size: 8–10 in (at the shoulder)

Weight: 6–12 lbs

Athletic Ability: Can jump 6 times body length

Diet: Cat food, local birds, lizards, and rodents

(Likely) Wild Origin:

FROM THE GRAIN HOUSE TO THE LAP

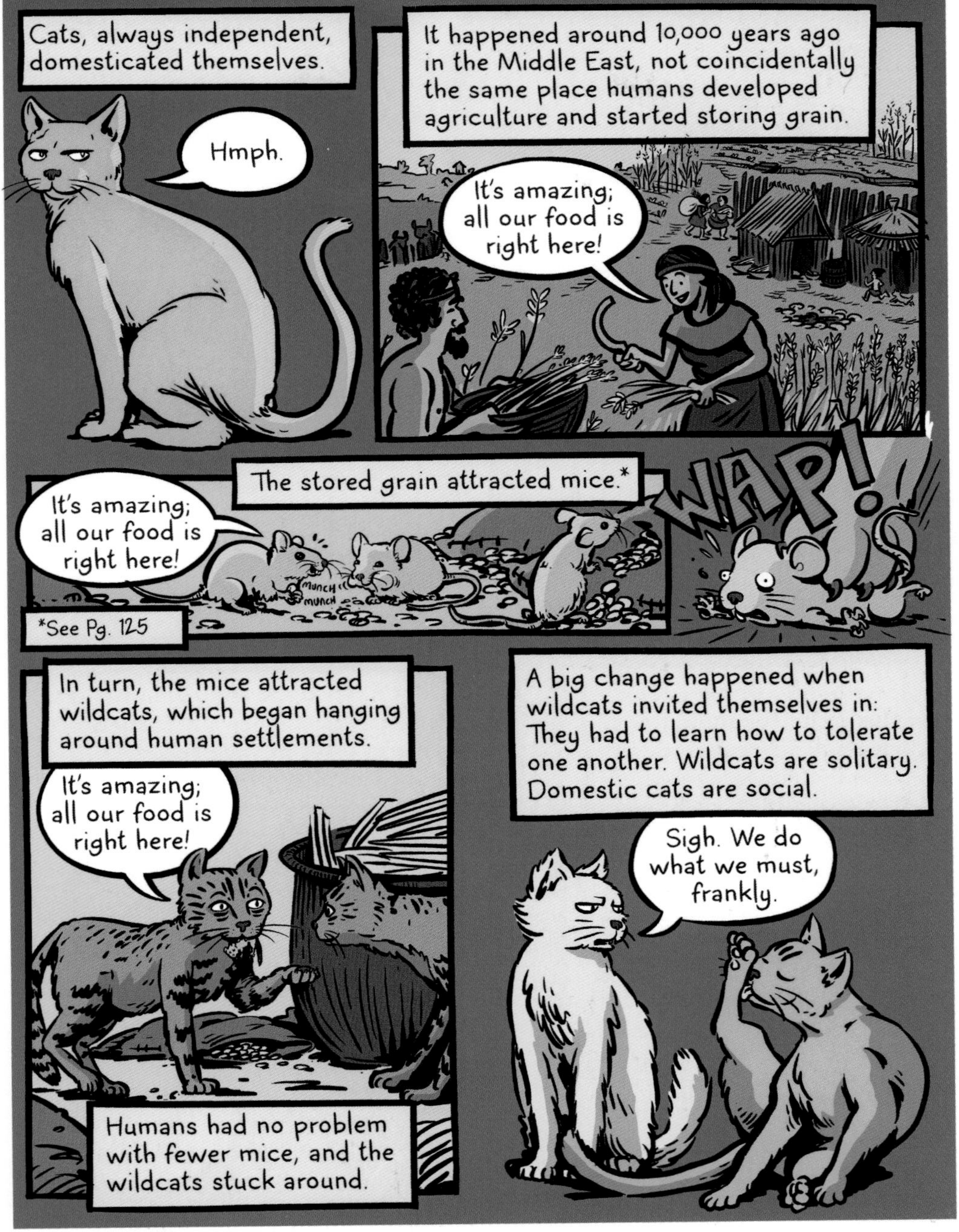
Cats, always independent, domesticated themselves.
Hmph.
It happened around 10,000 years ago in the Middle East, not coincidentally the same place humans developed agriculture and started storing grain.
It's amazing; all our food is right here!
The stored grain attracted mice.*
It's amazing; all our food is right here!
MUNCH MUNCH
*See Pg. 125
WAP!
In turn, the mice attracted wildcats, which began hanging around human settlements.
It's amazing; all our food is right here!
Humans had no problem with fewer mice, and the wildcats stuck around.
A big change happened when wildcats invited themselves in: They had to learn how to tolerate one another. Wildcats are solitary. Domestic cats are social.
Sigh. We do what we must, frankly.

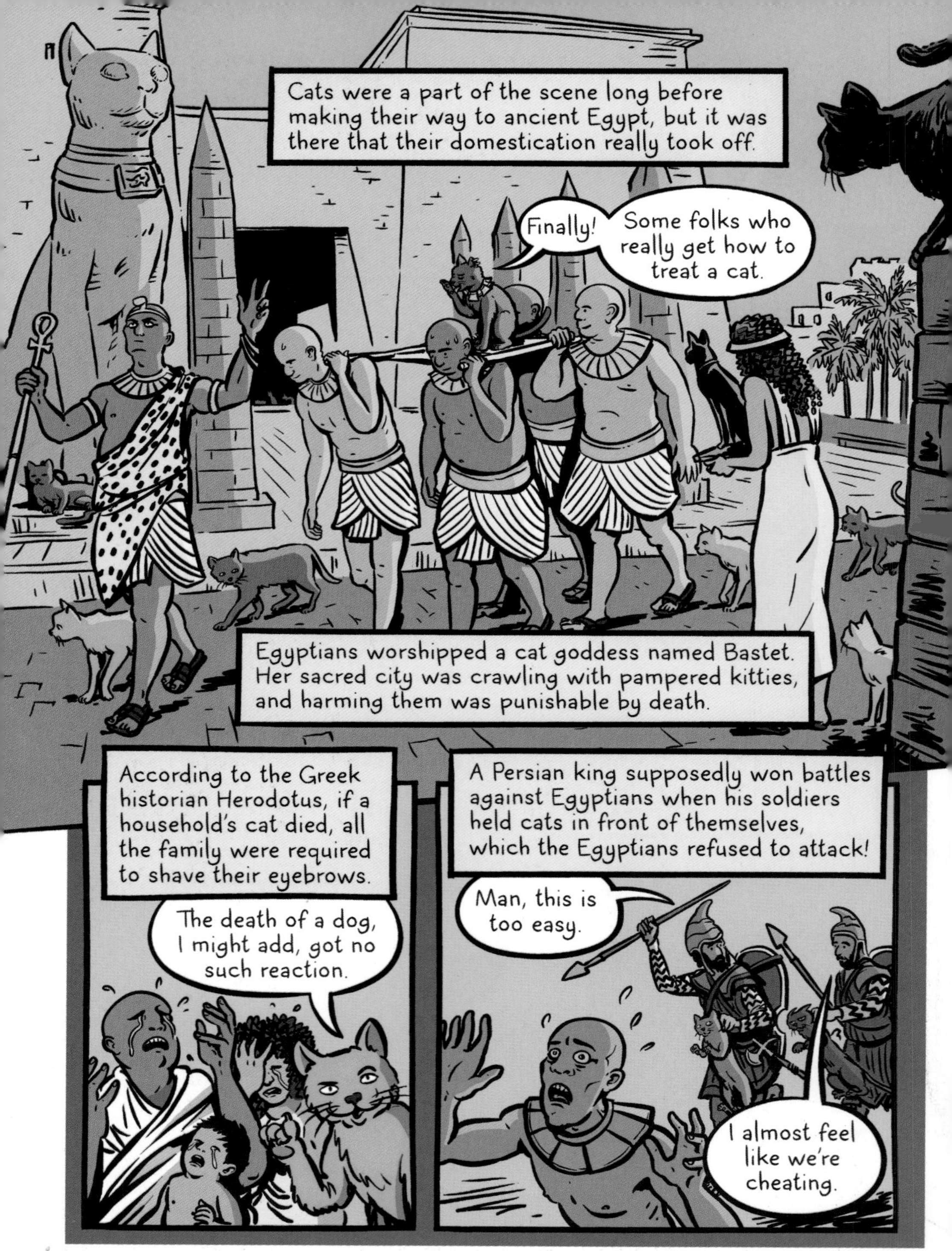
Cats were a part of the scene long before making their way to ancient Egypt, but it was there that their domestication really took off.
Finally!
Some folks who really get how to treat a cat.
Egyptians worshipped a cat goddess named Bastet. Her sacred city was crawling with pampered kitties, and harming them was punishable by death.
According to the Greek historian Herodotus, if a household's cat died, all the family were required to shave their eyebrows.
The death of a dog, I might add, got no such reaction.
A Persian king supposedly won battles against Egyptians when his soldiers held cats in front of themselves, which the Egyptians refused to attack!
Man, this is too easy.
I almost feel like we're cheating.

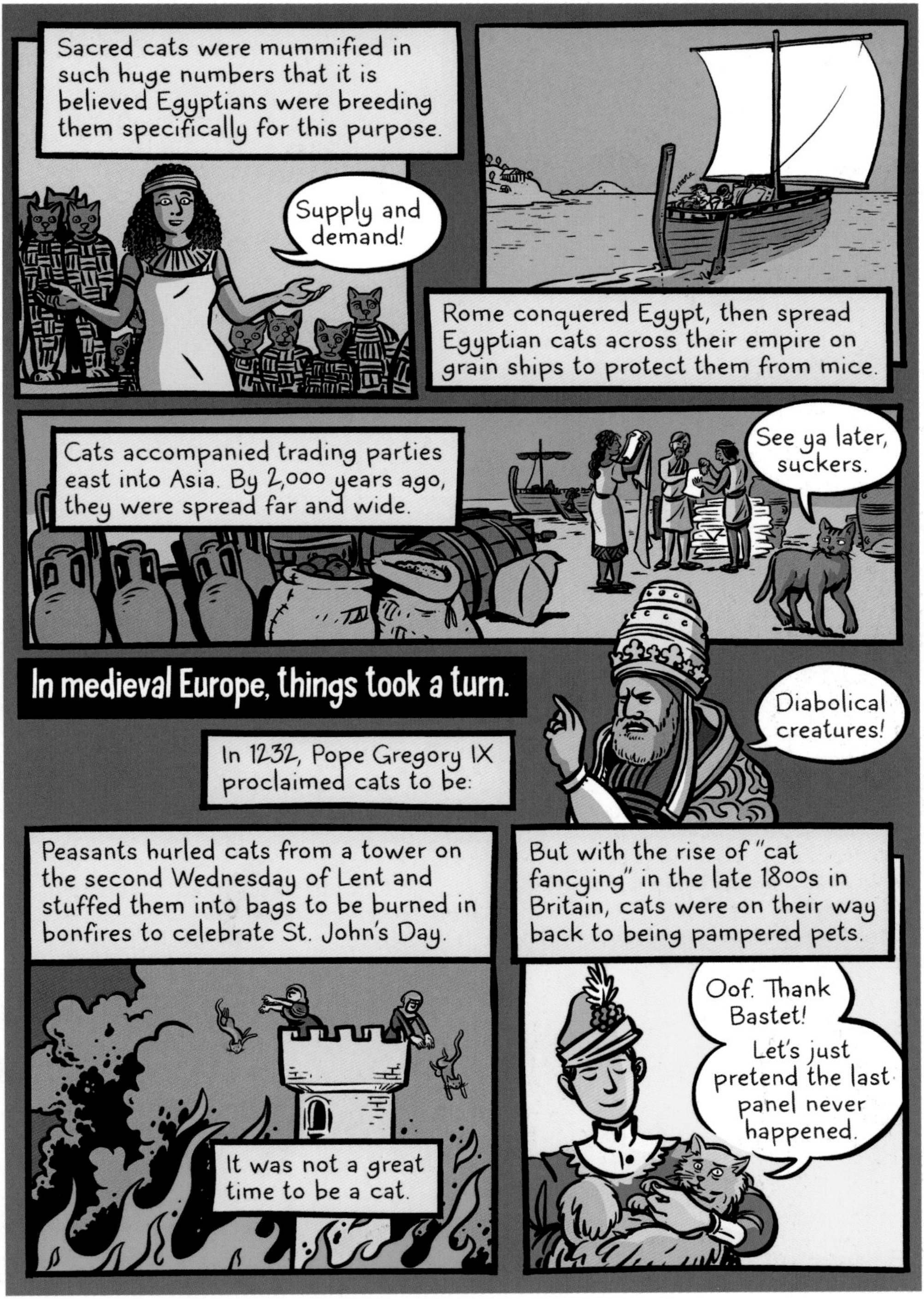

Sacred cats were mummified in such huge numbers that it is believed Egyptians were breeding them specifically for this purpose.
Supply and demand!
Rome conquered Egypt, then spread Egyptian cats across their empire on grain ships to protect them from mice.
Cats accompanied trading parties east into Asia. By 2,000 years ago, they were spread far and wide.
See ya later, suckers.
In medieval Europe, things took a turn.
In 1232, Pope Gregory IX proclaimed cats to be:
Diabolical creatures!
Peasants hurled cats from a tower on the second Wednesday of Lent and stuffed them into bags to be burned in bonfires to celebrate St. John's Day.
It was not a great time to be a cat.
But with the rise of "cat fancying" in the late 1800s in Britain, cats were on their way back to being pampered pets.
Oof. Thank Bastet!
Let's just pretend the last panel never happened.

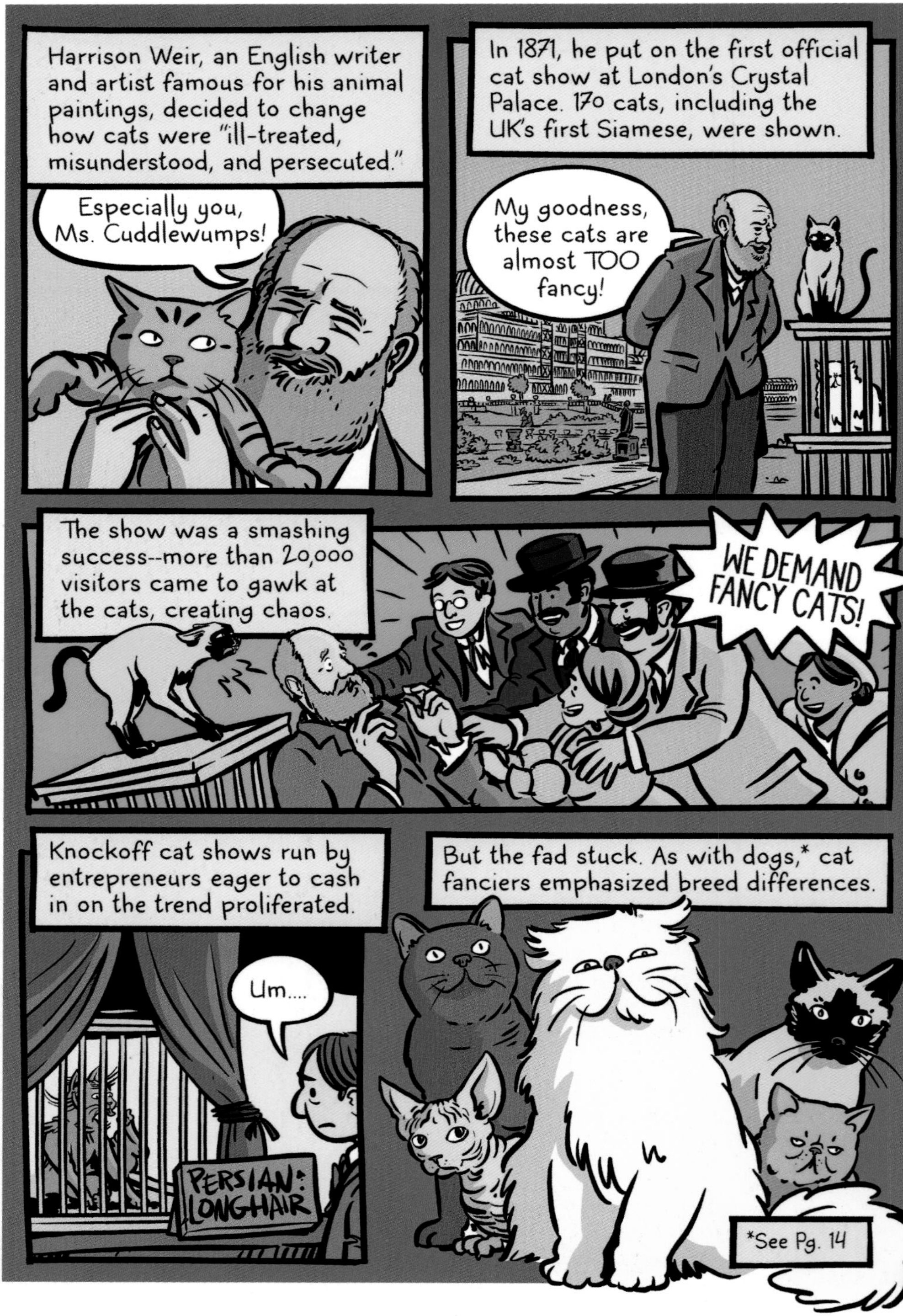
Harrison Weir, an English writer and artist famous for his animal paintings, decided to change how cats were "ill-treated, misunderstood, and persecuted."
Especially you, Ms. Cuddlewumps!
In 1871, he put on the first official cat show at London's Crystal Palace. 170 cats, including the UK's first Siamese, were shown.
My goodness, these cats are almost TOO fancy!
The show was a smashing success--more than 20,000 visitors came to gawk at the cats, creating chaos.
WE DEMAND FANCY CATS!
Knockoff cat shows run by entrepreneurs eager to cash in on the trend proliferated.
Um....
PERSIAN: LONGHAIR
But the fad stuck. As with dogs,* cat fanciers emphasized breed differences.
*See Pg. 14

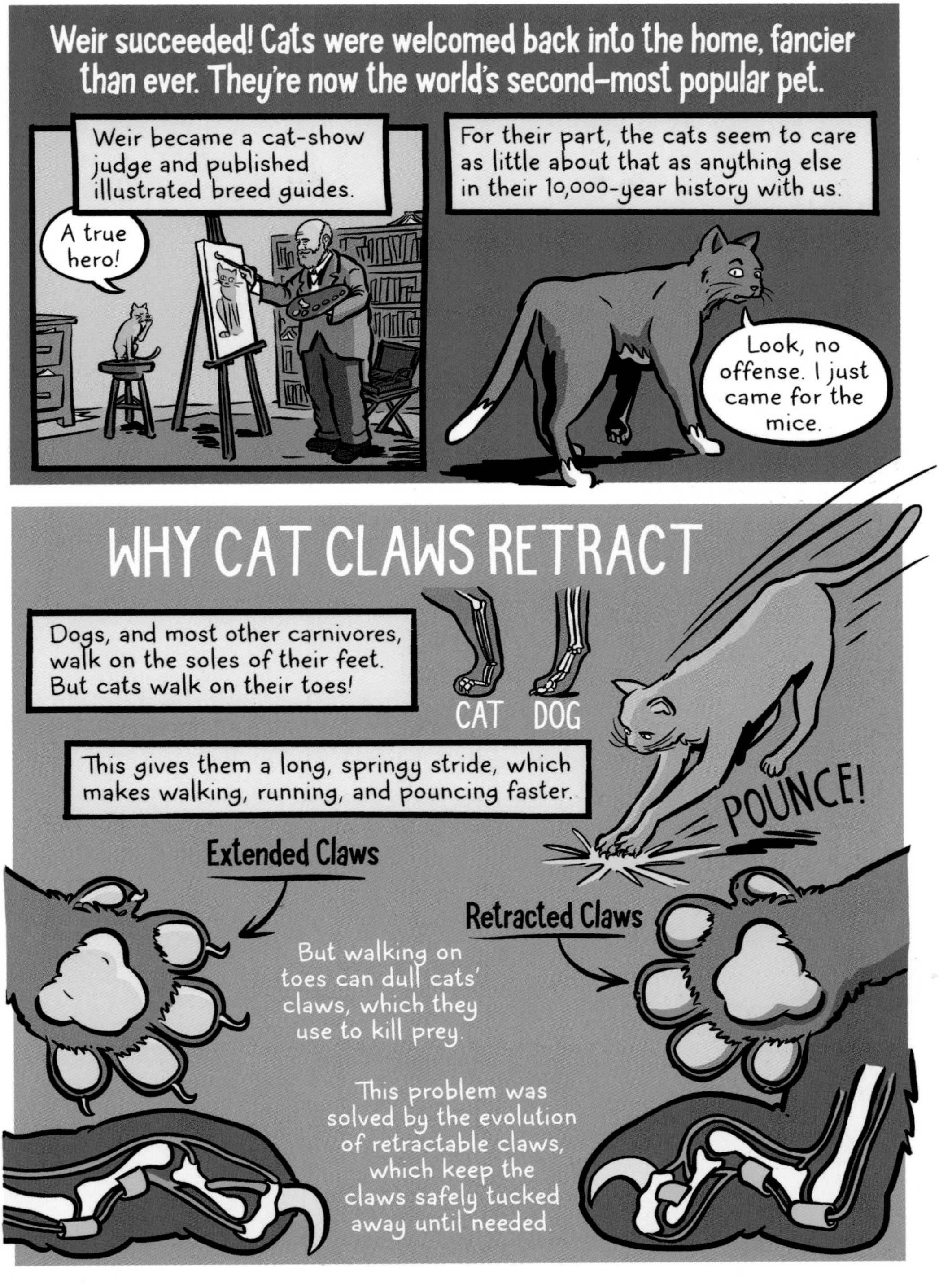

Weir succeeded! Cats were welcomed back into the home, fancier than ever. They're now the world's second-most popular pet.

Weir became a cat-show judge and published illustrated breed guides.

A true hero!

For their part, the cats seem to care as little about that as anything else in their 10,000-year history with us.

Look, no offense. I just came for the mice.

WHY CAT CLAWS RETRACT

Dogs, and most other carnivores, walk on the soles of their feet. But cats walk on their toes!

CAT DOG

This gives them a long, springy stride, which makes walking, running, and pouncing faster.

POUNCE!

Extended Claws

Retracted Claws

But walking on toes can dull cats' claws, which they use to kill prey.

This problem was solved by the evolution of retractable claws, which keep the claws safely tucked away until needed.

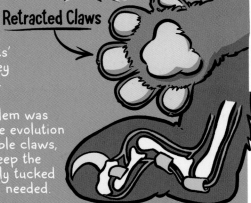

HORSES

Equus ferus caballus

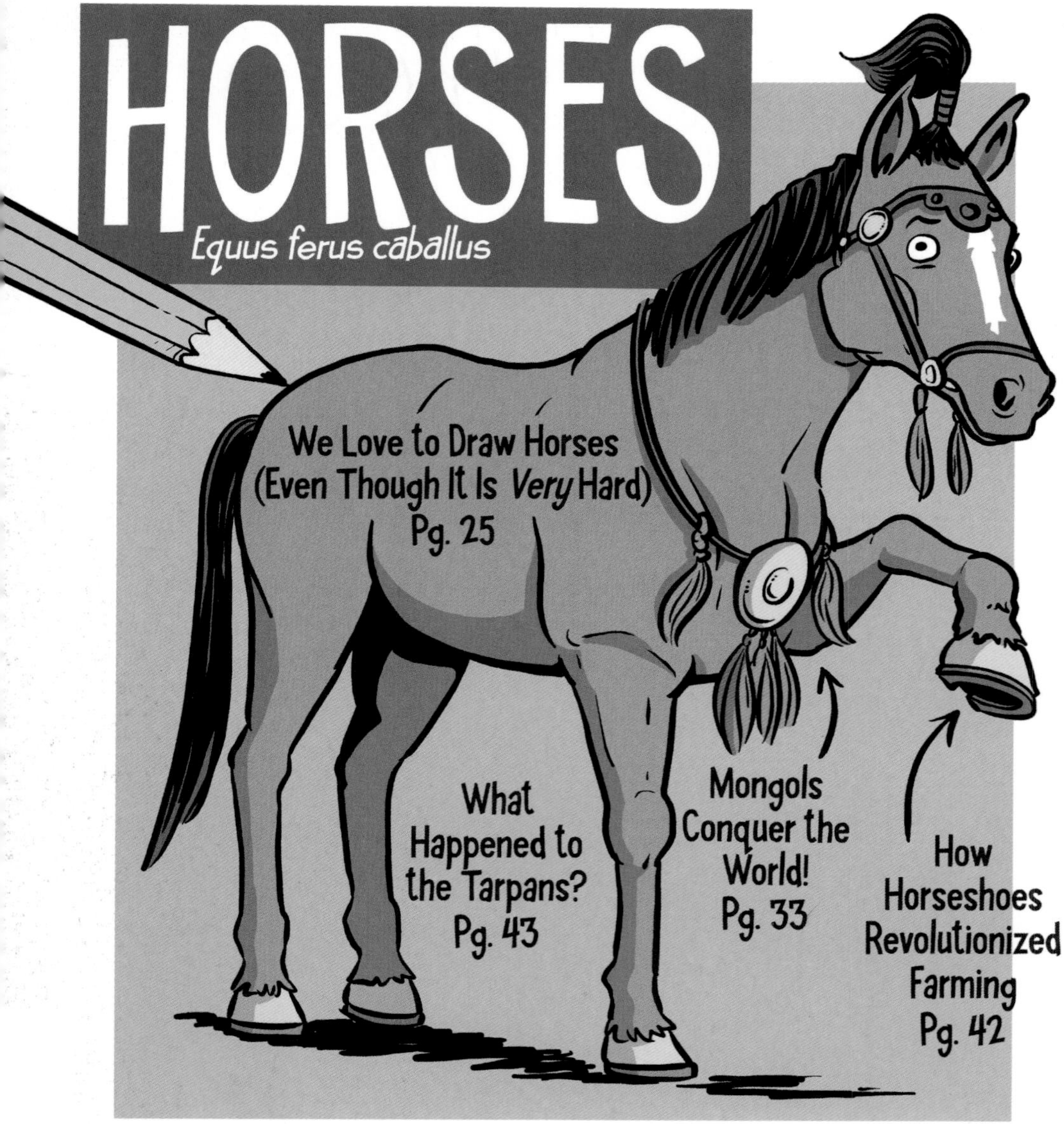
We Love to Draw Horses (Even Though It Is *Very* Hard) Pg. 25
What Happened to the Tarpans? Pg. 43
Mongols Conquer the World! Pg. 33
How Horseshoes Revolutionized Farming Pg. 42

Size: 34–72 in (at the shoulder)

Weight: 200–2,200 lbs

Oddball Anatomy: Largest eyes of any land mammal

Diet: Grass, alfalfa, hay, oats, an occasional apple treat

(Likely) **Wild Origin:**

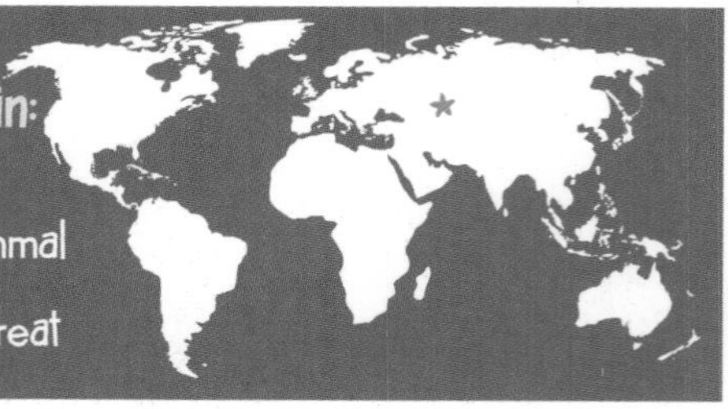

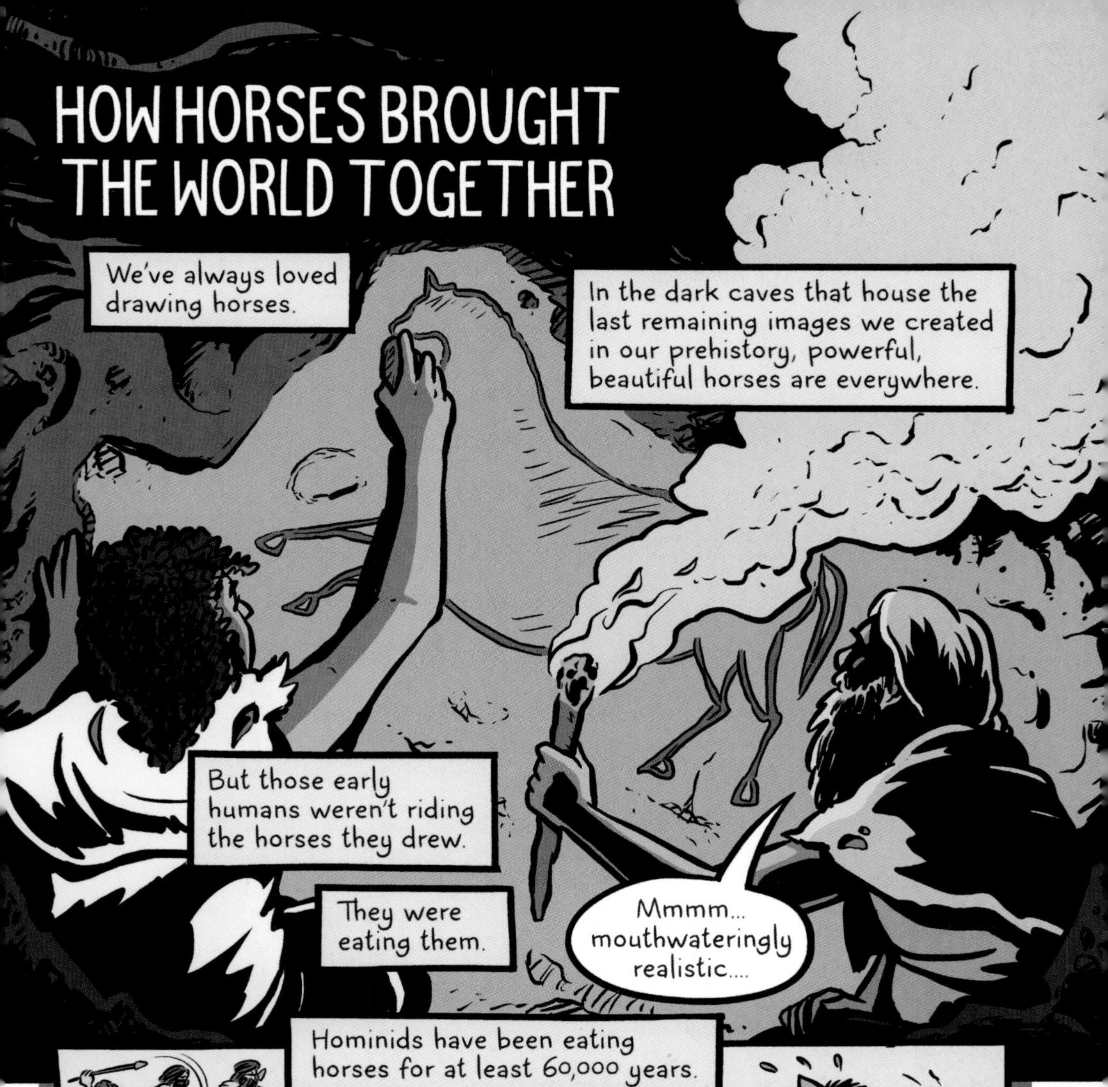
HOW HORSES BROUGHT THE WORLD TOGETHER
We've always loved drawing horses.
In the dark caves that house the last remaining images we created in our prehistory, powerful, beautiful horses are everywhere.
But those early humans weren't riding the horses they drew.
They were eating them.
Mmmm... mouthwateringly realistic....

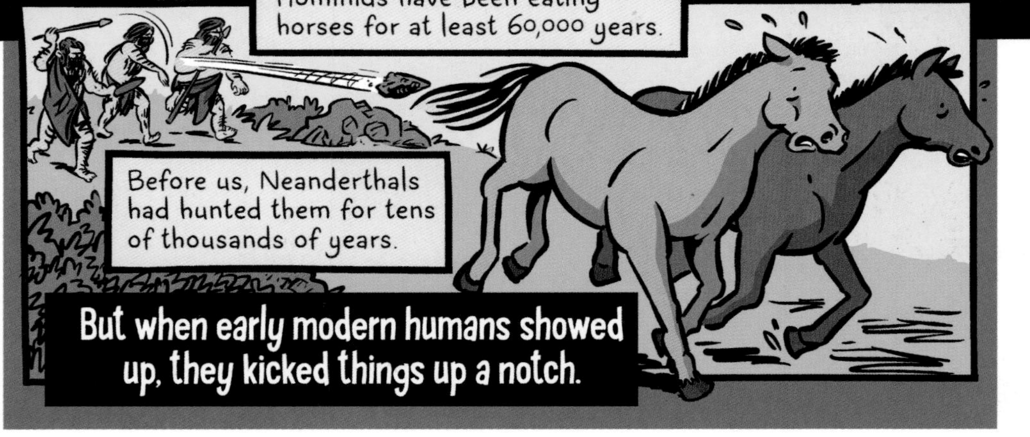
Hominids have been eating horses for at least 60,000 years.
Before us, Neanderthals had hunted them for tens of thousands of years.
But when early modern humans showed up, they kicked things up a notch.

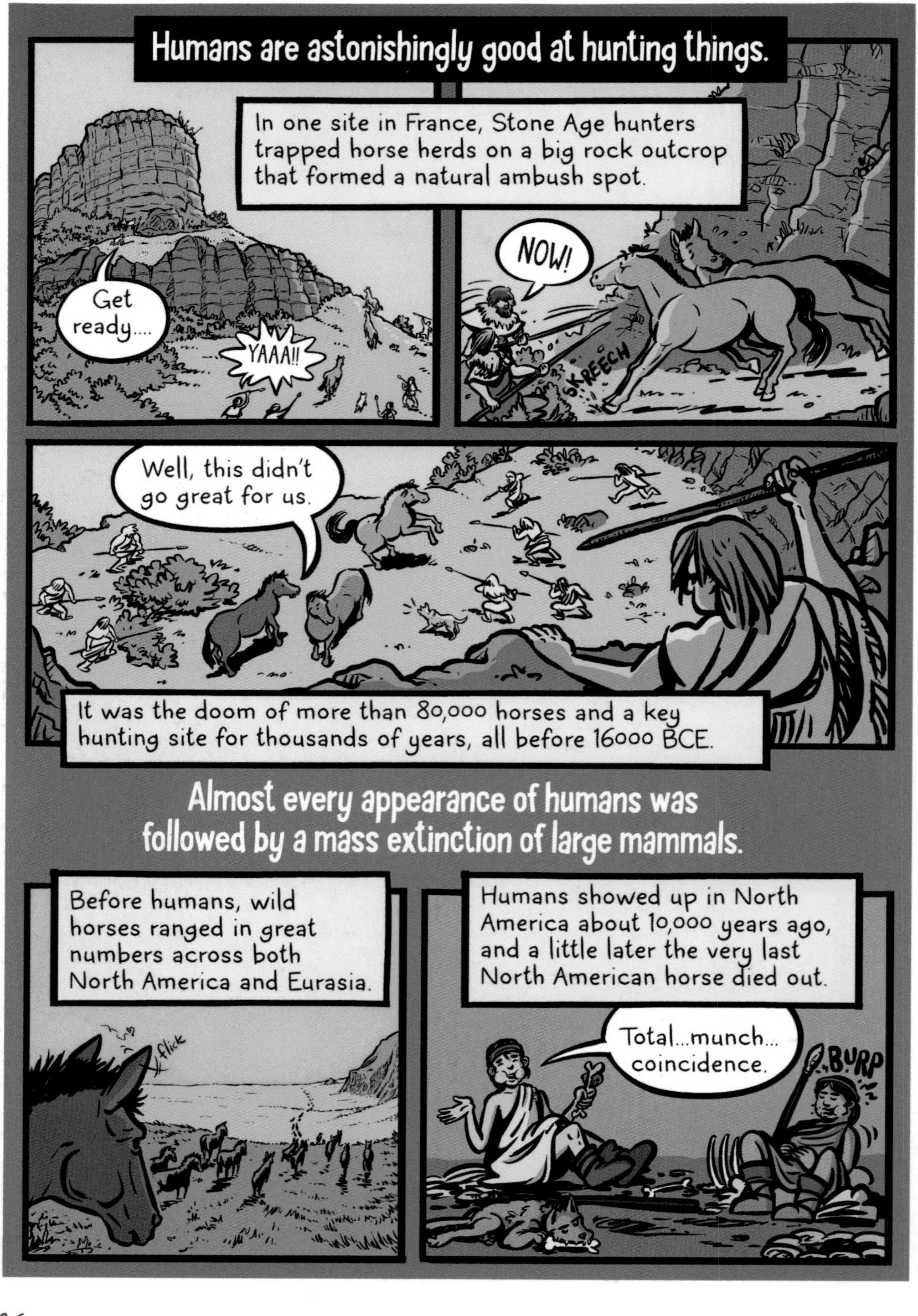
Humans are astonishingly good at hunting things.
In one site in France, Stone Age hunters trapped horse herds on a big rock outcrop that formed a natural ambush spot.
Get ready....
YAAA!!
NOW!
SKREECH
Well, this didn't go great for us.
It was the doom of more than 80,000 horses and a key hunting site for thousands of years, all before 16000 BCE.
Almost every appearance of humans was followed by a mass extinction of large mammals.
Before humans, wild horses ranged in great numbers across both North America and Eurasia.
flick
Humans showed up in North America about 10,000 years ago, and a little later the very last North American horse died out.
Total...munch... coincidence.
BURP

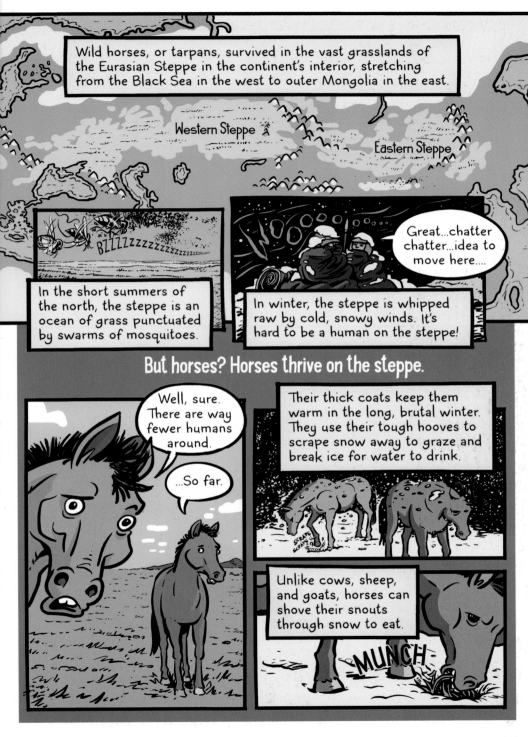

Wild horses, or tarpans, survived in the vast grasslands of the Eurasian Steppe in the continent's interior, stretching from the Black Sea in the west to outer Mongolia in the east.

Western Steppe

Eastern Steppe

BZZZZZZZZZZZZZZZZZZ

WOOOOOO

Great...chatter chatter...idea to move here....

In the short summers of the north, the steppe is an ocean of grass punctuated by swarms of mosquitoes.

In winter, the steppe is whipped raw by cold, snowy winds. It's hard to be a human on the steppe!

But horses? Horses thrive on the steppe.

Well, sure. There are way fewer humans around.

...So far.

Their thick coats keep them warm in the long, brutal winter. They use their tough hooves to scrape snow away to graze and break ice for water to drink.

Unlike cows, sheep, and goats, horses can shove their snouts through snow to eat.

MUNCH

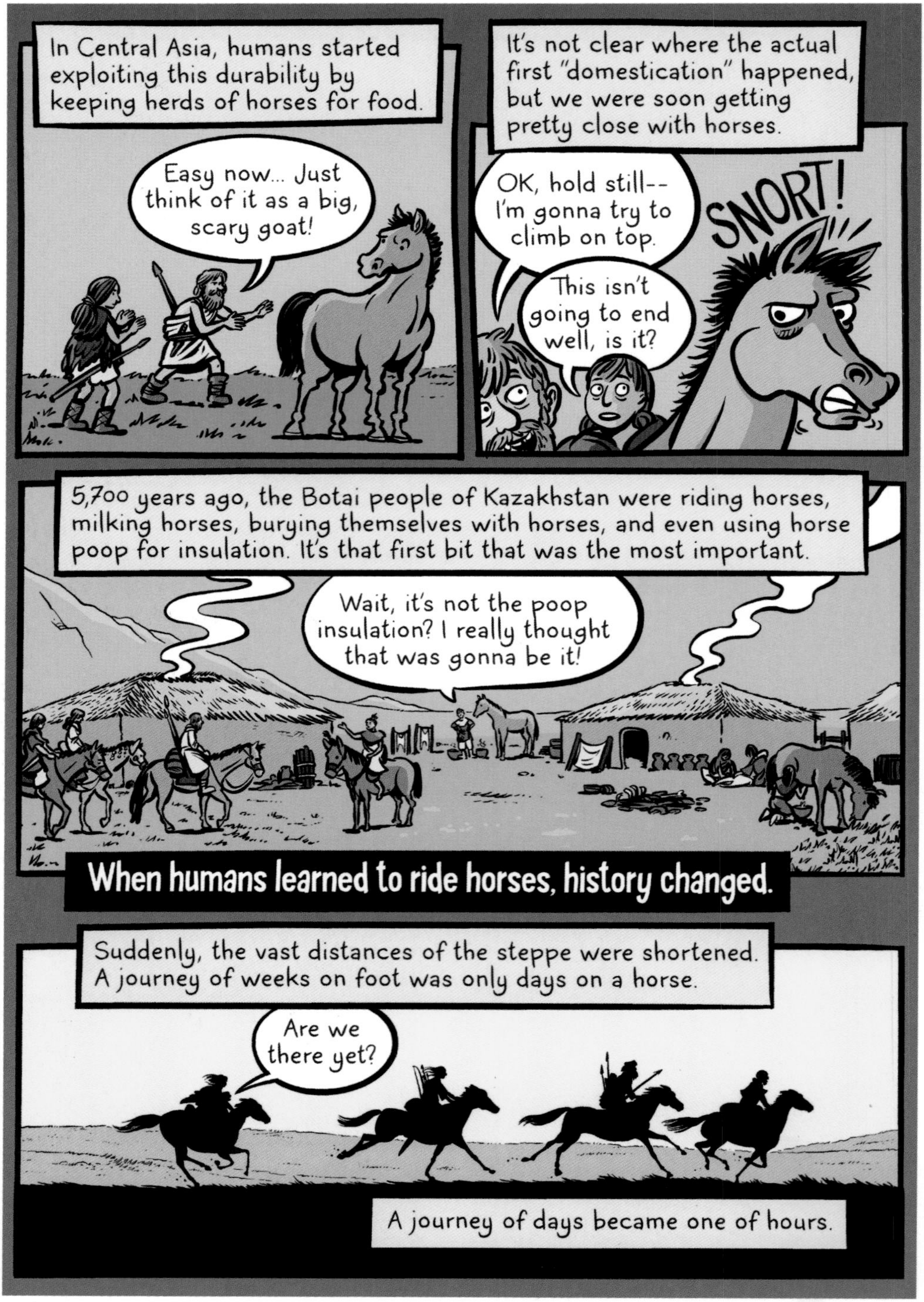

In Central Asia, humans started exploiting this durability by keeping herds of horses for food.
Easy now... Just think of it as a big, scary goat!
It's not clear where the actual first "domestication" happened, but we were soon getting pretty close with horses.
OK, hold still-- I'm gonna try to climb on top.
SNORT!
This isn't going to end well, is it?
5,700 years ago, the Botai people of Kazakhstan were riding horses, milking horses, burying themselves with horses, and even using horse poop for insulation. It's that first bit that was the most important.
Wait, it's not the poop insulation? I really thought that was gonna be it!
When humans learned to ride horses, history changed.
Suddenly, the vast distances of the steppe were shortened. A journey of weeks on foot was only days on a horse.
Are we there yet?
A journey of days became one of hours.

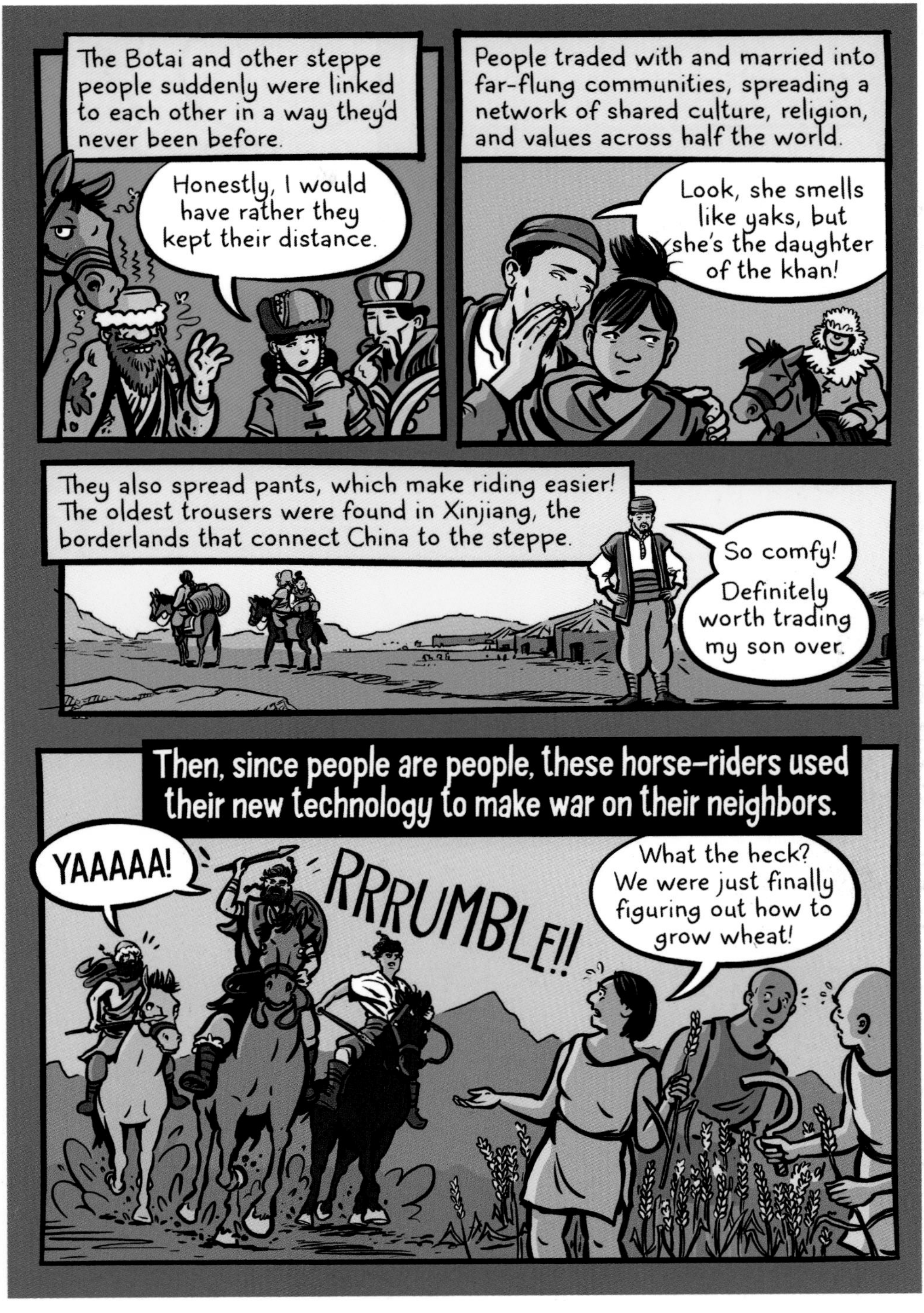
The Botai and other steppe people suddenly were linked to each other in a way they'd never been before.
Honestly, I would have rather they kept their distance.
People traded with and married into far-flung communities, spreading a network of shared culture, religion, and values across half the world.
Look, she smells like yaks, but she's the daughter of the khan!
They also spread pants, which make riding easier! The oldest trousers were found in Xinjiang, the borderlands that connect China to the steppe.
So comfy!
Definitely worth trading my son over.
Then, since people are people, these horse-riders used their new technology to make war on their neighbors.
YAAAAA!
RRRUMBLE!!
What the heck? We were just finally figuring out how to grow wheat!

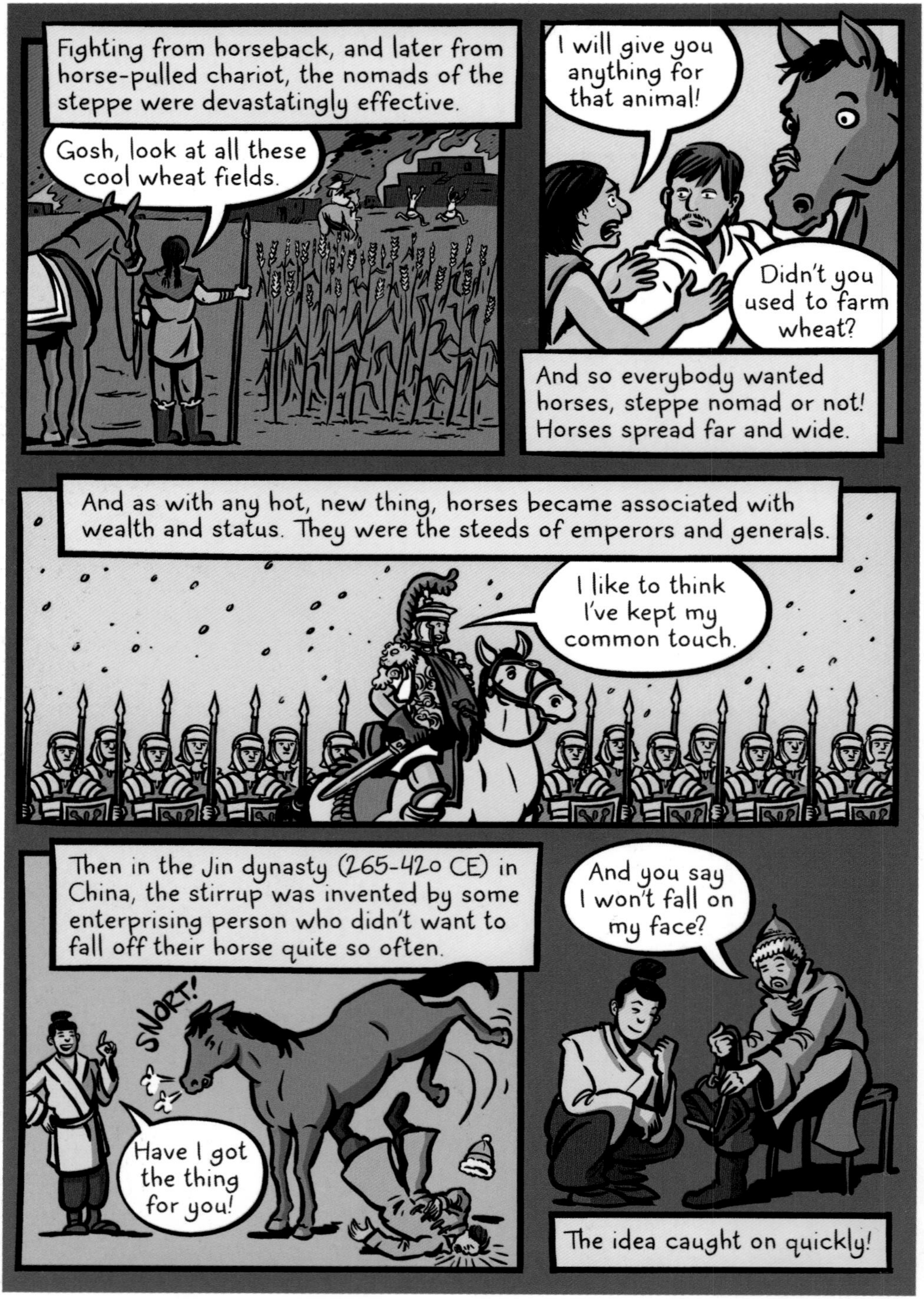

Fighting from horseback, and later from horse-pulled chariot, the nomads of the steppe were devastatingly effective.
Gosh, look at all these cool wheat fields.
I will give you anything for that animal!
Didn't you used to farm wheat?
And so everybody wanted horses, steppe nomad or not! Horses spread far and wide.
And as with any hot, new thing, horses became associated with wealth and status. They were the steeds of emperors and generals.
I like to think I've kept my common touch.
Then in the Jin dynasty (265-420 CE) in China, the stirrup was invented by some enterprising person who didn't want to fall off their horse quite so often.
SNORT!
Have I got the thing for you!
And you say I won't fall on my face?
The idea caught on quickly!

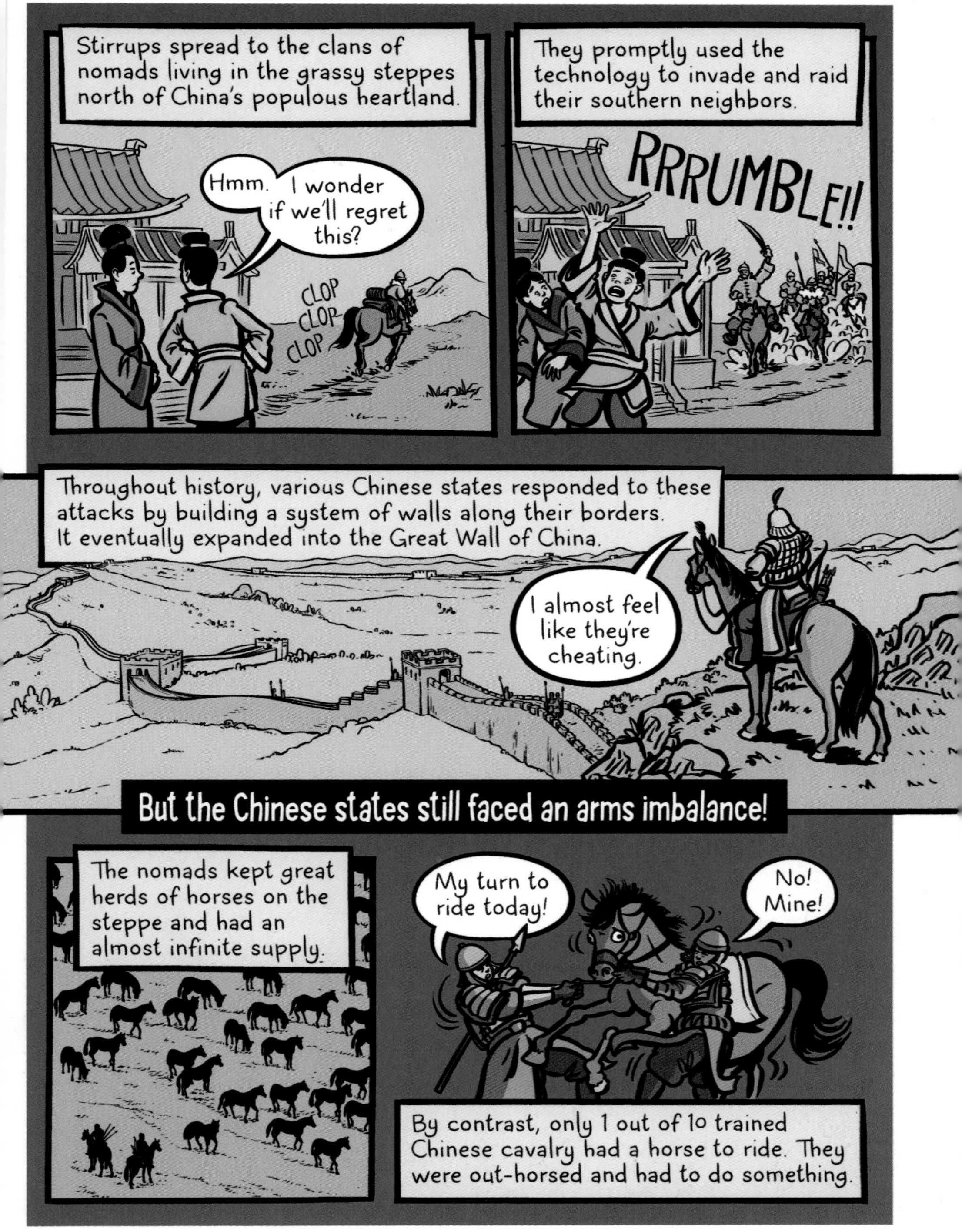

Stirrups spread to the clans of nomads living in the grassy steppes north of China's populous heartland.
Hmm. I wonder if we'll regret this?
CLOP CLOP CLOP
They promptly used the technology to invade and raid their southern neighbors.
RRRUMBLE!!
Throughout history, various Chinese states responded to these attacks by building a system of walls along their borders. It eventually expanded into the Great Wall of China.
I almost feel like they're cheating.
But the Chinese states still faced an arms imbalance!
The nomads kept great herds of horses on the steppe and had an almost infinite supply.
My turn to ride today!
No! Mine!
By contrast, only 1 out of 10 trained Chinese cavalry had a horse to ride. They were out-horsed and had to do something.

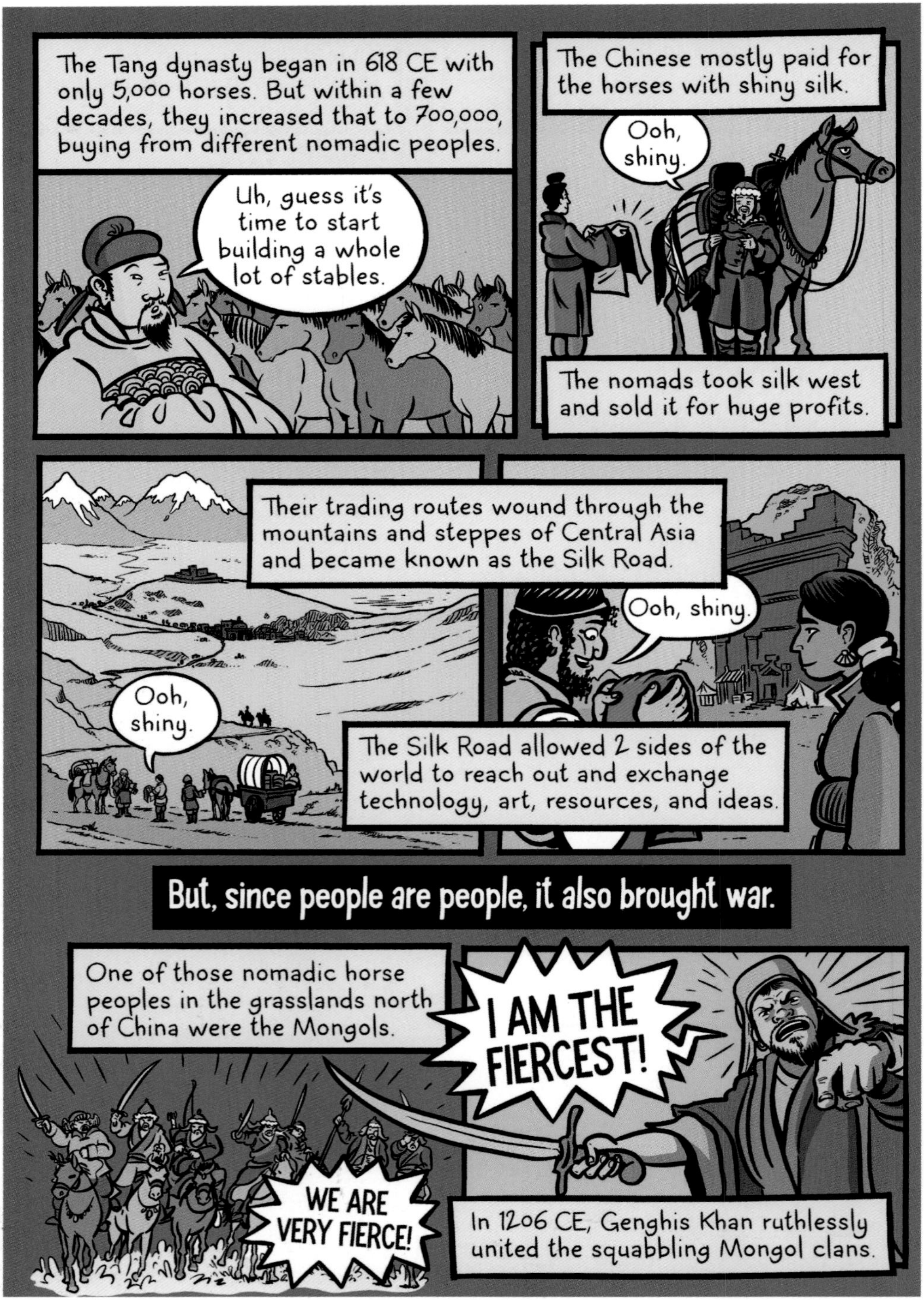

The Tang dynasty began in 618 CE with only 5,000 horses. But within a few decades, they increased that to 700,000, buying from different nomadic peoples.
Uh, guess it's time to start building a whole lot of stables.
The Chinese mostly paid for the horses with shiny silk.
Ooh, shiny.
The nomads took silk west and sold it for huge profits.
Their trading routes wound through the mountains and steppes of Central Asia and became known as the Silk Road.
Ooh, shiny.
Ooh, shiny.
The Silk Road allowed 2 sides of the world to reach out and exchange technology, art, resources, and ideas.
But, since people are people, it also brought war.
One of those nomadic horse peoples in the grasslands north of China were the Mongols.
I AM THE FIERCEST!
WE ARE VERY FIERCE!
In 1206 CE, Genghis Khan ruthlessly united the squabbling Mongol clans.

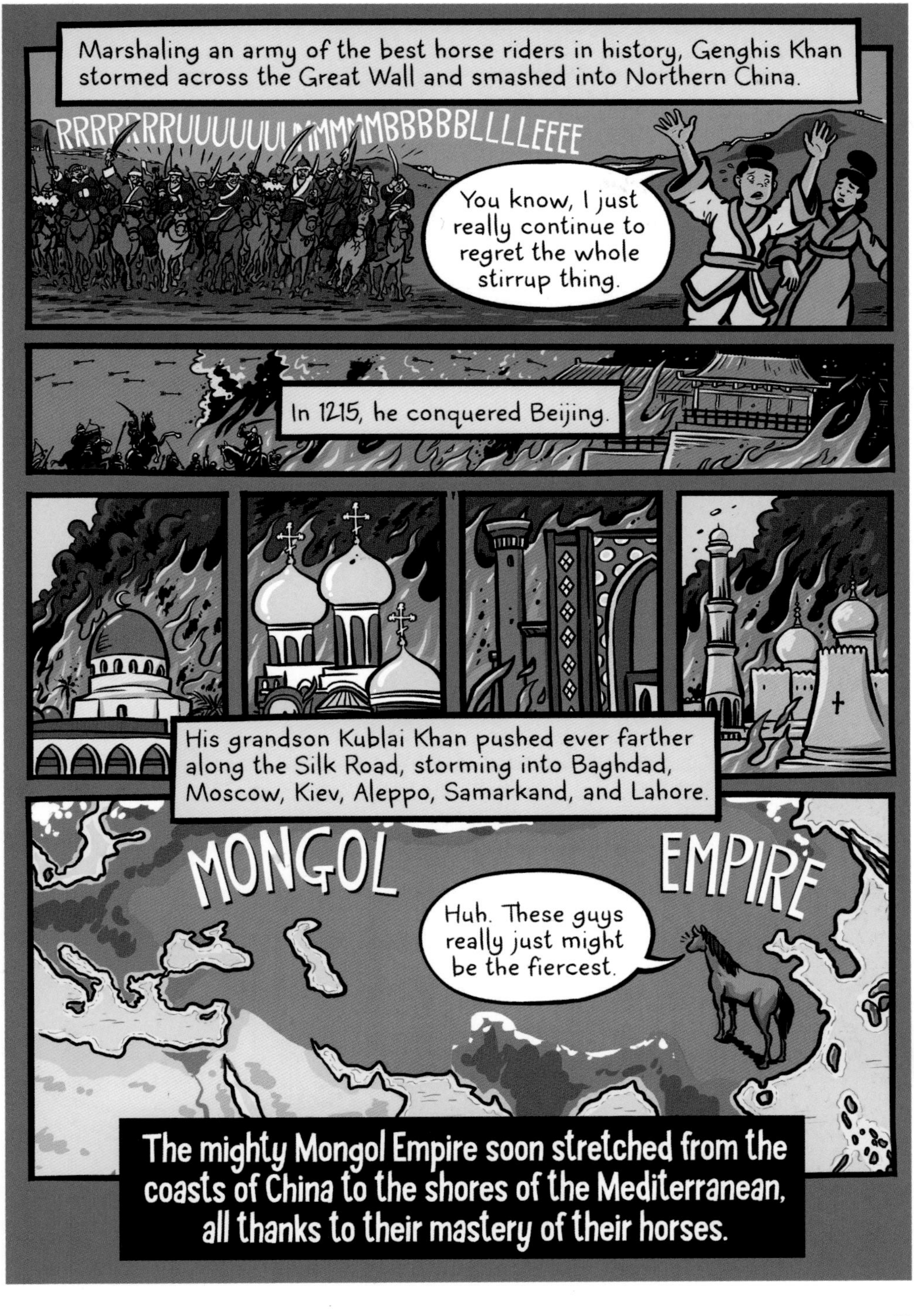
Marshaling an army of the best horse riders in history, Genghis Khan stormed across the Great Wall and smashed into Northern China.
RRRRRRRUUUUUUUMMMMMBBBBBLLLLEEEE
You know, I just really continue to regret the whole stirrup thing.
In 1215, he conquered Beijing.
His grandson Kublai Khan pushed ever farther along the Silk Road, storming into Baghdad, Moscow, Kiev, Aleppo, Samarkand, and Lahore.
MONGOL EMPIRE
Huh. These guys really just might be the fiercest.
The mighty Mongol Empire soon stretched from the coasts of China to the shores of the Mediterranean, all thanks to their mastery of their horses.

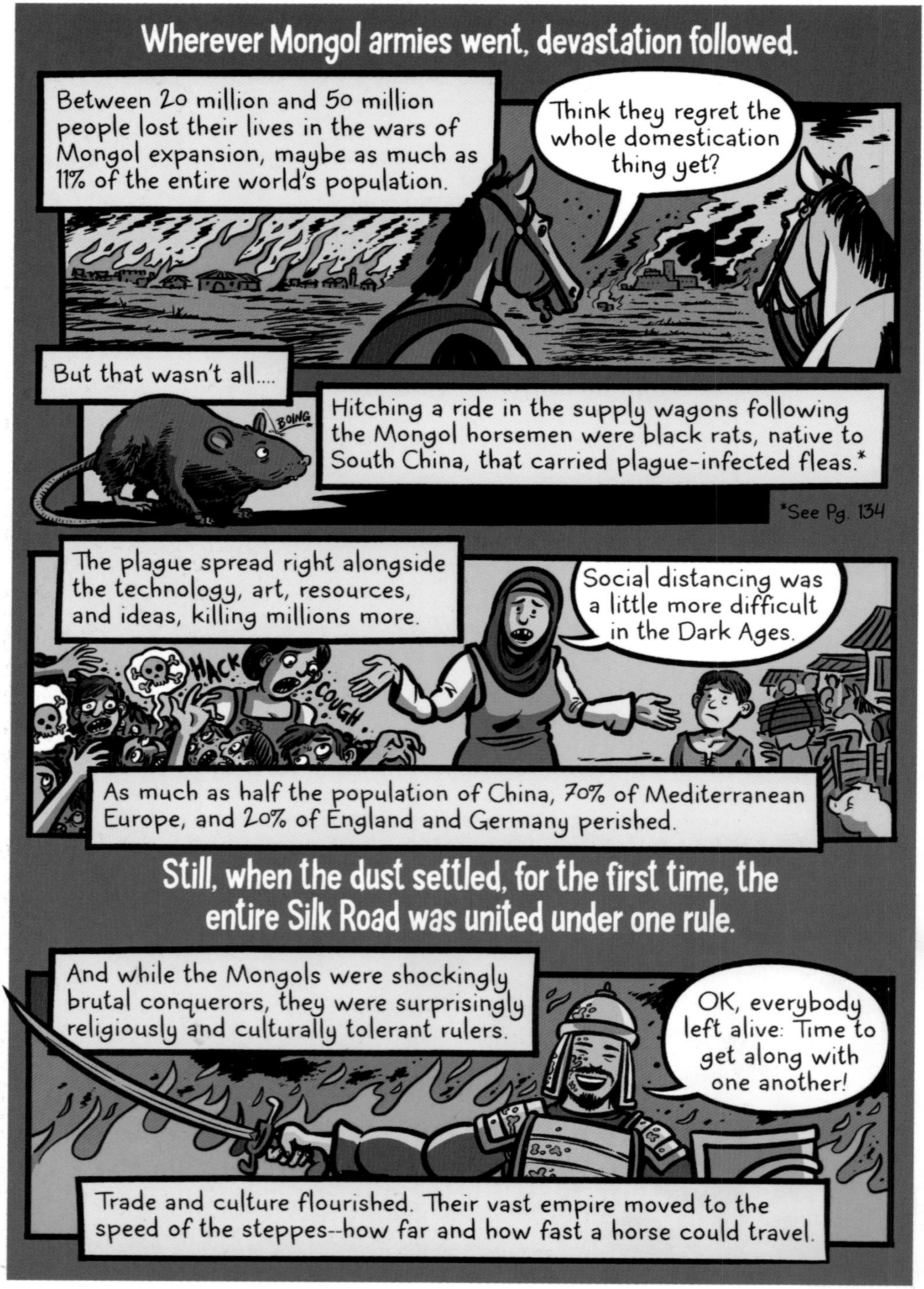
Wherever Mongol armies went, devastation followed.
Between 20 million and 50 million people lost their lives in the wars of Mongol expansion, maybe as much as 11% of the entire world's population.
Think they regret the whole domestication thing yet?
But that wasn't all....
BOING
Hitching a ride in the supply wagons following the Mongol horsemen were black rats, native to South China, that carried plague-infected fleas.*
*See Pg. 134
The plague spread right alongside the technology, art, resources, and ideas, killing millions more.
Social distancing was a little more difficult in the Dark Ages.
HACK
COUGH
As much as half the population of China, 70% of Mediterranean Europe, and 20% of England and Germany perished.
Still, when the dust settled, for the first time, the entire Silk Road was united under one rule.
And while the Mongols were shockingly brutal conquerors, they were surprisingly religiously and culturally tolerant rulers.
OK, everybody left alive: Time to get along with one another!
Trade and culture flourished. Their vast empire moved to the speed of the steppes--how far and how fast a horse could travel.

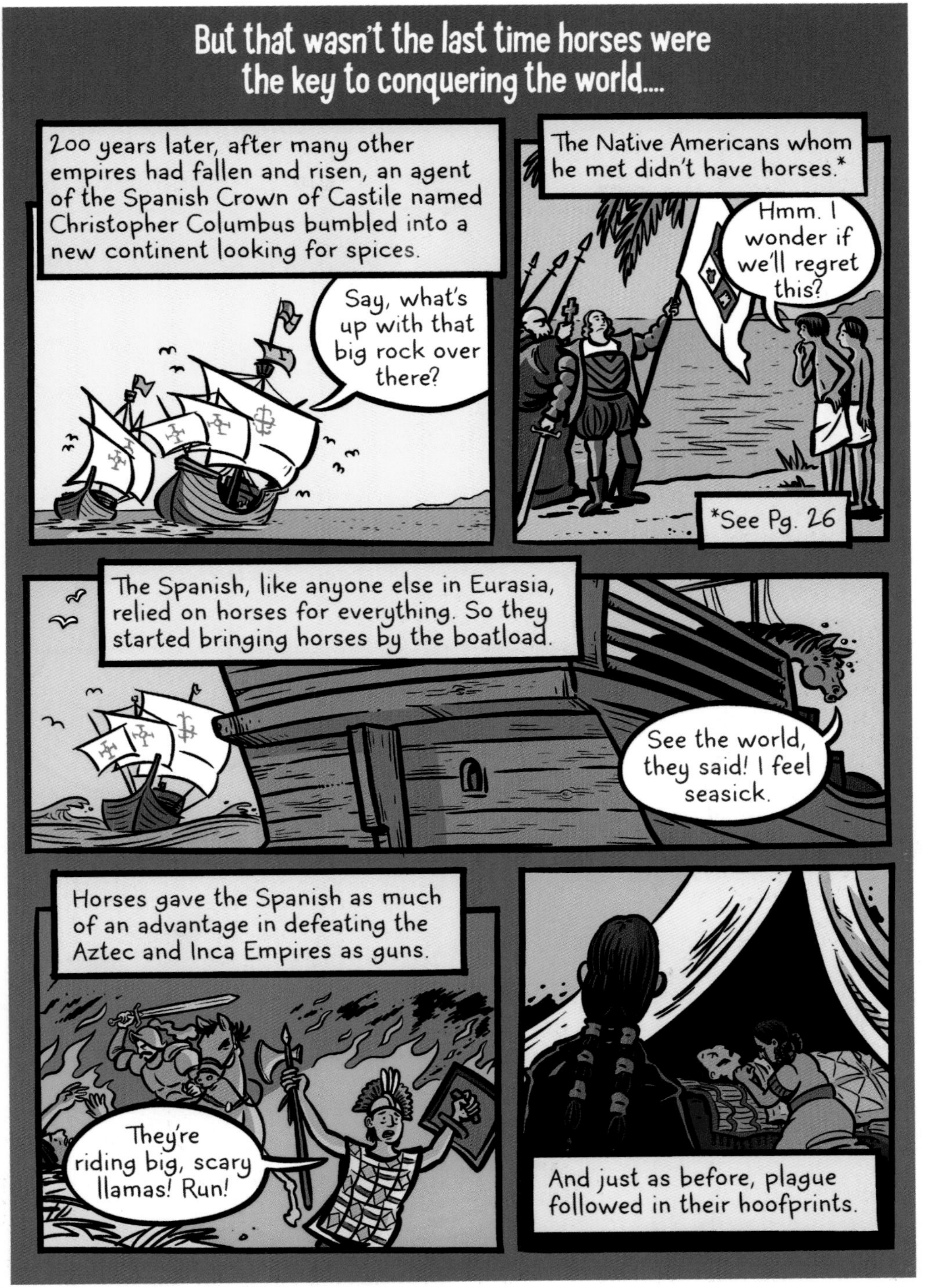
But that wasn't the last time horses were the key to conquering the world....
200 years later, after many other empires had fallen and risen, an agent of the Spanish Crown of Castile named Christopher Columbus bumbled into a new continent looking for spices.
Say, what's up with that big rock over there?
The Native Americans whom he met didn't have horses.*
Hmm. I wonder if we'll regret this?
*See Pg. 26
The Spanish, like anyone else in Eurasia, relied on horses for everything. So they started bringing horses by the boatload.
See the world, they said! I feel seasick.
Horses gave the Spanish as much of an advantage in defeating the Aztec and Inca Empires as guns.
They're riding big, scary llamas! Run!
And just as before, plague followed in their hoofprints.

Long before Europeans reached North America in any number, the effect of their contact was already being felt.

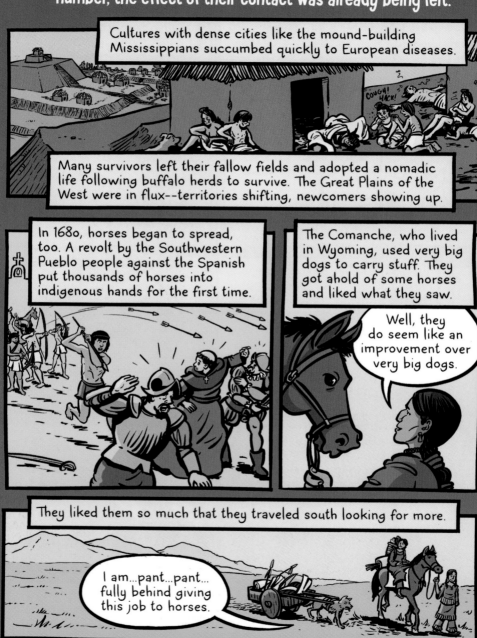

Cultures with dense cities like the mound-building Mississippians succumbed quickly to European diseases.

COUGH! HACK!

Many survivors left their fallow fields and adopted a nomadic life following buffalo herds to survive. The Great Plains of the West were in flux--territories shifting, newcomers showing up.

In 1680, horses began to spread, too. A revolt by the Southwestern Pueblo people against the Spanish put thousands of horses into indigenous hands for the first time.

The Comanche, who lived in Wyoming, used very big dogs to carry stuff. They got ahold of some horses and liked what they saw.

Well, they do seem like an improvement over very big dogs.

They liked them so much that they traveled south looking for more.

I am...pant...pant... fully behind giving this job to horses.

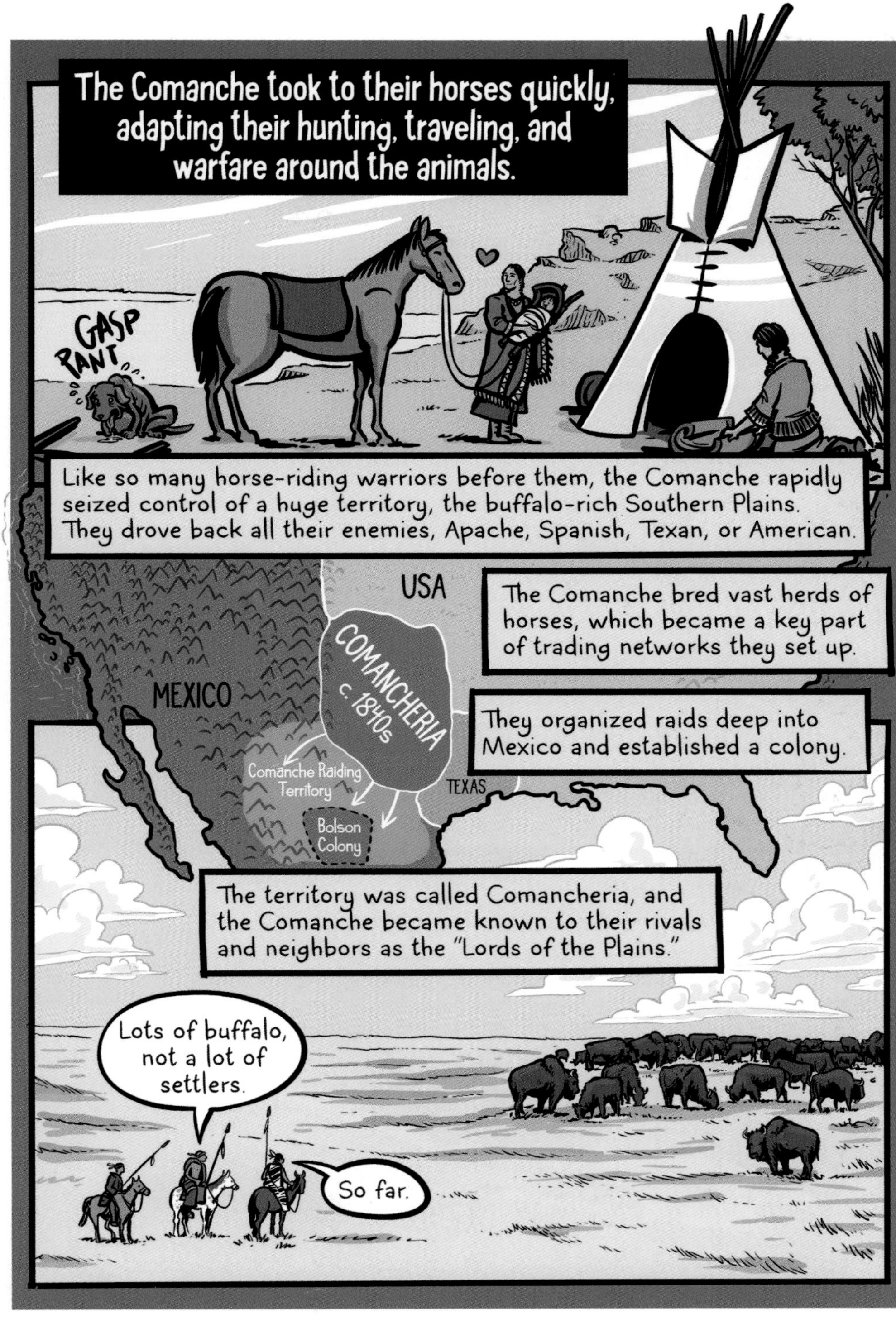

The Comanche took to their horses quickly, adapting their hunting, traveling, and warfare around the animals.
GASP
PANT
Like so many horse-riding warriors before them, the Comanche rapidly seized control of a huge territory, the buffalo-rich Southern Plains. They drove back all their enemies, Apache, Spanish, Texan, or American.
USA
The Comanche bred vast herds of horses, which became a key part of trading networks they set up.
MEXICO
COMANCHERIA c. 1840s
They organized raids deep into Mexico and established a colony.
Comanche Raiding Territory
TEXAS
Bolson Colony
The territory was called Comancheria, and the Comanche became known to their rivals and neighbors as the "Lords of the Plains."
Lots of buffalo, not a lot of settlers.
So far.

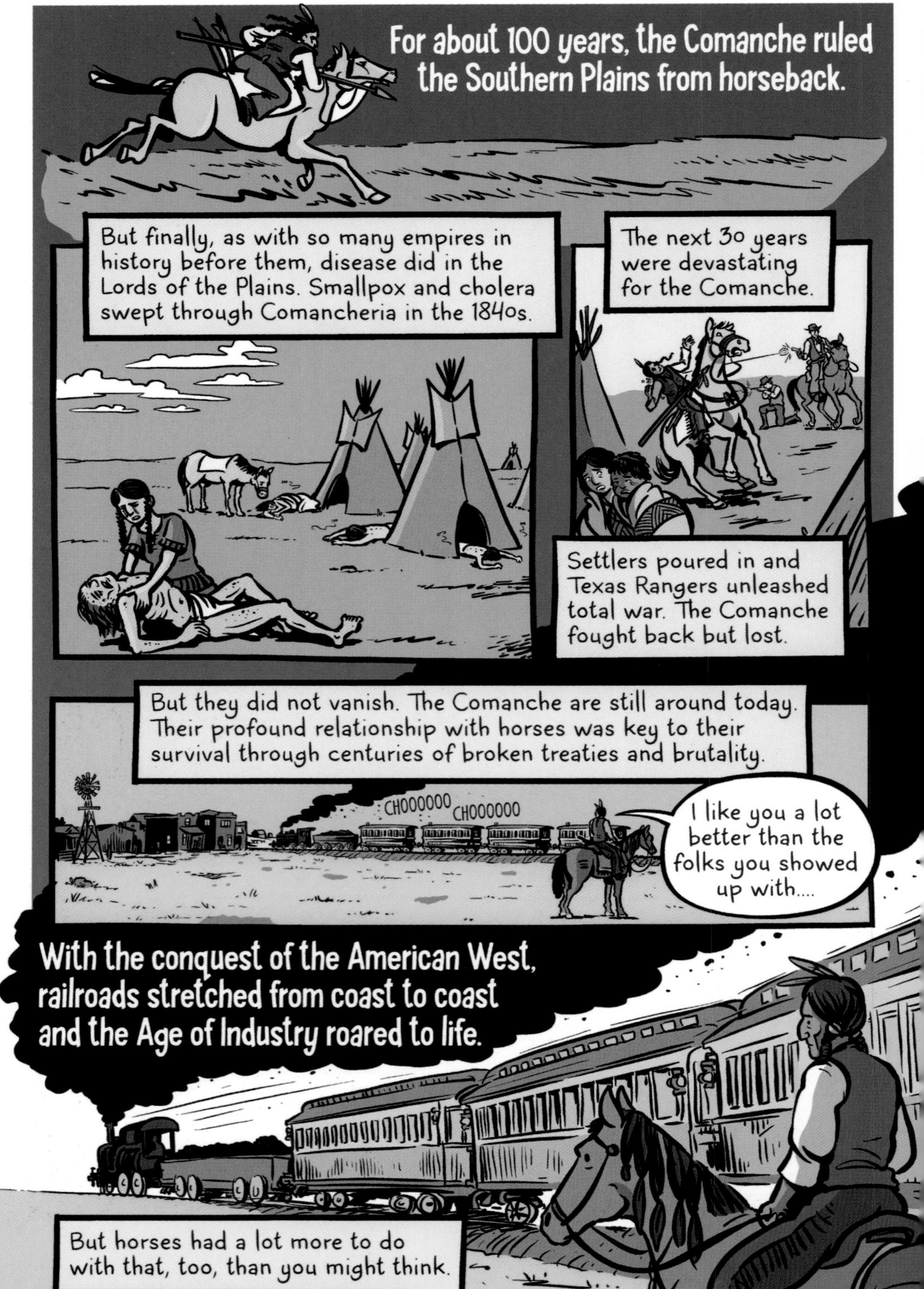

For about 100 years, the Comanche ruled the Southern Plains from horseback.
But finally, as with so many empires in history before them, disease did in the Lords of the Plains. Smallpox and cholera swept through Comancheria in the 1840s.
The next 30 years were devastating for the Comanche.
Settlers poured in and Texas Rangers unleashed total war. The Comanche fought back but lost.
But they did not vanish. The Comanche are still around today. Their profound relationship with horses was key to their survival through centuries of broken treaties and brutality.
CHOOOOOO
CHOOOOOO
I like you a lot better than the folks you showed up with....
With the conquest of the American West, railroads stretched from coast to coast and the Age of Industry roared to life.
But horses had a lot more to do with that, too, than you might think.

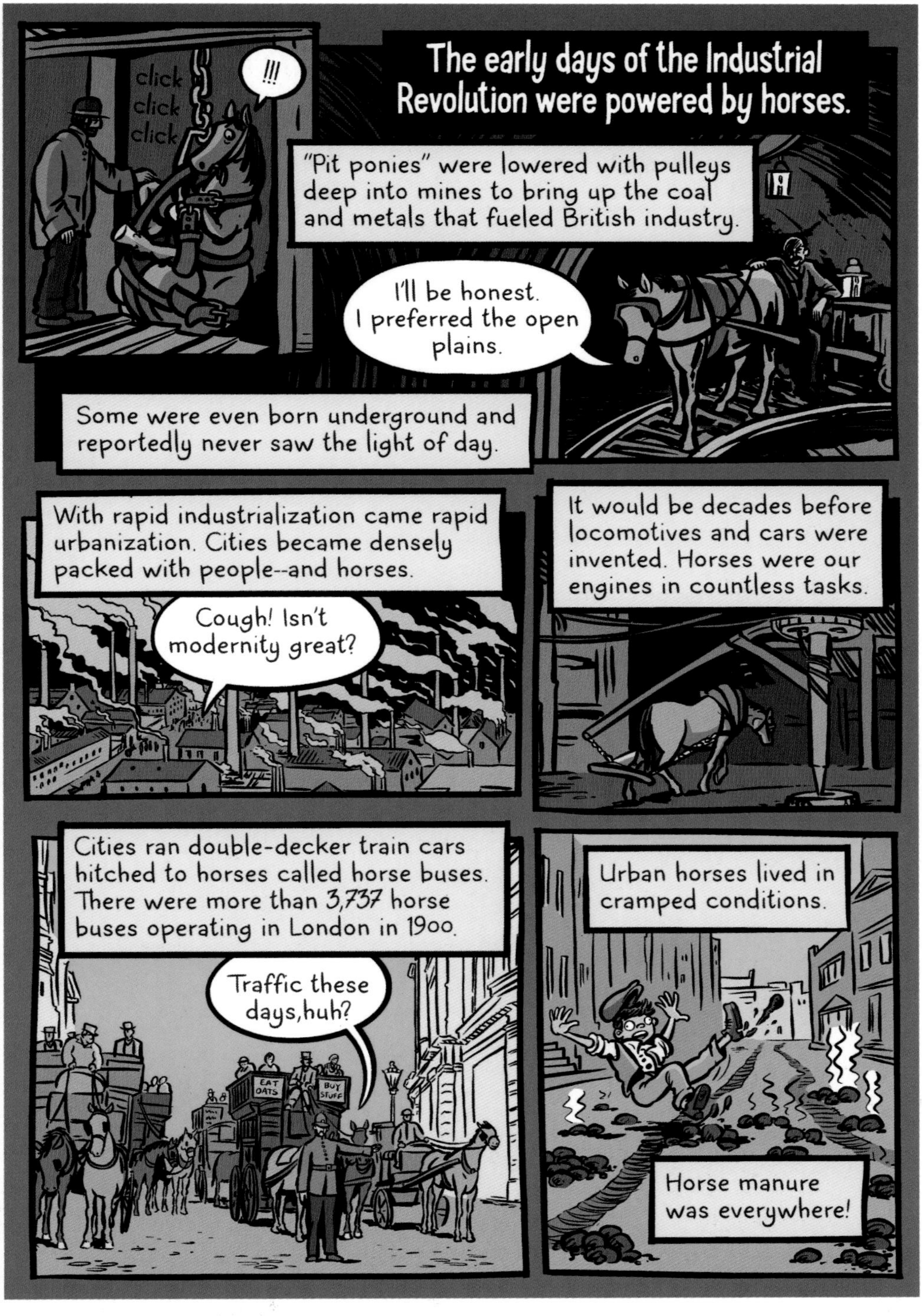
The early days of the Industrial Revolution were powered by horses.
click click click
!!!
"Pit ponies" were lowered with pulleys deep into mines to bring up the coal and metals that fueled British industry.
I'll be honest. I preferred the open plains.
Some were even born underground and reportedly never saw the light of day.
With rapid industrialization came rapid urbanization. Cities became densely packed with people--and horses.
Cough! Isn't modernity great?
It would be decades before locomotives and cars were invented. Horses were our engines in countless tasks.
Cities ran double-decker train cars hitched to horses called horse buses. There were more than 3,737 horse buses operating in London in 1900.
Traffic these days,huh?
EAT OATS
BUY STUFF
Urban horses lived in cramped conditions.
Horse manure was everywhere!

Meanwhile, horses continued to be key to human warfare.

But with the development of rapid-firing guns, horses and humans began to die in obscene numbers.

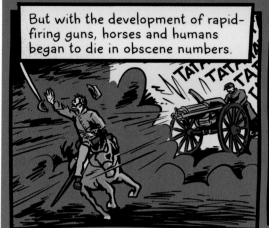

Half of the warhorses in the American Civil War died.

350,000 British horses died in the Boer War.

Then the invention of the automobile suddenly changed things. Jeeps replaced warhorses, sports cars replaced racehorses, trucks replaced mules, and actual buses replaced horse buses.

I'm OK with this; the war stuff was getting grim!

The global horse population collapsed in the 20th century, and they vanished from cities.

As farm equipment became increasingly industrialized, the number of horses dwindled in the countryside, too.

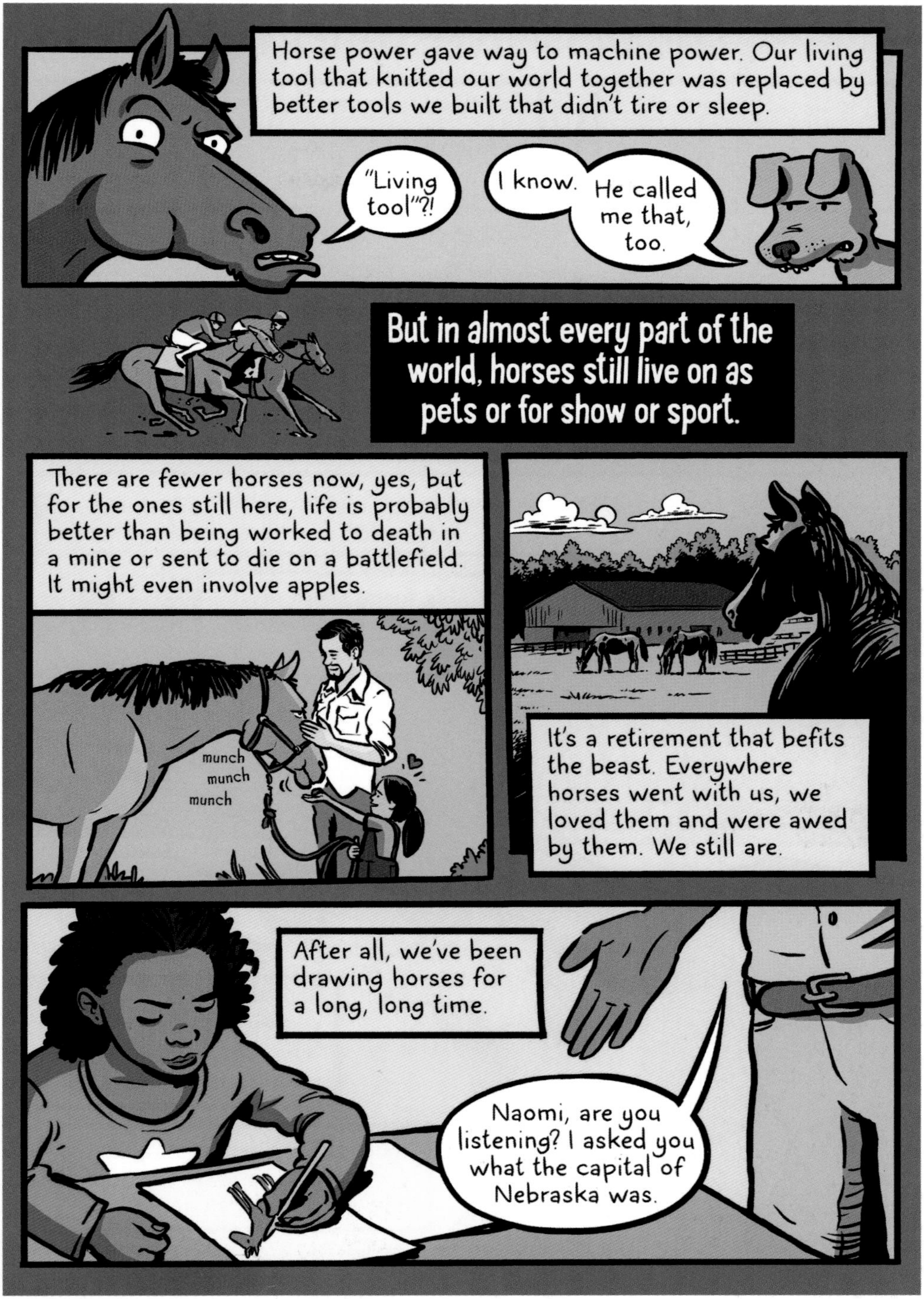
Horse power gave way to machine power. Our living tool that knitted our world together was replaced by better tools we built that didn't tire or sleep.
"Living tool"?!
I know.
He called me that, too.
But in almost every part of the world, horses still live on as pets or for show or sport.
There are fewer horses now, yes, but for the ones still here, life is probably better than being worked to death in a mine or sent to die on a battlefield. It might even involve apples.
munch munch munch
It's a retirement that befits the beast. Everywhere horses went with us, we loved them and were awed by them. We still are.
After all, we've been drawing horses for a long, long time.
Naomi, are you listening? I asked you what the capital of Nebraska was.

THE ULTIMATE FARM TOOL

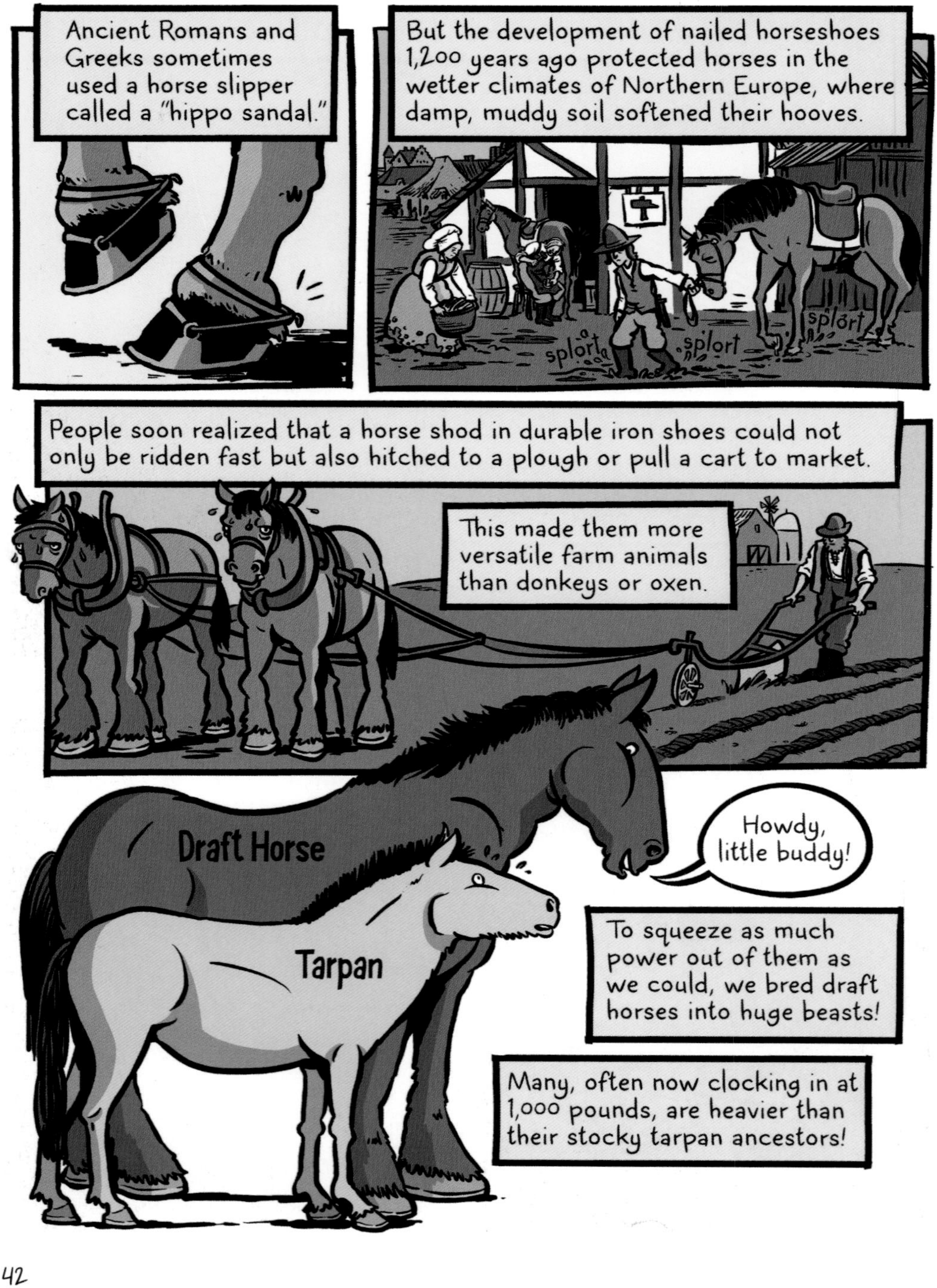
Ancient Romans and Greeks sometimes used a horse slipper called a "hippo sandal."
But the development of nailed horseshoes 1,200 years ago protected horses in the wetter climates of Northern Europe, where damp, muddy soil softened their hooves.
splort
splort
splort
People soon realized that a horse shod in durable iron shoes could not only be ridden fast but also hitched to a plough or pull a cart to market.
This made them more versatile farm animals than donkeys or oxen.
Draft Horse
Tarpan
Howdy, little buddy!
To squeeze as much power out of them as we could, we bred draft horses into huge beasts!
Many, often now clocking in at 1,000 pounds, are heavier than their stocky tarpan ancestors!

WHAT HAPPENED TO THE TARPANS?

One of the first things people used domestic horses for was hunting the Eurasian tarpan, the wild ancestor of the domesticated horse, into oblivion in most of their range.

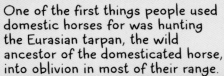

Dude! What is wrong with you?

The last reported wild tarpans were finally captured in Poland's Białowieża Forest in 1806 and interbred into local horse herds.

Do the new guys seem a little rough around the edges to you?

Sigh.

The last tarpan died at a Moscow zoo in 1909.

And with that, the wild horse was extinct in both Eurasia and North America.

Then, in the 1930s, a pair of Nazi zoologist brothers attempted to re-create tarpans by selectively breeding horses to fit their demented worldview of a "pure" premodern Europe.

I bet nobody expected Nazi zoologists!

The tarpan-like Nazi horses were sent to a game-hunting preserve in--you guessed it--Białowieża Forest.

"Tarpan-like Nazi horse"? My politics are quite progressive, I'll have you know!

A small population survived the war and still runs wild today.

HAMSTERS

(Golden/Syrian Hamster) *Mesocricetus auratus*

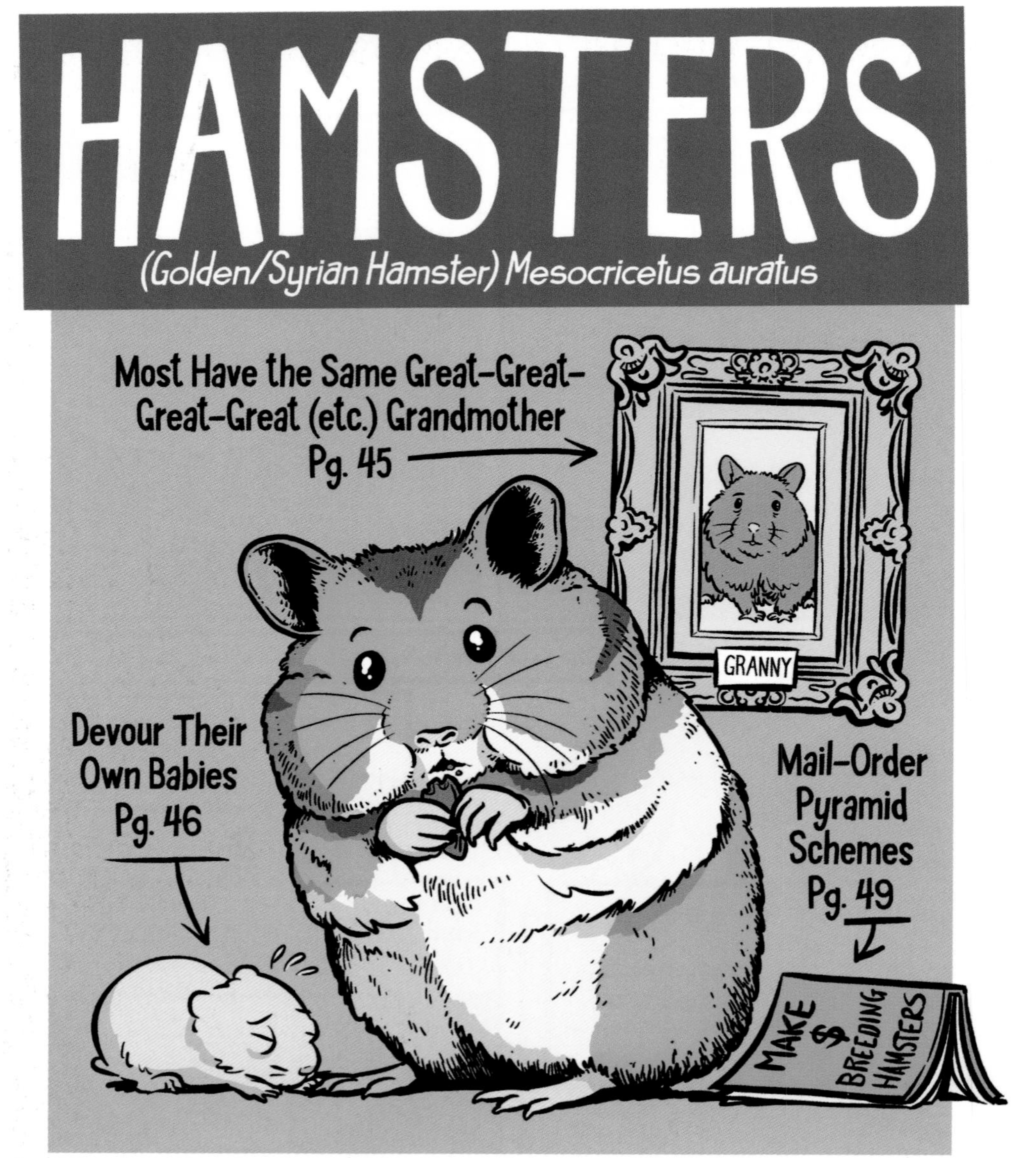

Most Have the Same Great-Great-Great-Great (etc.) Grandmother
Pg. 45
GRANNY
Devour Their Own Babies
Pg. 46
Mail-Order Pyramid Schemes
Pg. 49
MAKE $ BREEDING HAMSTERS

Size: 6 in (length)

Weight: 3-5 oz

Maximum Cheek Capacity: 20% of body weight

Diet: Seeds, plants, pellets, the occasional insect

(Likely) **Wild Origin:**

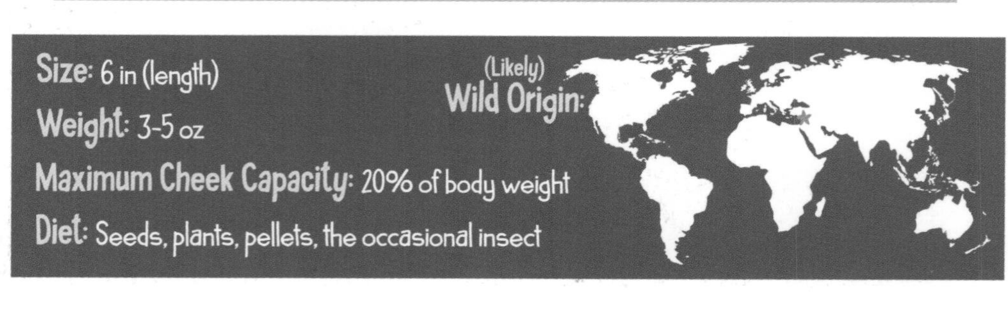

ALMOST ALL PET HAMSTERS DESCEND FROM THE SAME MOM

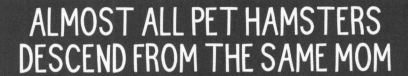

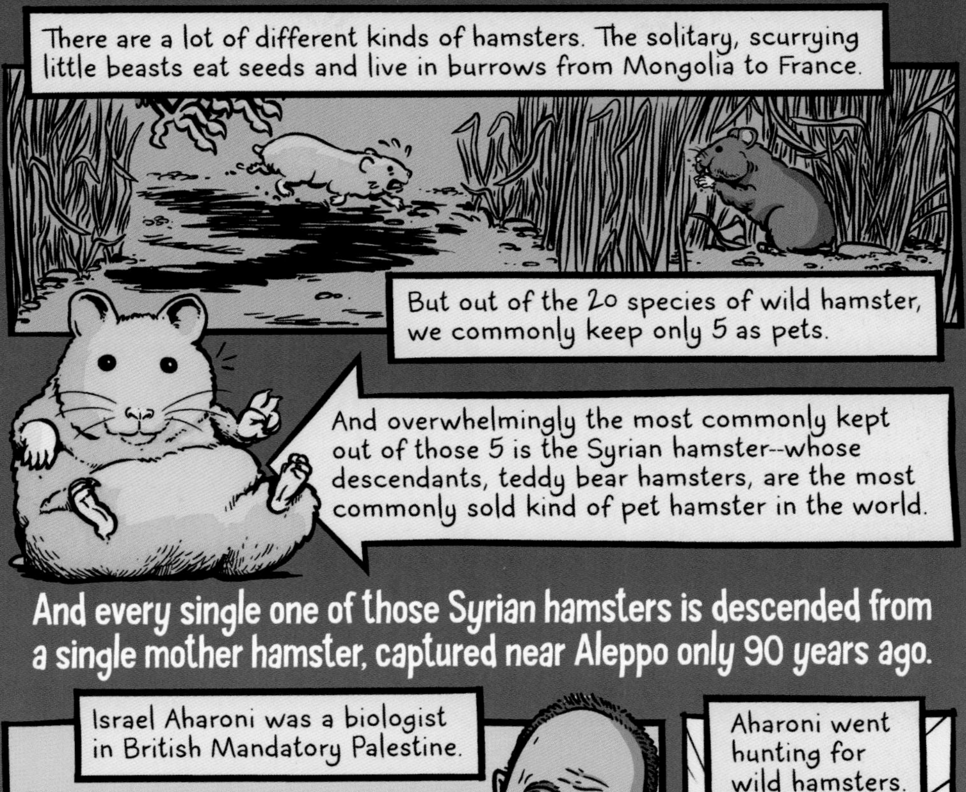
There are a lot of different kinds of hamsters. The solitary, scurrying little beasts eat seeds and live in burrows from Mongolia to France.
But out of the 20 species of wild hamster, we commonly keep only 5 as pets.
And overwhelmingly the most commonly kept out of those 5 is the Syrian hamster--whose descendants, teddy bear hamsters, are the most commonly sold kind of pet hamster in the world.

And every single one of those Syrian hamsters is descended from a single mother hamster, captured near Aleppo only 90 years ago.

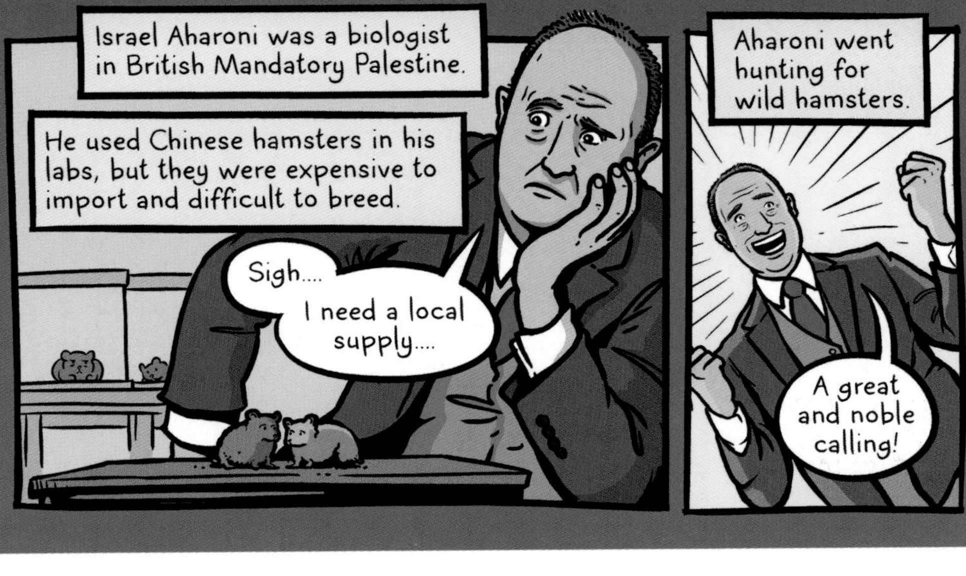
Israel Aharoni was a biologist in British Mandatory Palestine.
He used Chinese hamsters in his labs, but they were expensive to import and difficult to breed.
Sigh.... I need a local supply....
Aharoni went hunting for wild hamsters.
A great and noble calling!

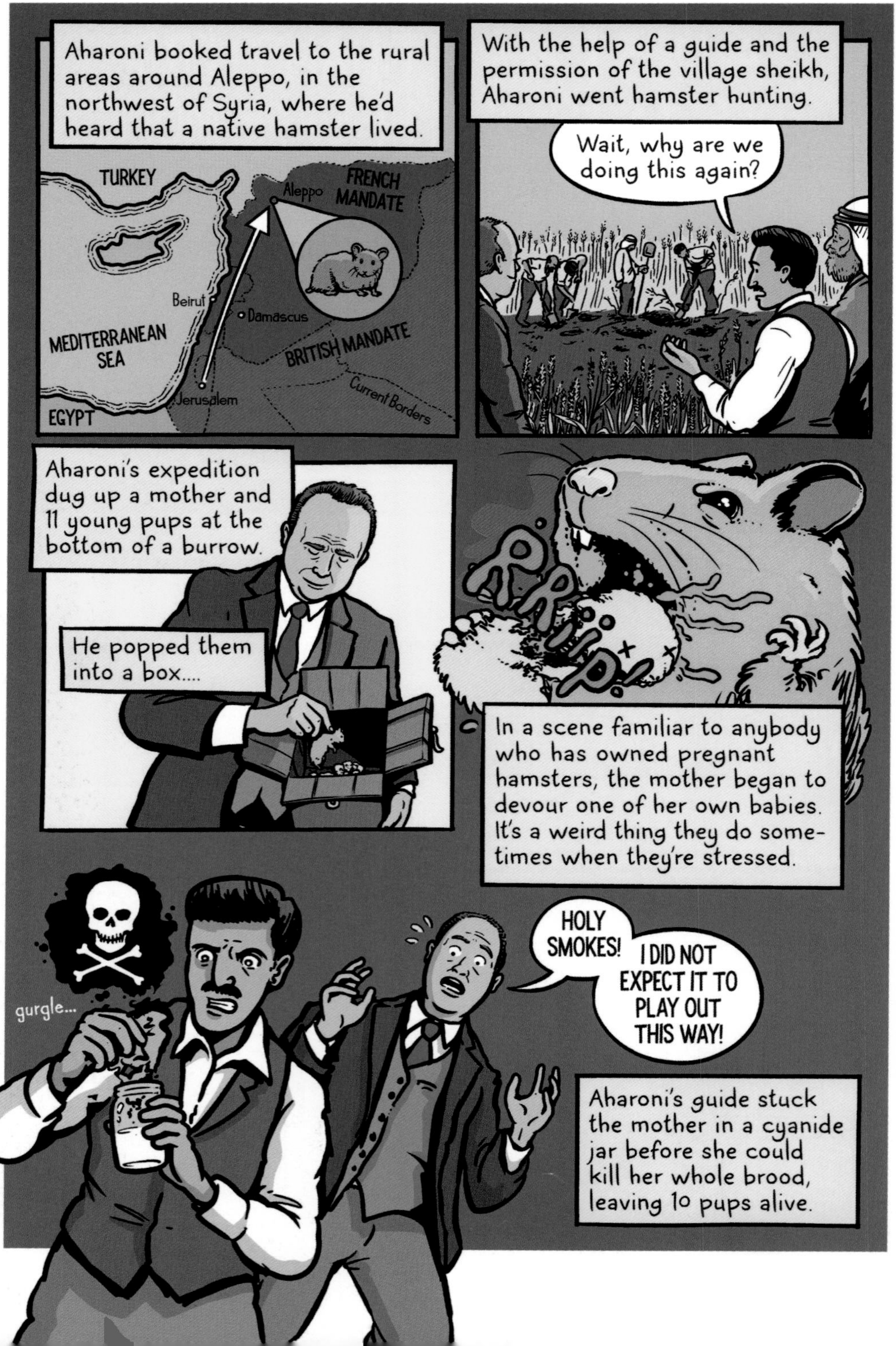
Aharoni booked travel to the rural areas around Aleppo, in the northwest of Syria, where he'd heard that a native hamster lived.
TURKEY
FRENCH MANDATE
Aleppo
Beirut
Damascus
MEDITERRANEAN SEA
BRITISH MANDATE
Jerusalem
Current Borders
EGYPT
With the help of a guide and the permission of the village sheikh, Aharoni went hamster hunting.
Wait, why are we doing this again?
Aharoni's expedition dug up a mother and 11 young pups at the bottom of a burrow.
He popped them into a box....
RRRIP!
In a scene familiar to anybody who has owned pregnant hamsters, the mother began to devour one of her own babies. It's a weird thing they do sometimes when they're stressed.
gurgle...
HOLY SMOKES!
I DID NOT EXPECT IT TO PLAY OUT THIS WAY!
Aharoni's guide stuck the mother in a cyanide jar before she could kill her whole brood, leaving 10 pups alive.

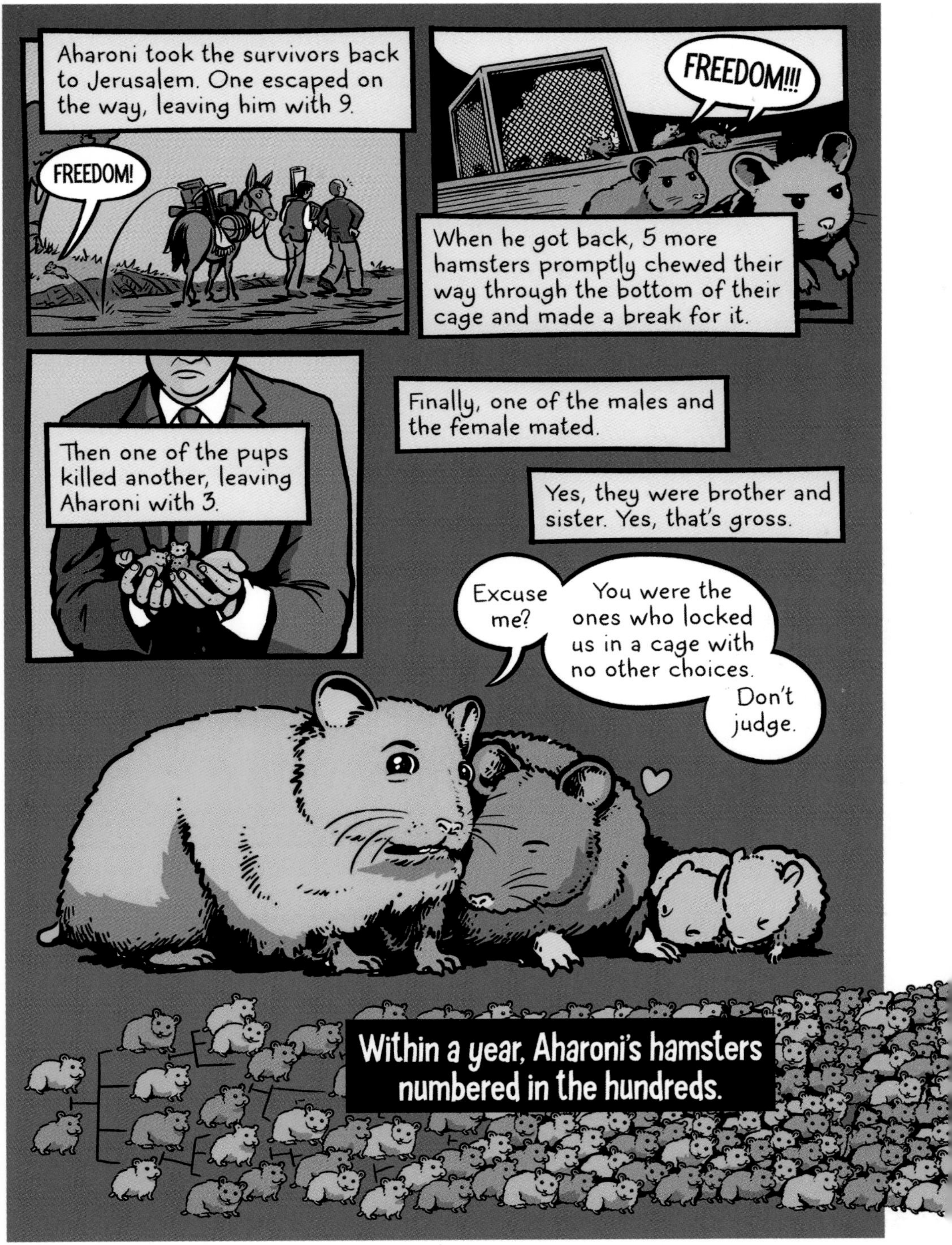

Aharoni took the survivors back to Jerusalem. One escaped on the way, leaving him with 9.
FREEDOM!
FREEDOM!!!
When he got back, 5 more hamsters promptly chewed their way through the bottom of their cage and made a break for it.
Then one of the pups killed another, leaving Aharoni with 3.
Finally, one of the males and the female mated.
Yes, they were brother and sister. Yes, that's gross.
Excuse me?
You were the ones who locked us in a cage with no other choices.
Don't judge.
Within a year, Aharoni's hamsters numbered in the hundreds.

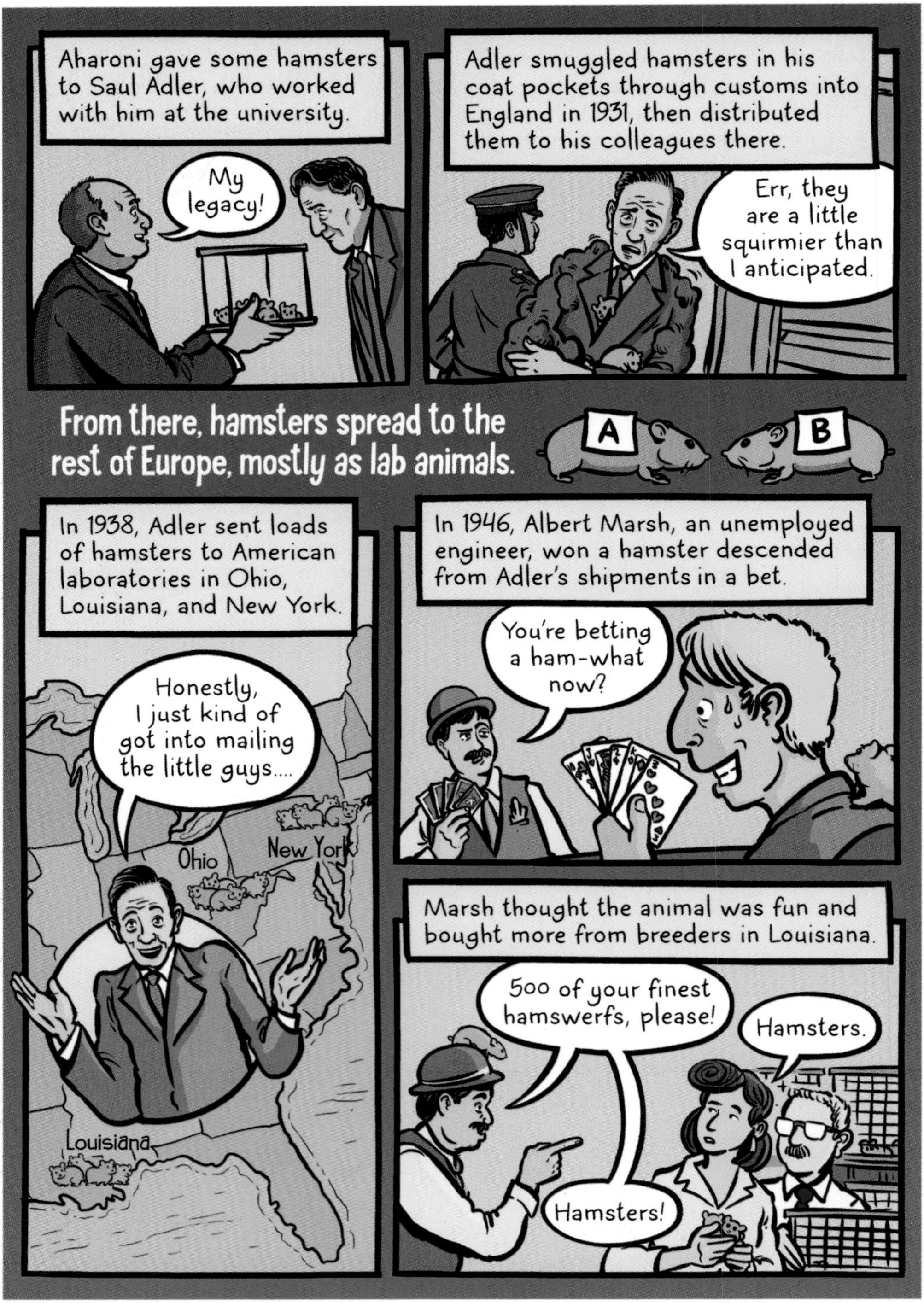
Aharoni gave some hamsters to Saul Adler, who worked with him at the university.
My legacy!
Adler smuggled hamsters in his coat pockets through customs into England in 1931, then distributed them to his colleagues there.
Err, they are a little squirmier than I anticipated.
From there, hamsters spread to the rest of Europe, mostly as lab animals.
A
B
In 1938, Adler sent loads of hamsters to American laboratories in Ohio, Louisiana, and New York.
Honestly, I just kind of got into mailing the little guys....
Ohio
New York
Louisiana
In 1946, Albert Marsh, an unemployed engineer, won a hamster descended from Adler's shipments in a bet.
You're betting a ham-what now?
Marsh thought the animal was fun and bought more from breeders in Louisiana.
500 of your finest hamswerfs, please!
Hamsters.
Hamsters!

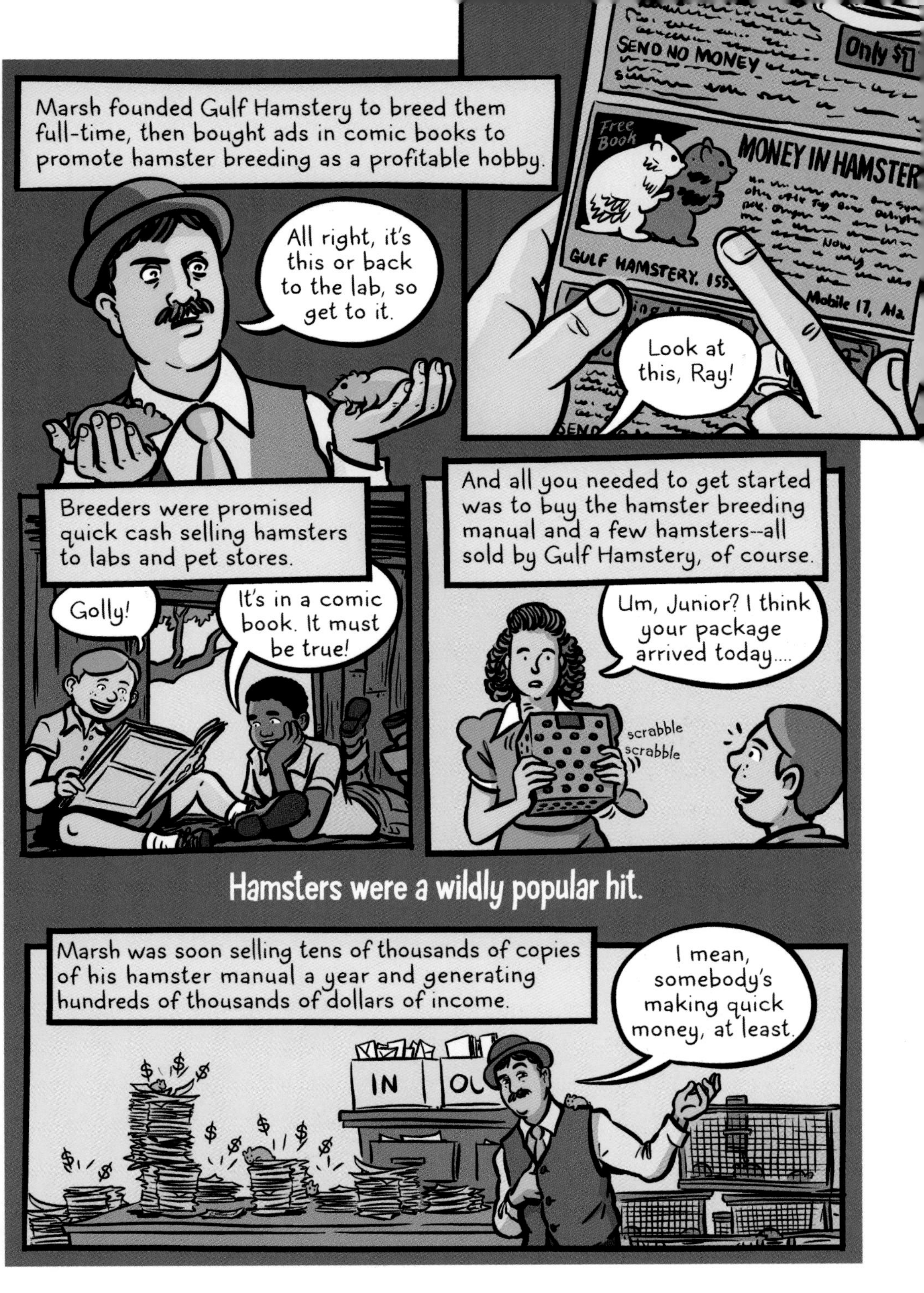
Marsh founded Gulf Hamstery to breed them full-time, then bought ads in comic books to promote hamster breeding as a profitable hobby.
All right, it's this or back to the lab, so get to it.
SEND NO MONEY
Only $1
Free Book
MONEY IN HAMSTER
GULF HAMSTERY.
Mobile 17, Ala.
Look at this, Ray!
Breeders were promised quick cash selling hamsters to labs and pet stores.
Golly!
It's in a comic book. It must be true!
And all you needed to get started was to buy the hamster breeding manual and a few hamsters--all sold by Gulf Hamstery, of course.
Um, Junior? I think your package arrived today....
scrabble scrabble
Hamsters were a wildly popular hit.
Marsh was soon selling tens of thousands of copies of his hamster manual a year and generating hundreds of thousands of dollars of income.
I mean, somebody's making quick money, at least.
IN

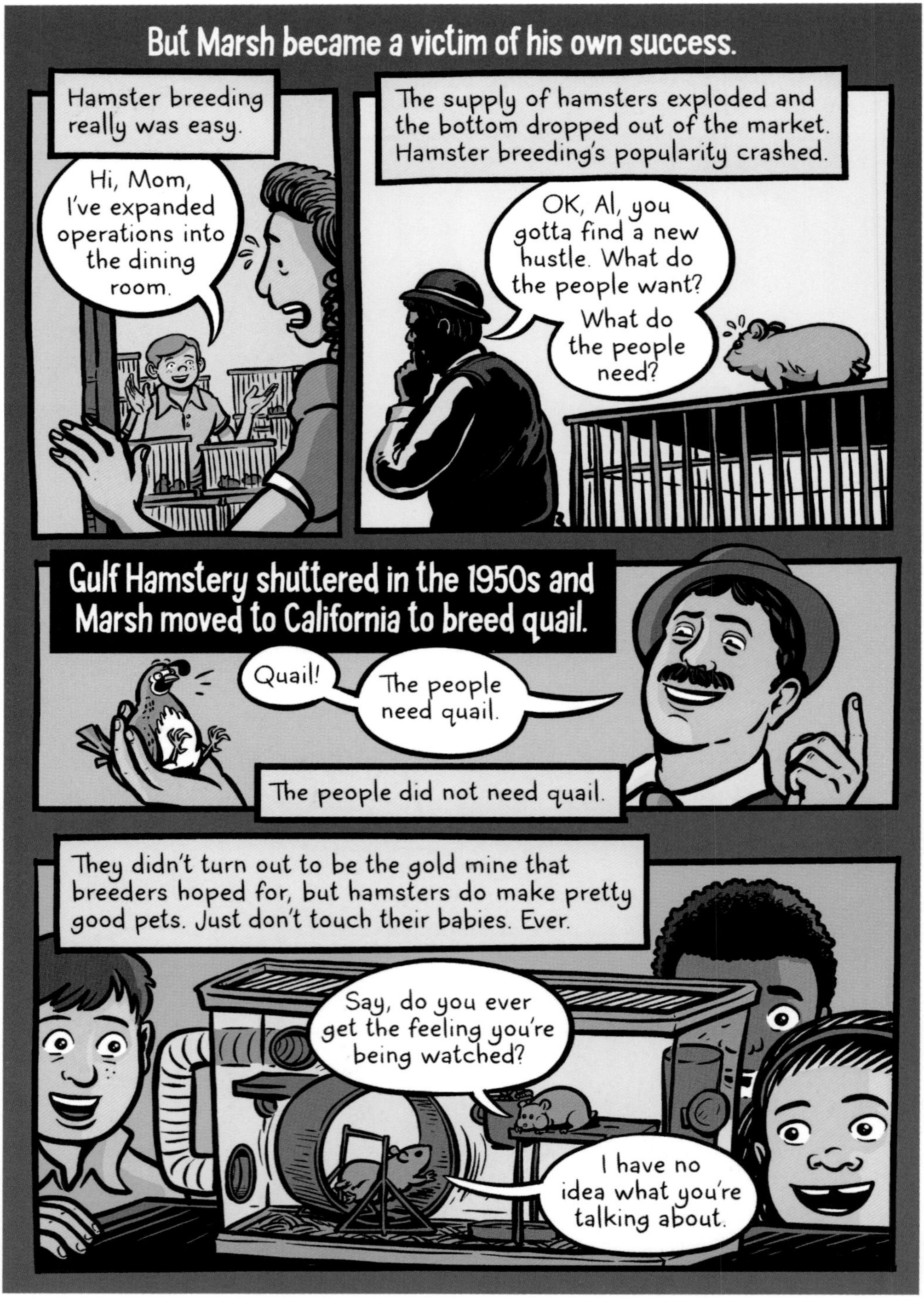
But Marsh became a victim of his own success.
Hamster breeding really was easy.
Hi, Mom, I've expanded operations into the dining room.
The supply of hamsters exploded and the bottom dropped out of the market. Hamster breeding's popularity crashed.
OK, Al, you gotta find a new hustle. What do the people want?
What do the people need?
Gulf Hamstery shuttered in the 1950s and Marsh moved to California to breed quail.
Quail!
The people need quail.
The people did not need quail.
They didn't turn out to be the gold mine that breeders hoped for, but hamsters do make pretty good pets. Just don't touch their babies. Ever.
Say, do you ever get the feeling you're being watched?
I have no idea what you're talking about.

OTHER HAMSTER VARIETIES

While "golden" Syrian hamsters are the most commonly found pet species, they're not the only domesticated hamsters.

We also keep 4 unrelated species of hamsters as pets.

Campbell's Dwarf Hamster

Phodopus campbelli

The most common dwarf hamster.
Originally from the Central Asian Steppe.

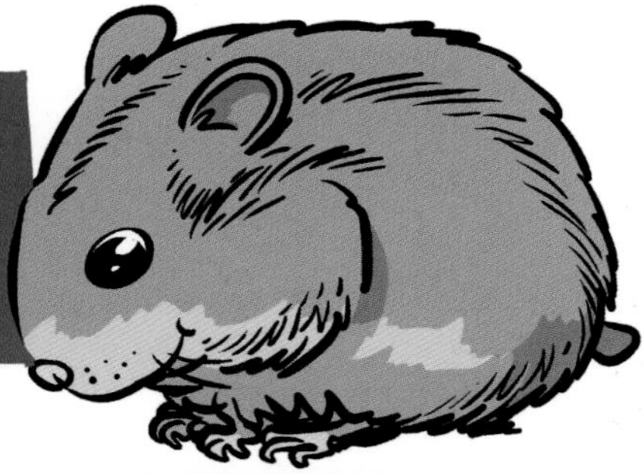

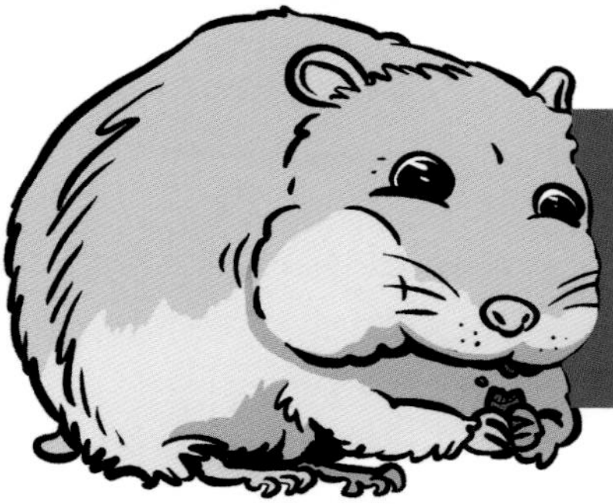

Winter White Dwarf Hamster

Phodopus sungorus

Coat changes color in the winter.
Originally from the Central Asian meadows.

Roborovski Dwarf Hamster

Phodopus roborovskii

Small, quick, and easily stressed.
Originally from the Central Asian deserts.

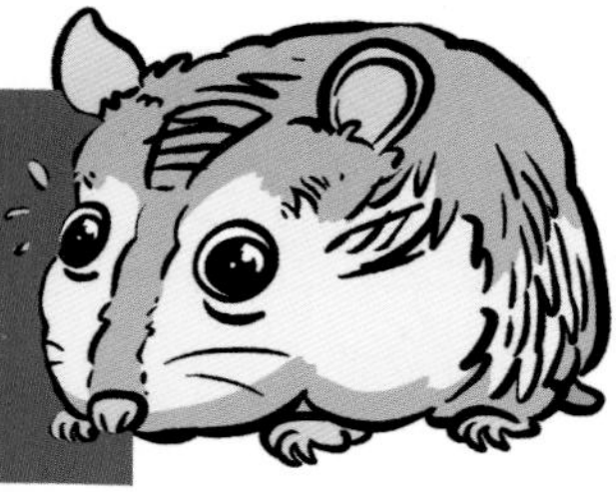

Chinese Hamster

Cricetulus griseus

Longer tail than other pet hamsters but can be nippy.

Originally from deserts in Mongolia and Northern China.

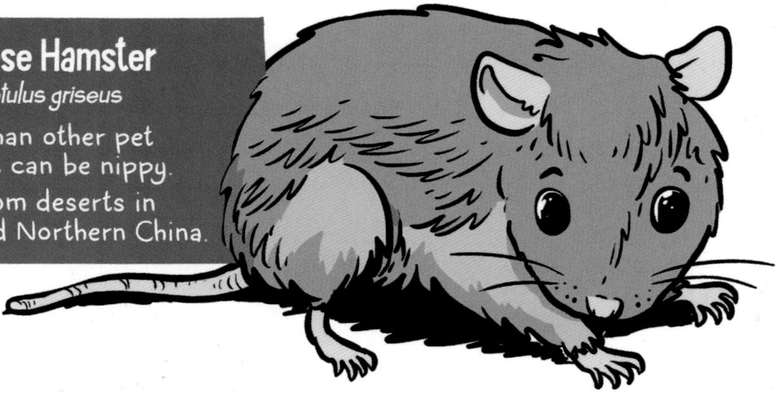

GOLDFISH

Carassius auratus

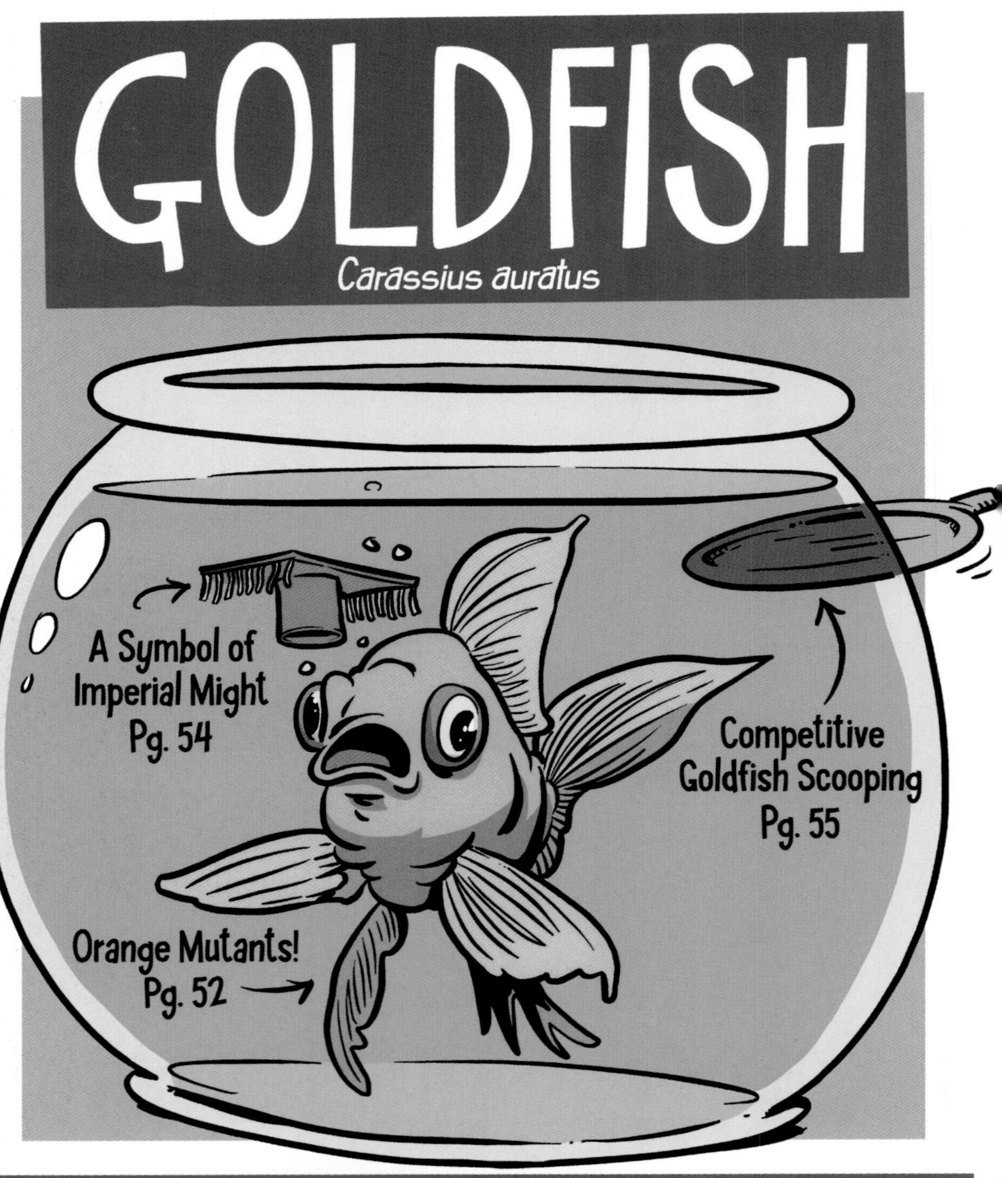

A Symbol of Imperial Might Pg. 54
Competitive Goldfish Scooping Pg. 55
Orange Mutants! Pg. 52

Size: 4-18 in

Weight: 8-154 oz

Oddball Anatomy: Set of teeth at back of mouth

Diet: Strange little flakes that smell weird

(Likely) **Wild Origin:**

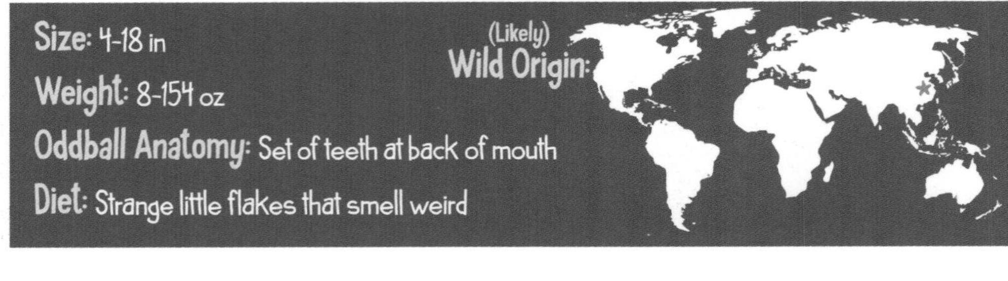

THE JOURNEY OF THE GOLDFISH FROM THE DINNER PLATE TO THE FISH TANK

Chinese people have been raising and farming Asian carp as food for thousands of years.

It's great! Once you get them in a pond, there's no way for them to get back out again!

The carp were naturally brown and gray, good camouflage for riverbeds.

Lot of good camouflage does you in a carp pond.

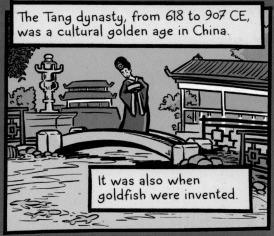

The Tang dynasty, from 618 to 907 CE, was a cultural golden age in China.

It was also when goldfish were invented.

Sometime during the Tang, someone noticed some yellow and orange mutants swimming among their usual boring gray-brown carp.

Ooh, shiny.

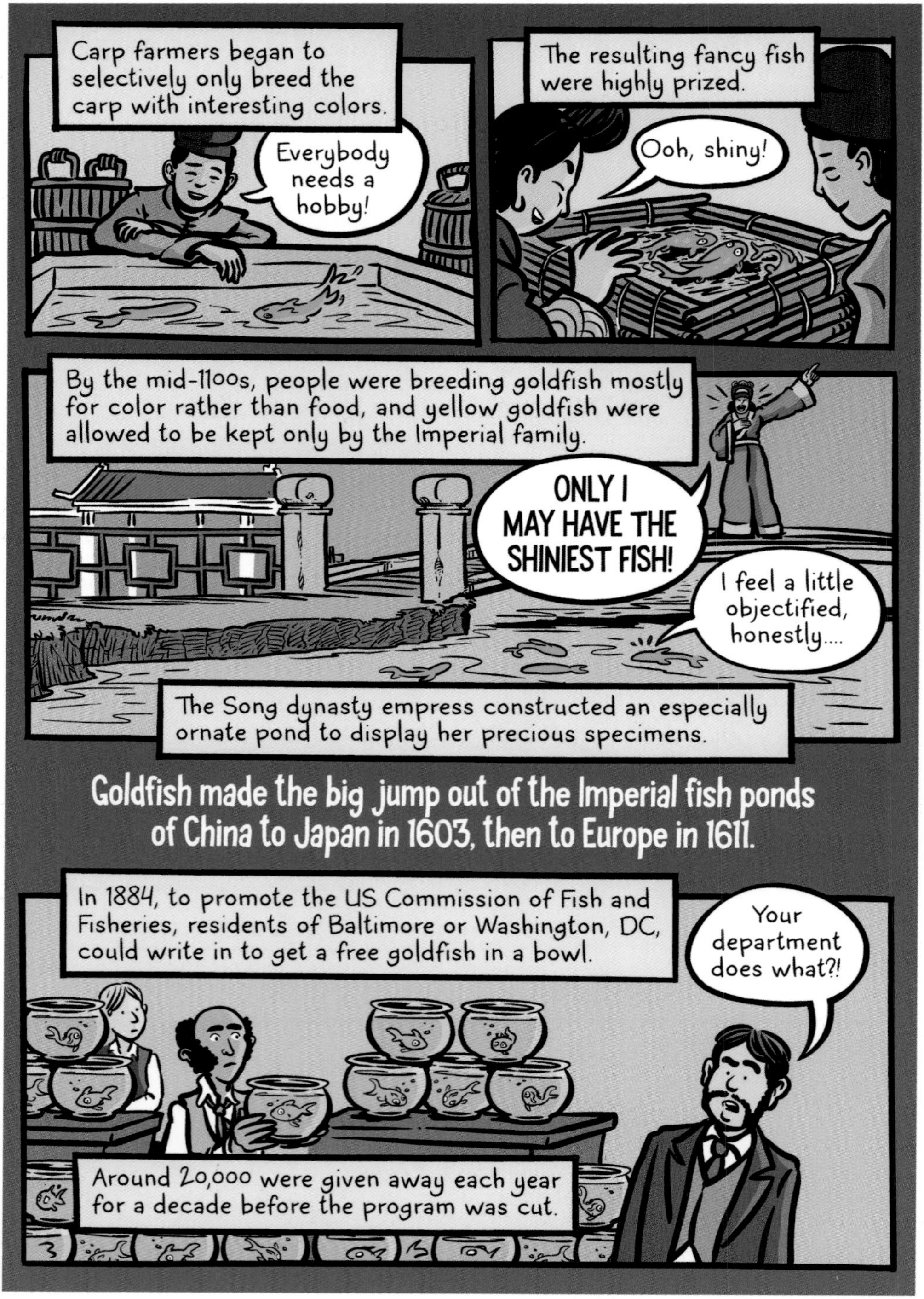
Carp farmers began to selectively only breed the carp with interesting colors.
Everybody needs a hobby!
The resulting fancy fish were highly prized.
Ooh, shiny!
By the mid-1100s, people were breeding goldfish mostly for color rather than food, and yellow goldfish were allowed to be kept only by the Imperial family.
ONLY I MAY HAVE THE SHINIEST FISH!
I feel a little objectified, honestly....
The Song dynasty empress constructed an especially ornate pond to display her precious specimens.
Goldfish made the big jump out of the Imperial fish ponds of China to Japan in 1603, then to Europe in 1611.
In 1884, to promote the US Commission of Fish and Fisheries, residents of Baltimore or Washington, DC, could write in to get a free goldfish in a bowl.
Your department does what?!
Around 20,000 were given away each year for a decade before the program was cut.

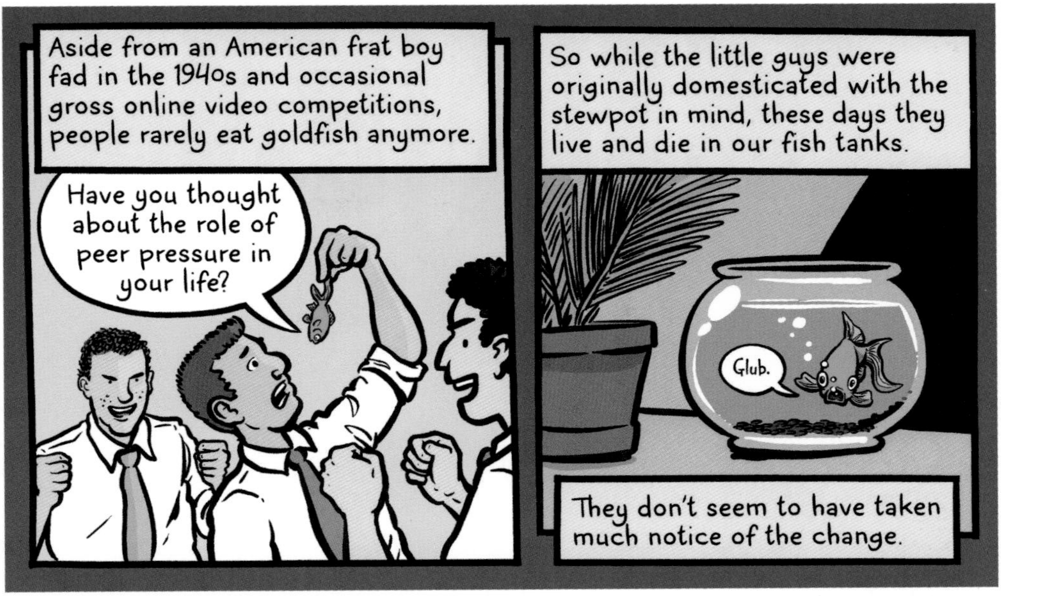
Aside from an American frat boy fad in the 1940s and occasional gross online video competitions, people rarely eat goldfish anymore.
Have you thought about the role of peer pressure in your life?
So while the little guys were originally domesticated with the stewpot in mind, these days they live and die in our fish tanks.
Glub.
They don't seem to have taken much notice of the change.

JAPAN'S LOVE OF GOLDFISH SCOOPING

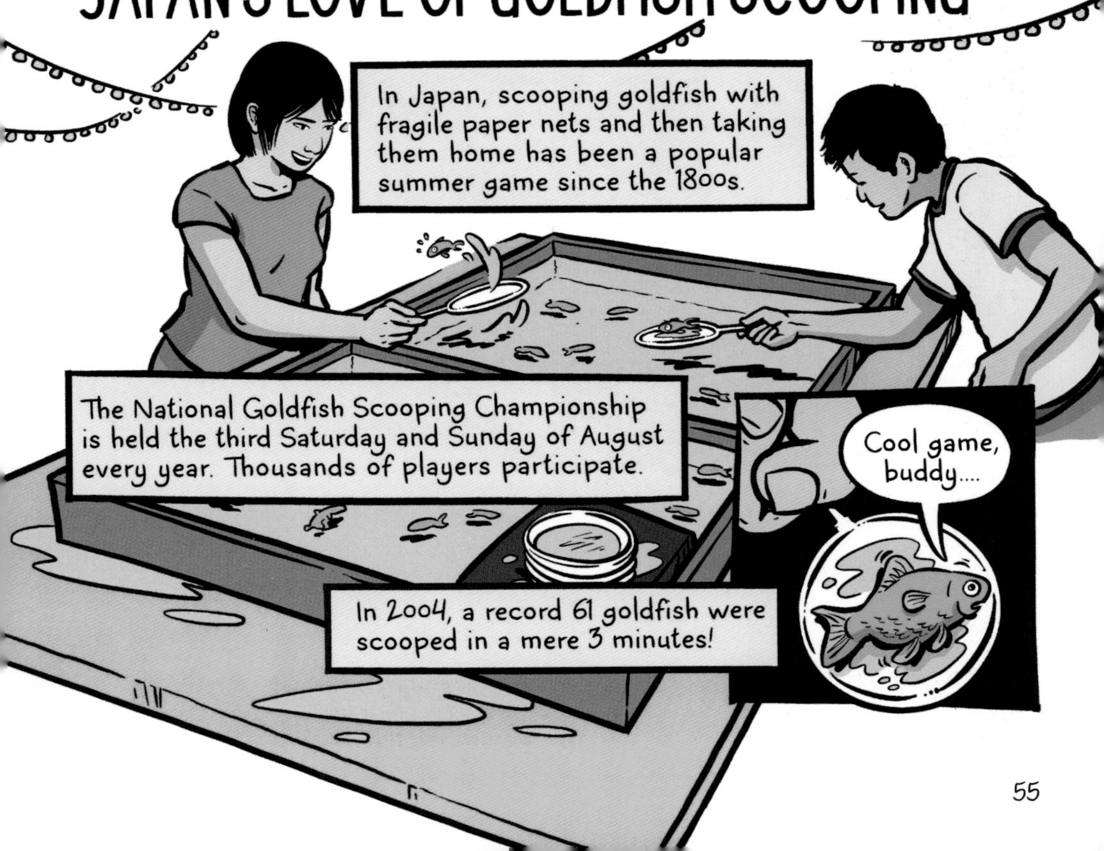
In Japan, scooping goldfish with fragile paper nets and then taking them home has been a popular summer game since the 1800s.
The National Goldfish Scooping Championship is held the third Saturday and Sunday of August every year. Thousands of players participate.
Cool game, buddy....
In 2004, a record 61 goldfish were scooped in a mere 3 minutes!

GUINEA PIGS

Cavia porcellus

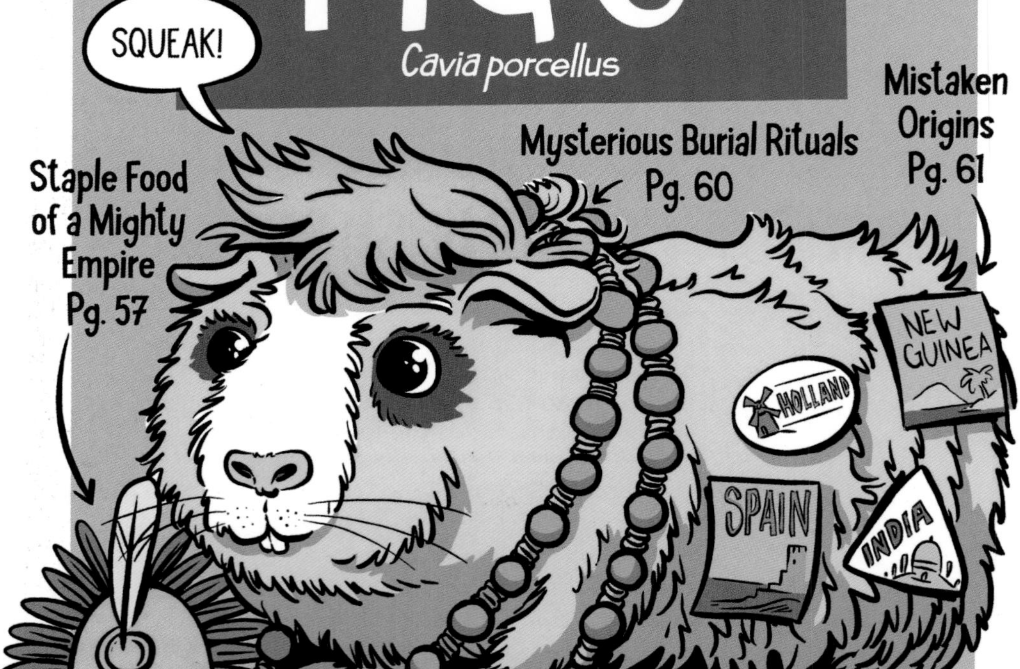
SQUEAK!
Staple Food of a Mighty Empire Pg. 57
Mysterious Burial Rituals Pg. 60
Mistaken Origins Pg. 61
NEW GUINEA
HOLLAND
SPAIN
INDIA

Size: 7.9–9.8 in (length)

Weight: 1.5–2.6 lbs

Unusual Use: Disease diagnosis in traditional medicine

Diet: Alfalfa, hay, vegetables, their own poop pellets

(Likely) Wild Origin:

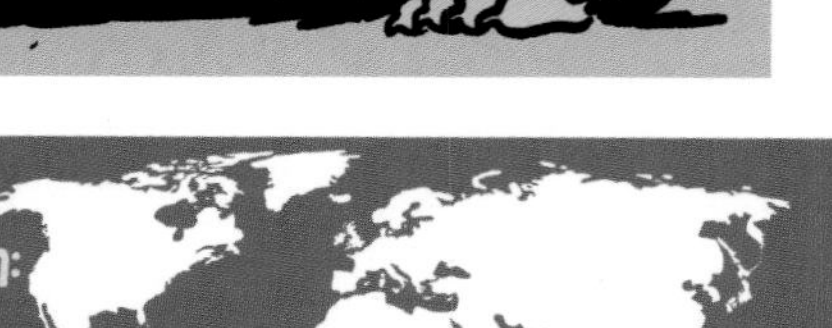

HOW GUINEA PIGS FED THE LARGEST NATIVE-AMERICAN EMPIRE

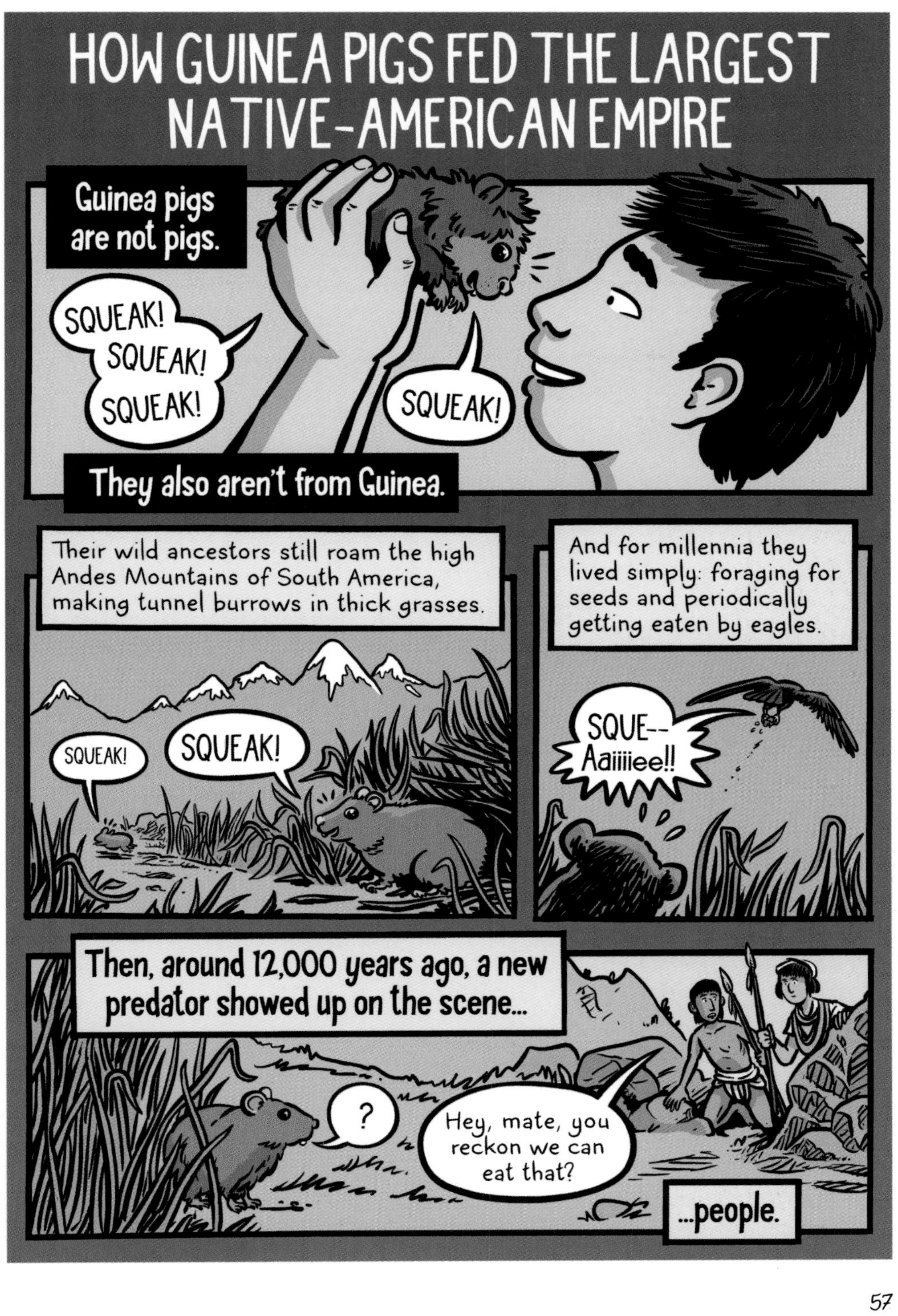
Guinea pigs are not pigs.
SQUEAK! SQUEAK! SQUEAK!
SQUEAK!
They also aren't from Guinea.
Their wild ancestors still roam the high Andes Mountains of South America, making tunnel burrows in thick grasses.
SQUEAK!
SQUEAK!
And for millennia they lived simply: foraging for seeds and periodically getting eaten by eagles.
SQUE-- Aaiiiiee!!
Then, around 12,000 years ago, a new predator showed up on the scene...
?
Hey, mate, you reckon we can eat that?
...people.

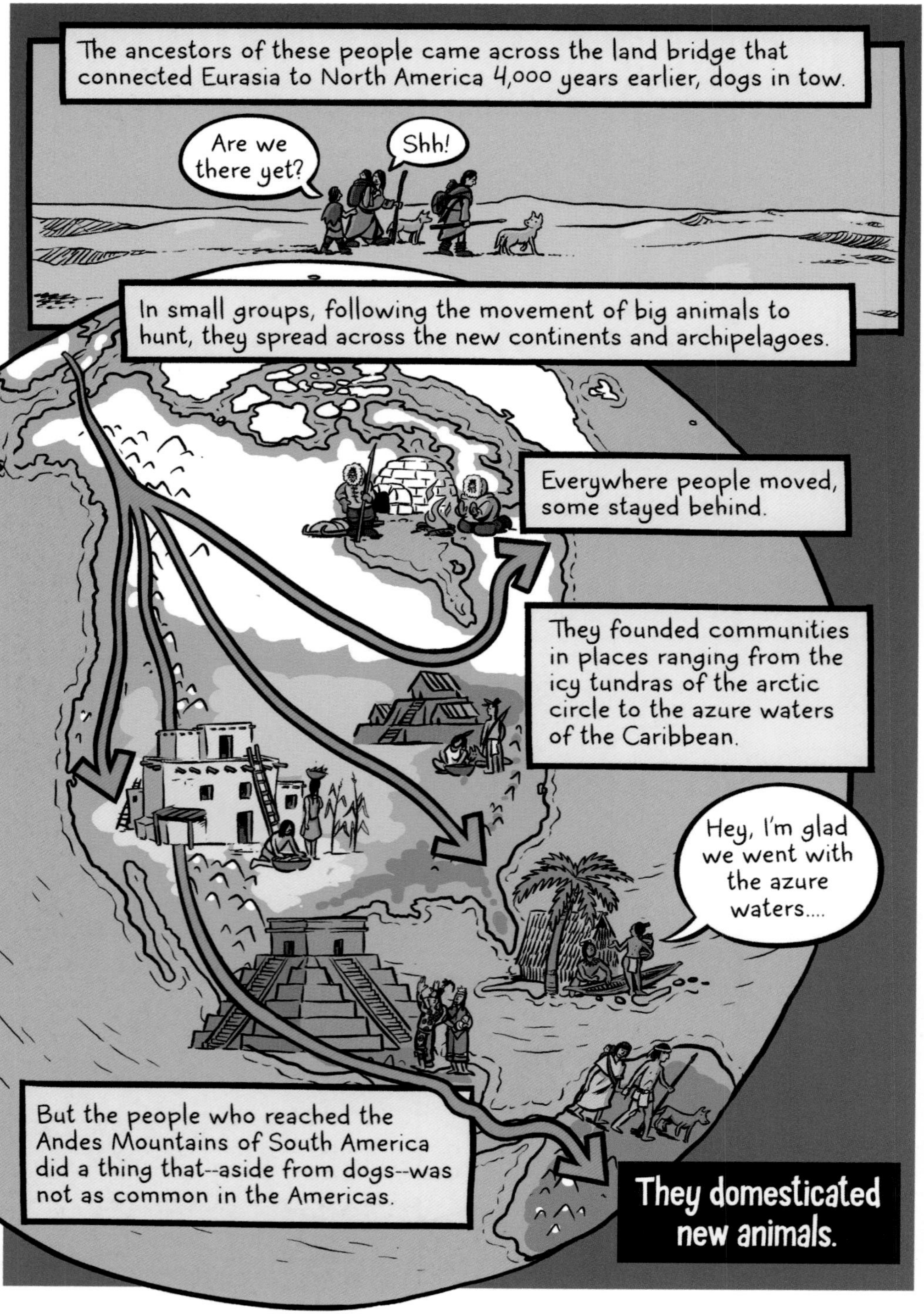
The ancestors of these people came across the land bridge that connected Eurasia to North America 4,000 years earlier, dogs in tow.
Are we there yet?
Shh!
In small groups, following the movement of big animals to hunt, they spread across the new continents and archipelagoes.
Everywhere people moved, some stayed behind.
They founded communities in places ranging from the icy tundras of the arctic circle to the azure waters of the Caribbean.
Hey, I'm glad we went with the azure waters....
But the people who reached the Andes Mountains of South America did a thing that--aside from dogs--was not as common in the Americas.
They domesticated new animals.

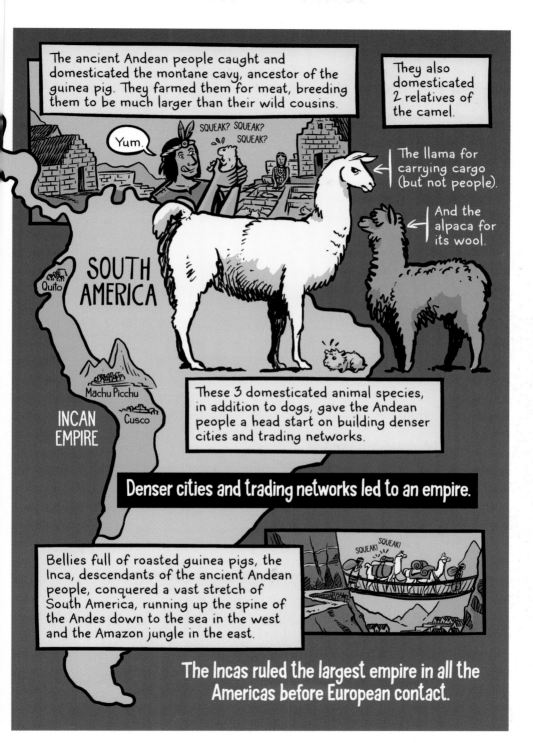

The ancient Andean people caught and domesticated the montane cavy, ancestor of the guinea pig. They farmed them for meat, breeding them to be much larger than their wild cousins.

They also domesticated 2 relatives of the camel.

Yum.

SQUEAK? SQUEAK? SQUEAK?

SOUTH AMERICA

Quito

The llama for carrying cargo (but not people).

And the alpaca for its wool.

Machu Picchu

Cusco

INCAN EMPIRE

These 3 domesticated animal species, in addition to dogs, gave the Andean people a head start on building denser cities and trading networks.

Denser cities and trading networks led to an empire.

Bellies full of roasted guinea pigs, the Inca, descendants of the ancient Andean people, conquered a vast stretch of South America, running up the spine of the Andes down to the sea in the west and the Amazon jungle in the east.

SQUEAK! SQUEAK!

The Incas ruled the largest empire in all the Americas before European contact.

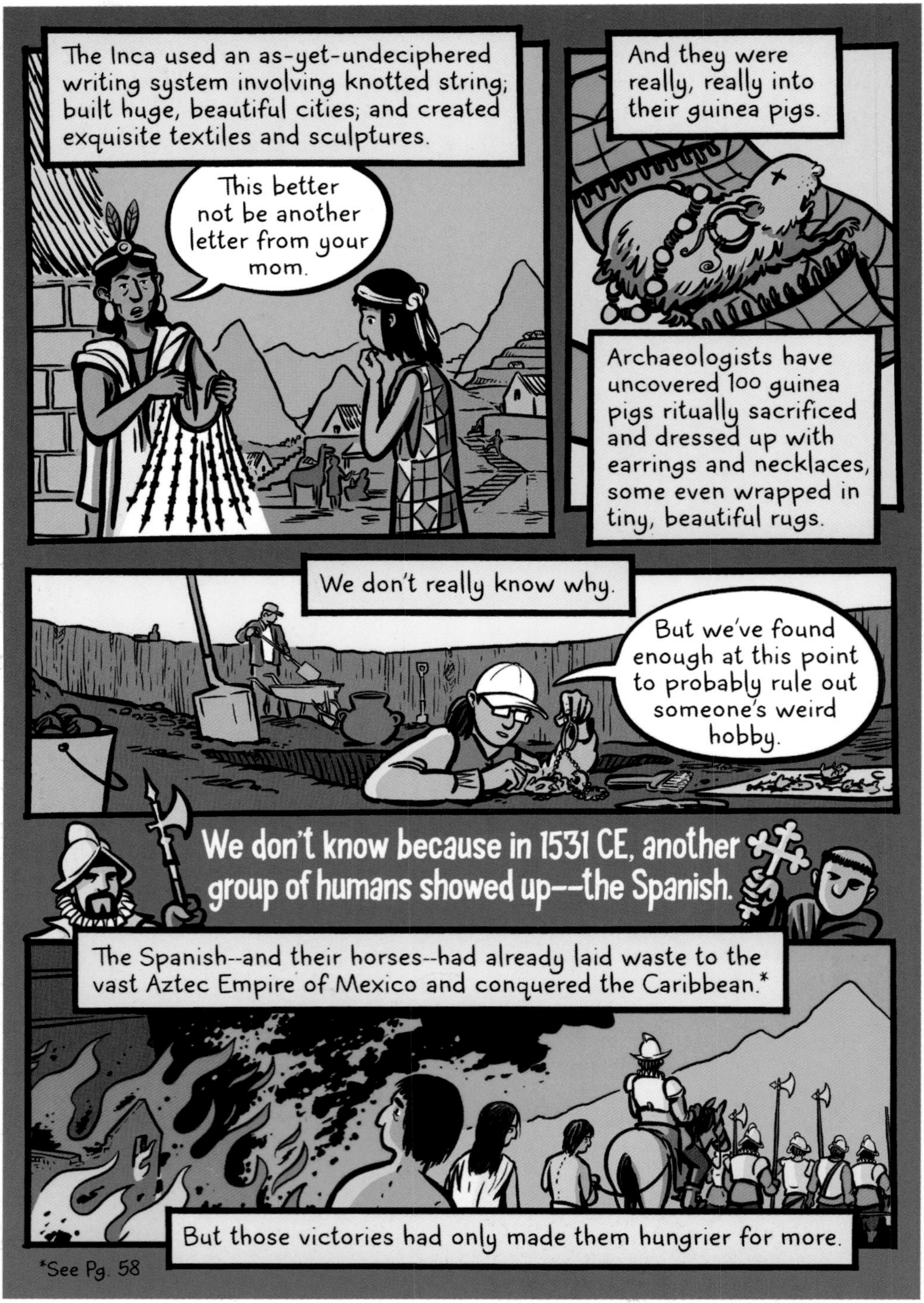
The Inca used an as-yet-undeciphered writing system involving knotted string; built huge, beautiful cities; and created exquisite textiles and sculptures.
This better not be another letter from your mom.
And they were really, really into their guinea pigs.
Archaeologists have uncovered 100 guinea pigs ritually sacrificed and dressed up with earrings and necklaces, some even wrapped in tiny, beautiful rugs.
We don't really know why.
But we've found enough at this point to probably rule out someone's weird hobby.
We don't know because in 1531 CE, another group of humans showed up--the Spanish.
The Spanish--and their horses--had already laid waste to the vast Aztec Empire of Mexico and conquered the Caribbean.*
But those victories had only made them hungrier for more.
*See Pg. 58

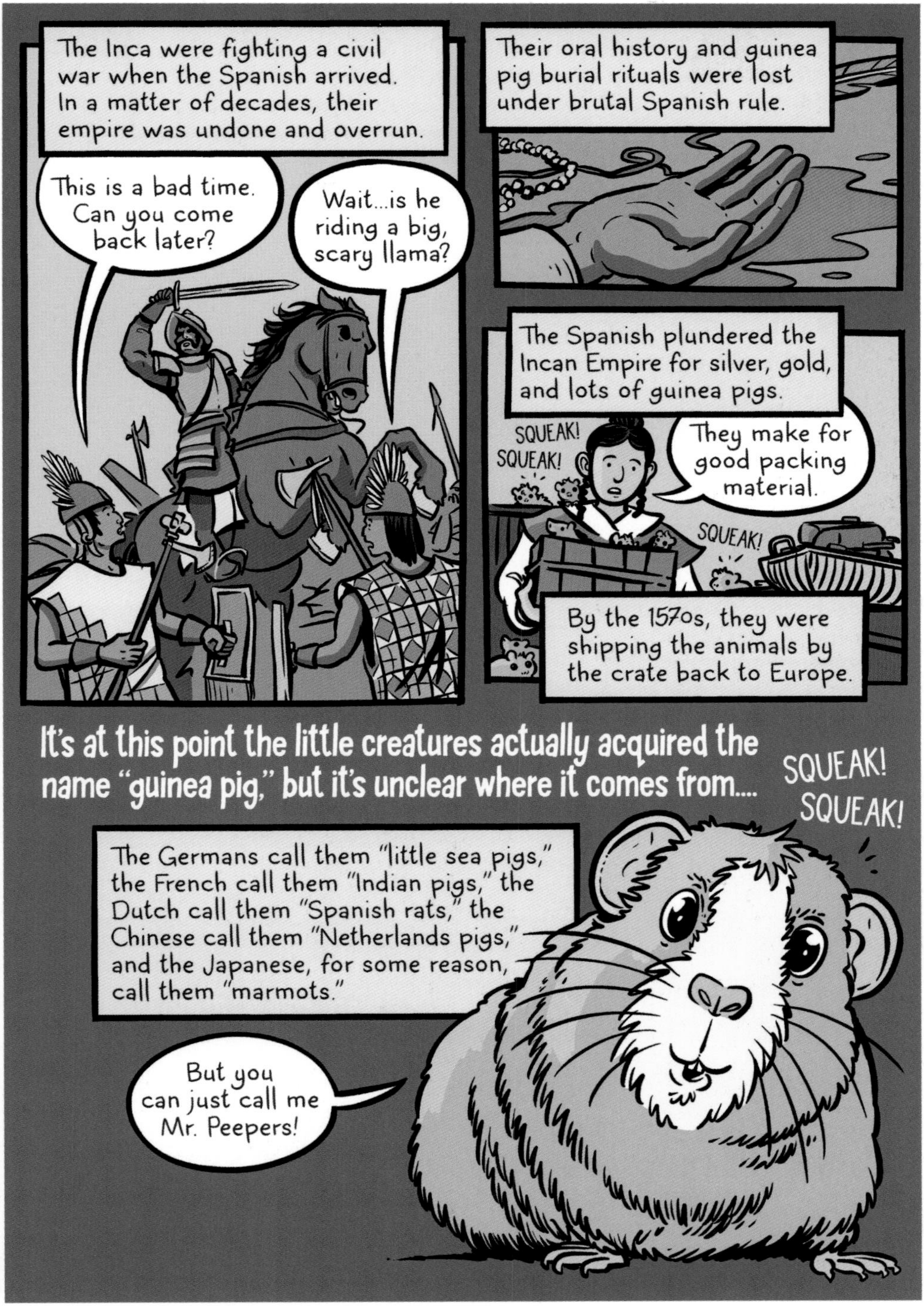
The Inca were fighting a civil war when the Spanish arrived. In a matter of decades, their empire was undone and overrun.
This is a bad time. Can you come back later?
Wait...is he riding a big, scary llama?
Their oral history and guinea pig burial rituals were lost under brutal Spanish rule.
The Spanish plundered the Incan Empire for silver, gold, and lots of guinea pigs.
SQUEAK! SQUEAK!
They make for good packing material.
SQUEAK!
By the 1570s, they were shipping the animals by the crate back to Europe.
It's at this point the little creatures actually acquired the name "guinea pig," but it's unclear where it comes from....
SQUEAK! SQUEAK!
The Germans call them "little sea pigs," the French call them "Indian pigs," the Dutch call them "Spanish rats," the Chinese call them "Netherlands pigs," and the Japanese, for some reason, call them "marmots."
But you can just call me Mr. Peepers!

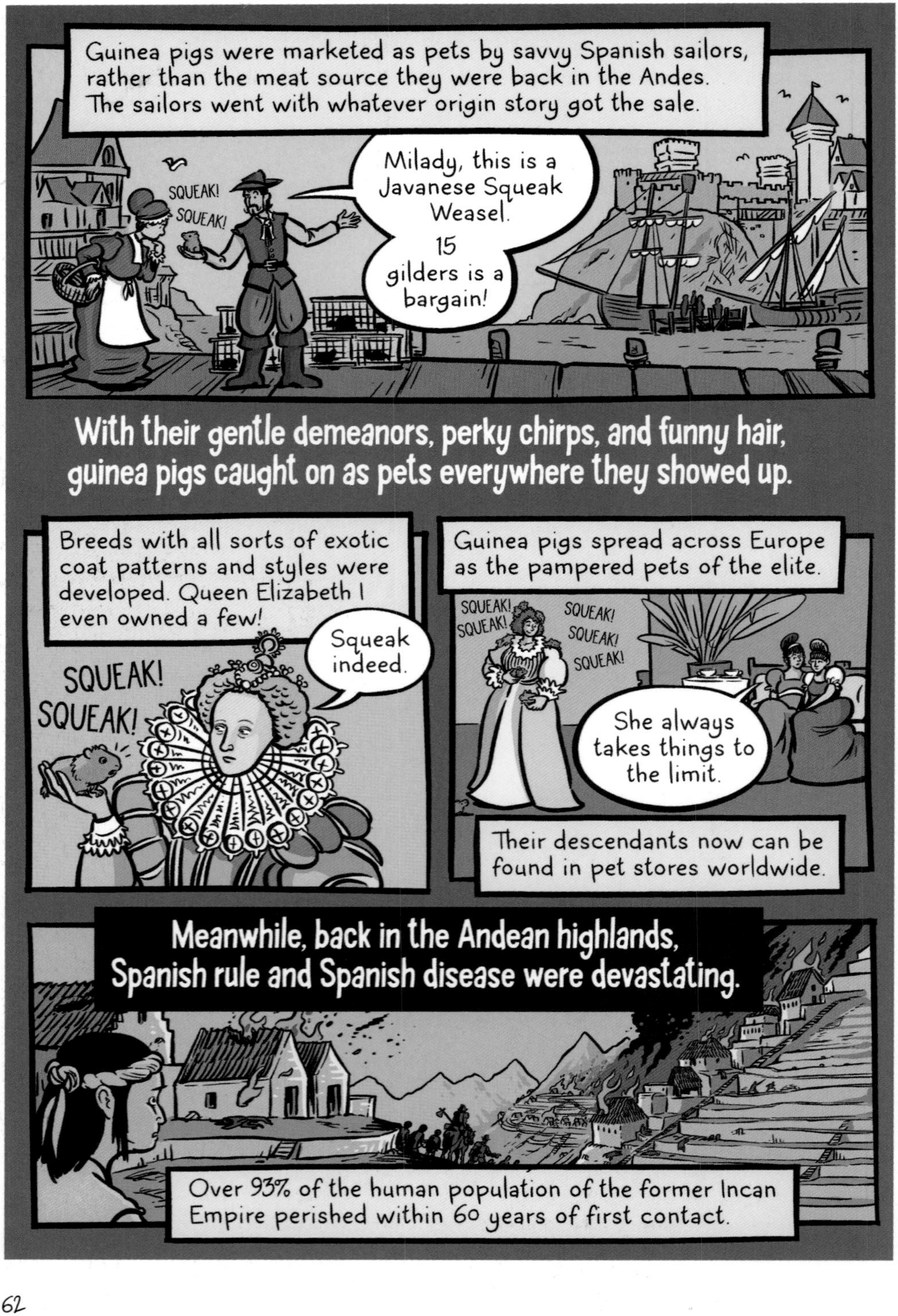
Guinea pigs were marketed as pets by savvy Spanish sailors, rather than the meat source they were back in the Andes. The sailors went with whatever origin story got the sale.
SQUEAK!
SQUEAK!
Milady, this is a Javanese Squeak Weasel.
15 gilders is a bargain!
With their gentle demeanors, perky chirps, and funny hair, guinea pigs caught on as pets everywhere they showed up.
Breeds with all sorts of exotic coat patterns and styles were developed. Queen Elizabeth I even owned a few!
SQUEAK!
SQUEAK!
Squeak indeed.
Guinea pigs spread across Europe as the pampered pets of the elite.
SQUEAK!
SQUEAK!
SQUEAK!
SQUEAK!
SQUEAK!
She always takes things to the limit.
Their descendants now can be found in pet stores worldwide.
Meanwhile, back in the Andean highlands, Spanish rule and Spanish disease were devastating.
Over 93% of the human population of the former Incan Empire perished within 60 years of first contact.

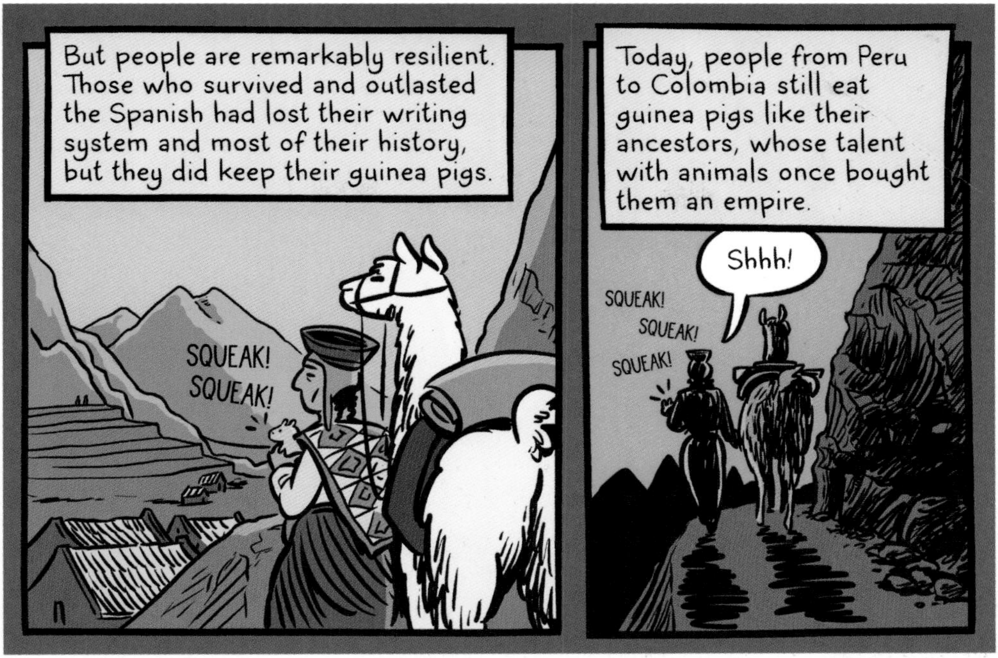

But people are remarkably resilient. Those who survived and outlasted the Spanish had lost their writing system and most of their history, but they did keep their guinea pigs.
SQUEAK! SQUEAK!
Today, people from Peru to Colombia still eat guinea pigs like their ancestors, whose talent with animals once bought them an empire.
Shhh!
SQUEAK!
SQUEAK!
SQUEAK!

GUINEA PIG EXPERIMENTS

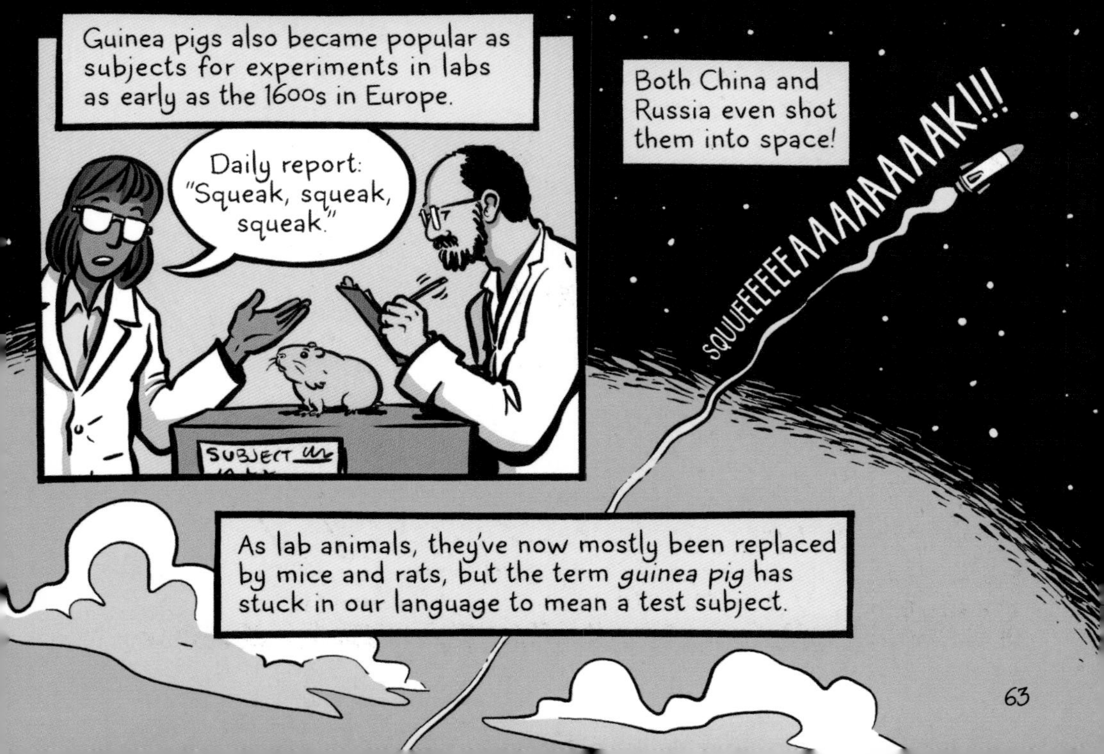

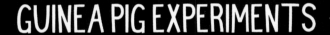

Guinea pigs also became popular as subjects for experiments in labs as early as the 1600s in Europe.
Daily report: "Squeak, squeak, squeak."
SUBJECT
Both China and Russia even shot them into space!
SQUEEEEEAAAAAAAAK!!!
As lab animals, they've now mostly been replaced by mice and rats, but the term *guinea pig* has stuck in our language to mean a test subject.

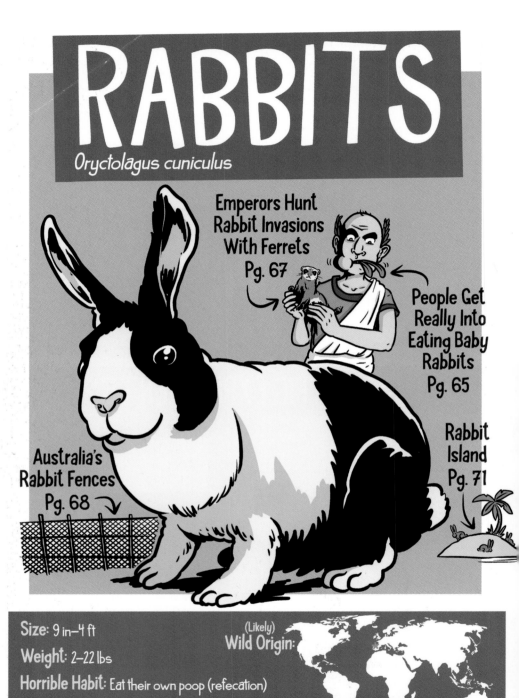

RABBITS

Oryctolagus cuniculus

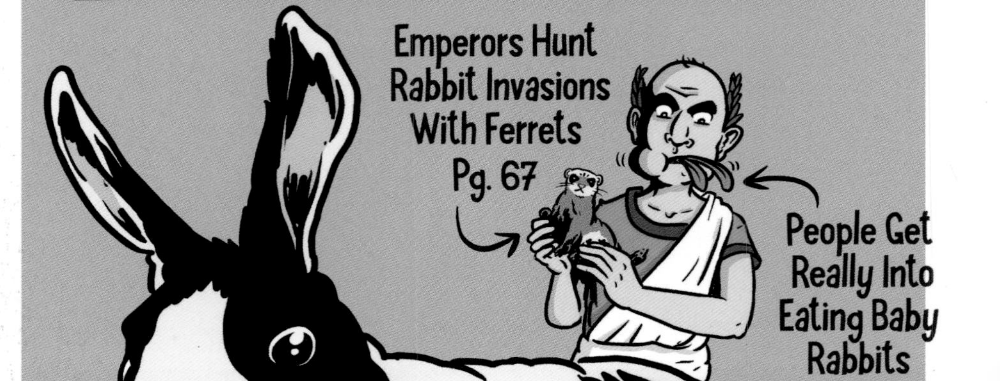
Emperors Hunt Rabbit Invasions With Ferrets Pg. 67
People Get Really Into Eating Baby Rabbits Pg. 65
Rabbit Island Pg. 71
Australia's Rabbit Fences Pg. 68

Size: 9 in–4 ft

Weight: 2–22 lbs

Horrible Habit: Eat their own poop (refecation)

Diet: Hay, vegetables, occasional fruits

(Likely) Wild Origin:

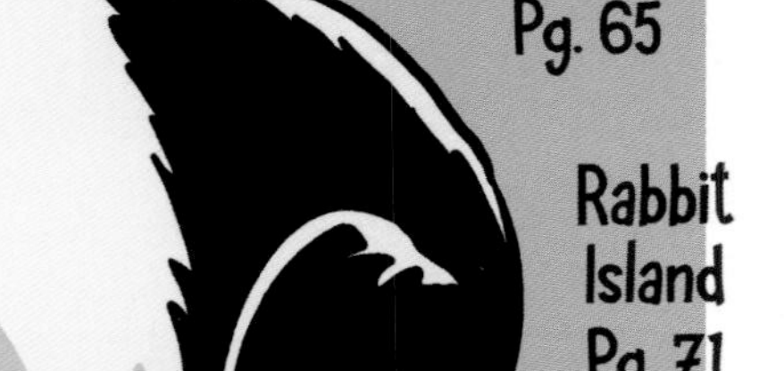

HOW RABBITS ATE AUSTRALIA

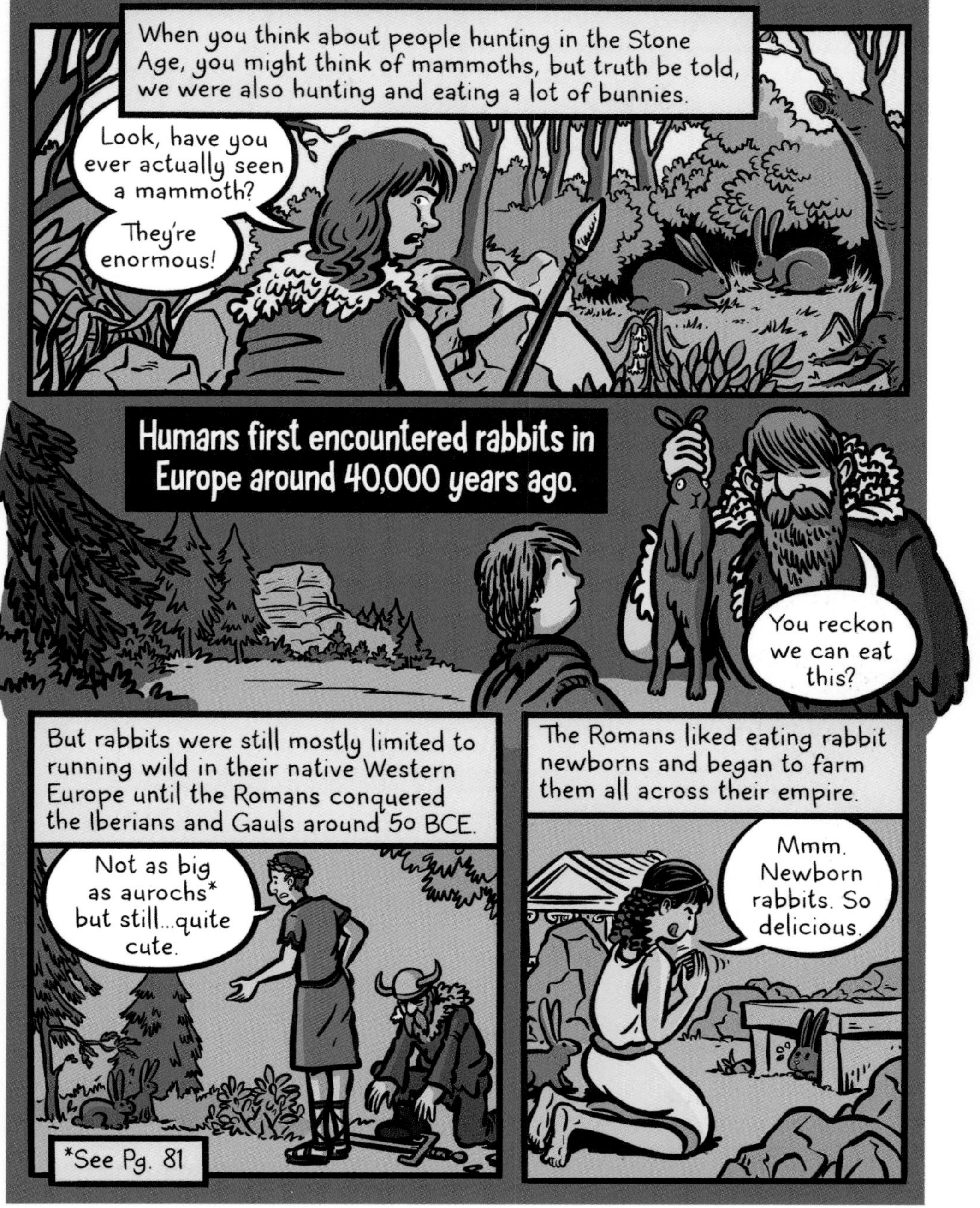
When you think about people hunting in the Stone Age, you might think of mammoths, but truth be told, we were also hunting and eating a lot of bunnies.
Look, have you ever actually seen a mammoth?
They're enormous!
Humans first encountered rabbits in Europe around 40,000 years ago.
You reckon we can eat this?
But rabbits were still mostly limited to running wild in their native Western Europe until the Romans conquered the Iberians and Gauls around 50 BCE.
Not as big as aurochs* but still...quite cute.
*See Pg. 81
The Romans liked eating rabbit newborns and began to farm them all across their empire.
Mmm. Newborn rabbits. So delicious.

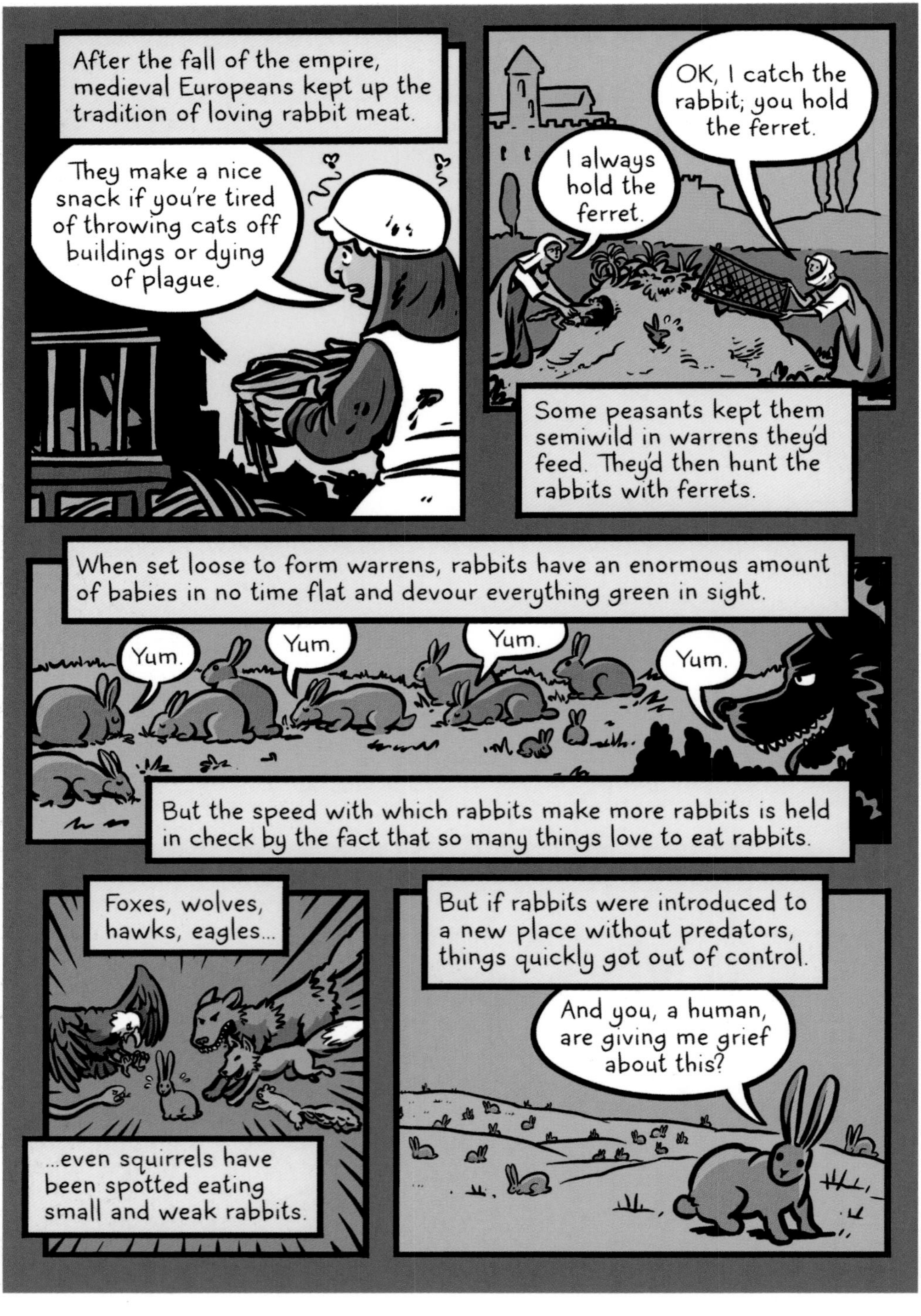
After the fall of the empire, medieval Europeans kept up the tradition of loving rabbit meat.
They make a nice snack if you're tired of throwing cats off buildings or dying of plague.
OK, I catch the rabbit; you hold the ferret.
I always hold the ferret.
Some peasants kept them semiwild in warrens they'd feed. They'd then hunt the rabbits with ferrets.
When set loose to form warrens, rabbits have an enormous amount of babies in no time flat and devour everything green in sight.
Yum.
Yum.
Yum.
Yum.
But the speed with which rabbits make more rabbits is held in check by the fact that so many things love to eat rabbits.
Foxes, wolves, hawks, eagles...
...even squirrels have been spotted eating small and weak rabbits.
But if rabbits were introduced to a new place without predators, things quickly got out of control.
And you, a human, are giving me grief about this?

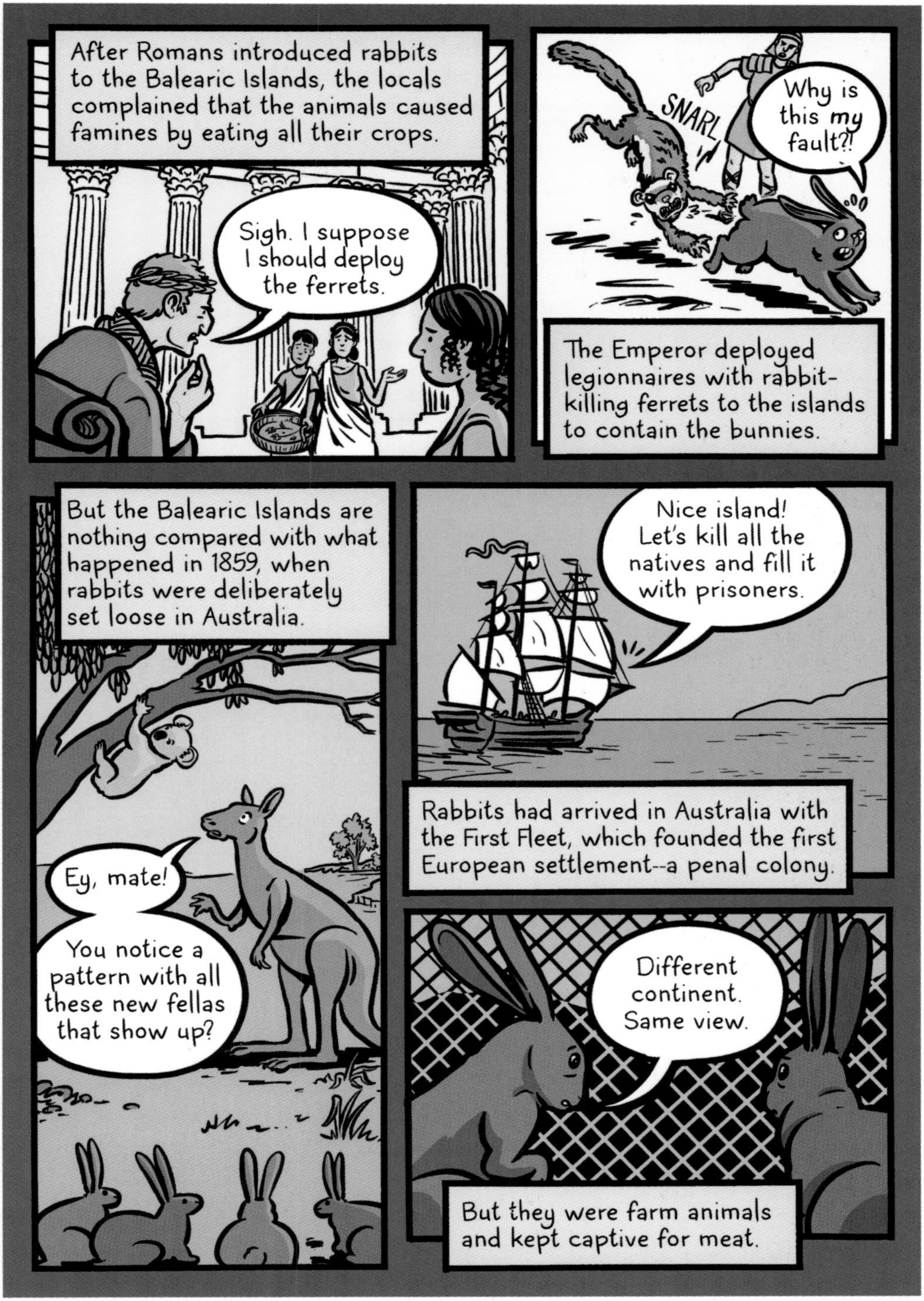

After Romans introduced rabbits to the Balearic Islands, the locals complained that the animals caused famines by eating all their crops.
Sigh. I suppose I should deploy the ferrets.
SNARL
Why is this my fault?!
The Emperor deployed legionnaires with rabbit-killing ferrets to the islands to contain the bunnies.
But the Balearic Islands are nothing compared with what happened in 1859, when rabbits were deliberately set loose in Australia.
Ey, mate!
You notice a pattern with all these new fellas that show up?
Nice island! Let's kill all the natives and fill it with prisoners.
Rabbits had arrived in Australia with the First Fleet, which founded the first European settlement--a penal colony.
Different continent. Same view.
But they were farm animals and kept captive for meat.

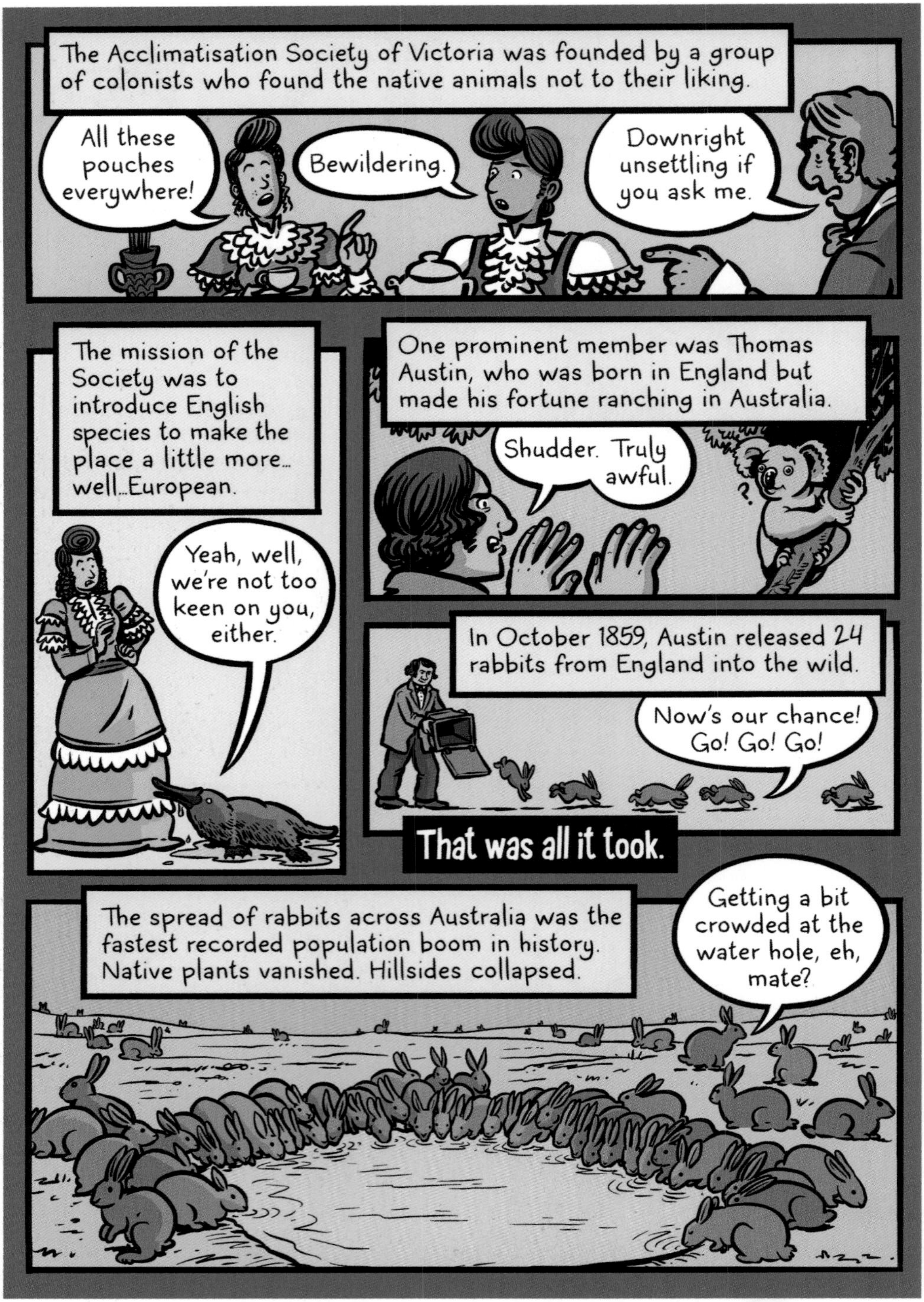

The Acclimatisation Society of Victoria was founded by a group of colonists who found the native animals not to their liking.
All these pouches everywhere!
Bewildering.
Downright unsettling if you ask me.
The mission of the Society was to introduce English species to make the place a little more... well...European.
Yeah, well, we're not too keen on you, either.
One prominent member was Thomas Austin, who was born in England but made his fortune ranching in Australia.
Shudder. Truly awful.
?
In October 1859, Austin released 24 rabbits from England into the wild.
Now's our chance! Go! Go! Go!
That was all it took.
The spread of rabbits across Australia was the fastest recorded population boom in history. Native plants vanished. Hillsides collapsed.
Getting a bit crowded at the water hole, eh, mate?

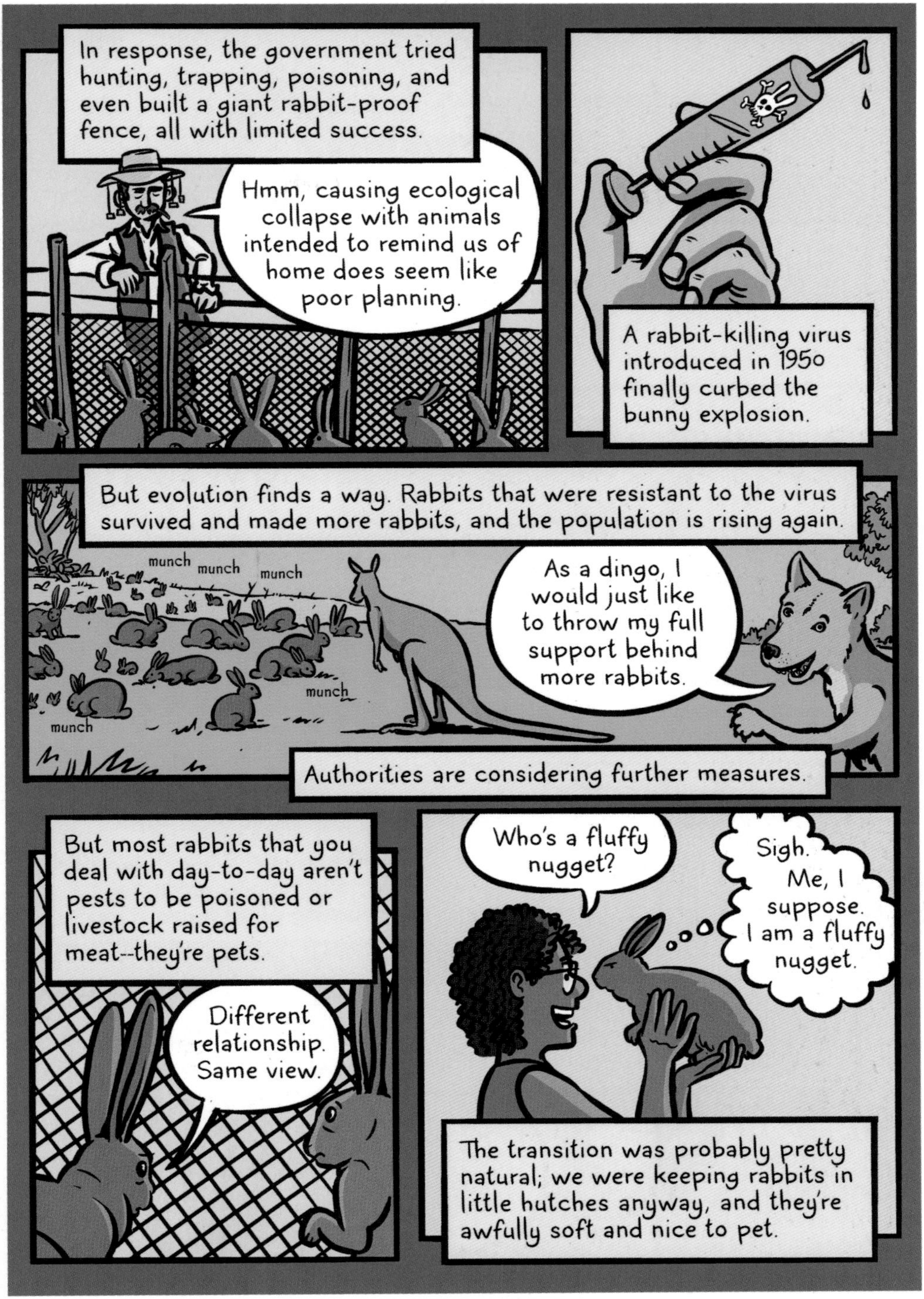
In response, the government tried hunting, trapping, poisoning, and even built a giant rabbit-proof fence, all with limited success.
Hmm, causing ecological collapse with animals intended to remind us of home does seem like poor planning.
A rabbit-killing virus introduced in 1950 finally curbed the bunny explosion.
But evolution finds a way. Rabbits that were resistant to the virus survived and made more rabbits, and the population is rising again.
munch munch munch
munch
munch
As a dingo, I would just like to throw my full support behind more rabbits.
Authorities are considering further measures.
But most rabbits that you deal with day-to-day aren't pests to be poisoned or livestock raised for meat--they're pets.
Different relationship. Same view.
Who's a fluffy nugget?
Sigh. Me, I suppose. I am a fluffy nugget.
The transition was probably pretty natural; we were keeping rabbits in little hutches anyway, and they're awfully soft and nice to pet.

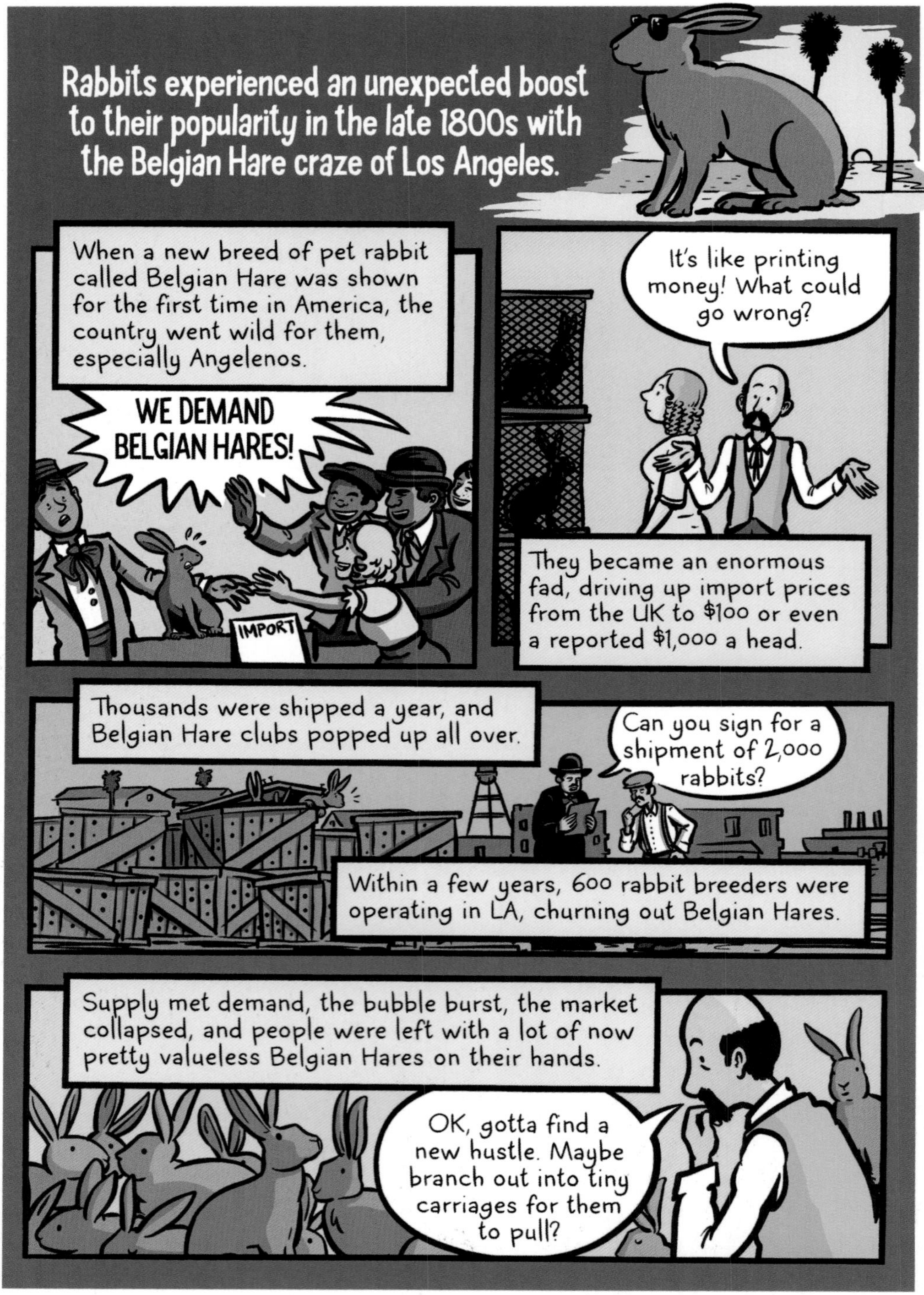
Rabbits experienced an unexpected boost to their popularity in the late 1800s with the Belgian Hare craze of Los Angeles.
When a new breed of pet rabbit called Belgian Hare was shown for the first time in America, the country went wild for them, especially Angelenos.
WE DEMAND BELGIAN HARES!
IMPORT
It's like printing money! What could go wrong?
They became an enormous fad, driving up import prices from the UK to $100 or even a reported $1,000 a head.
Thousands were shipped a year, and Belgian Hare clubs popped up all over.
Can you sign for a shipment of 2,000 rabbits?
Within a few years, 600 rabbit breeders were operating in LA, churning out Belgian Hares.
Supply met demand, the bubble burst, the market collapsed, and people were left with a lot of now pretty valueless Belgian Hares on their hands.
OK, gotta find a new hustle. Maybe branch out into tiny carriages for them to pull?

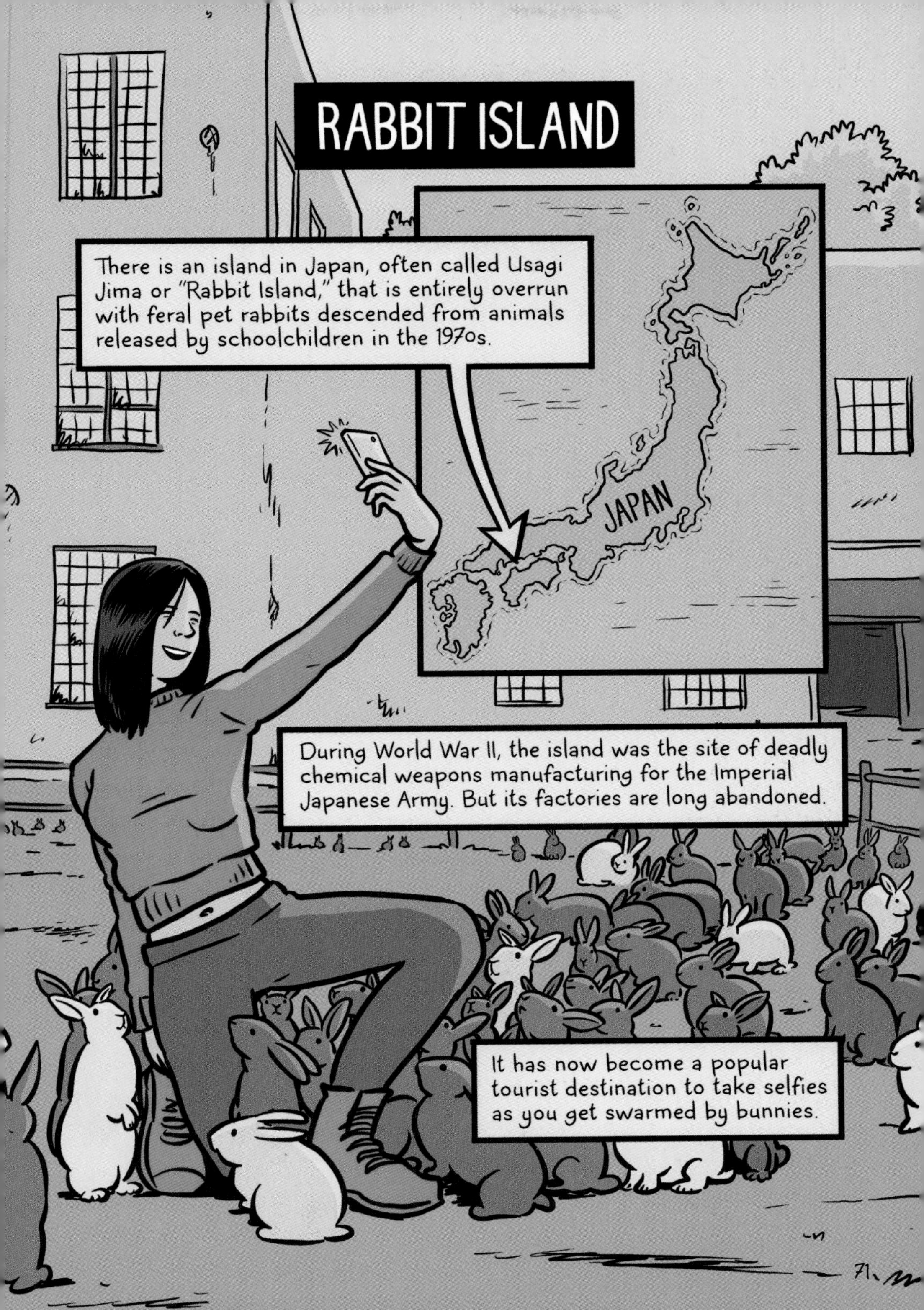
RABBIT ISLAND
There is an island in Japan, often called Usagi Jima or "Rabbit Island," that is entirely overrun with feral pet rabbits descended from animals released by schoolchildren in the 1970s.
JAPAN
During World War II, the island was the site of deadly chemical weapons manufacturing for the Imperial Japanese Army. But its factories are long abandoned.
It has now become a popular tourist destination to take selfies as you get swarmed by bunnies.

CREATURES WE FIND USEFUL

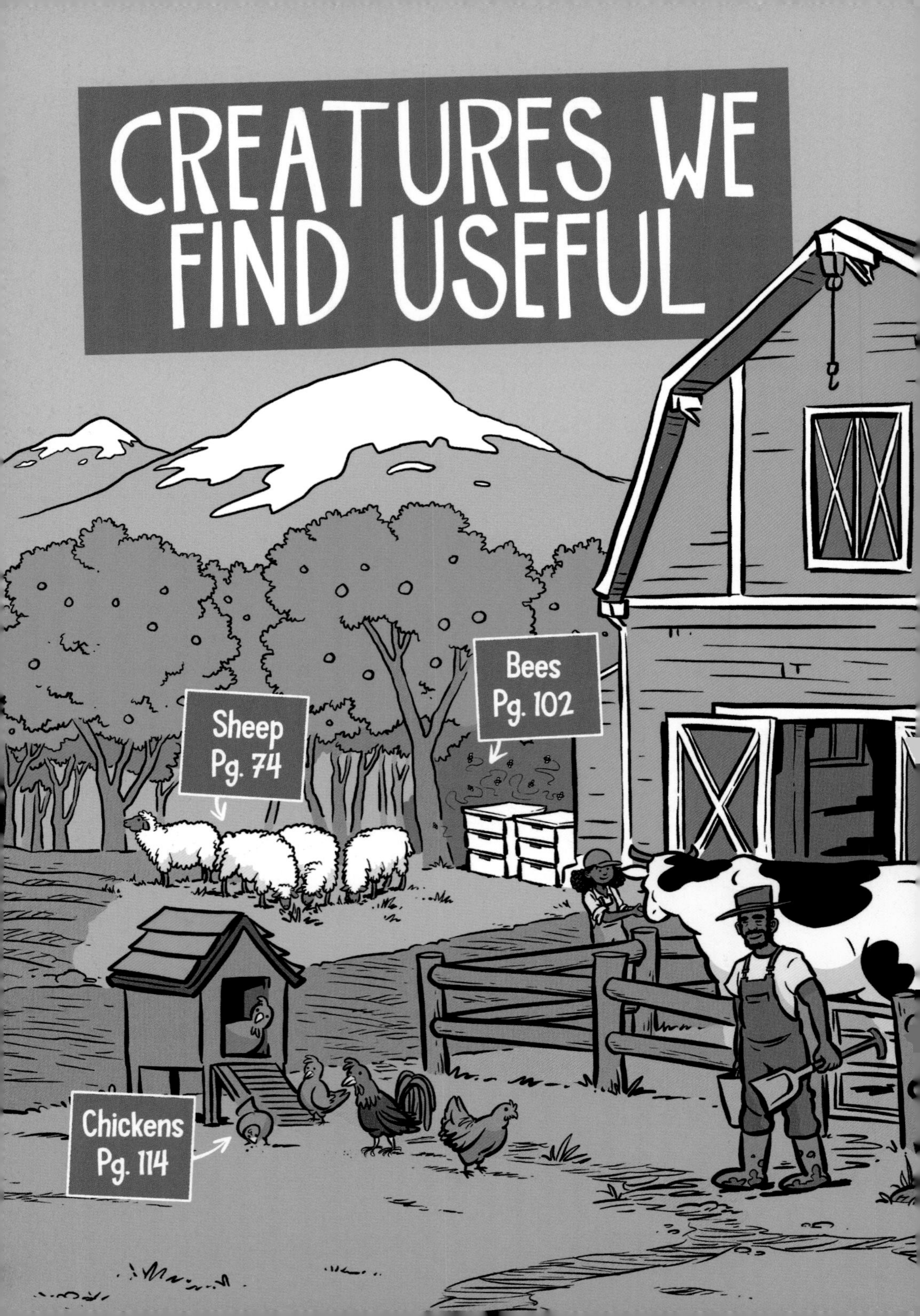
Sheep
Pg. 74
Bees
Pg. 102
Chickens
Pg. 114

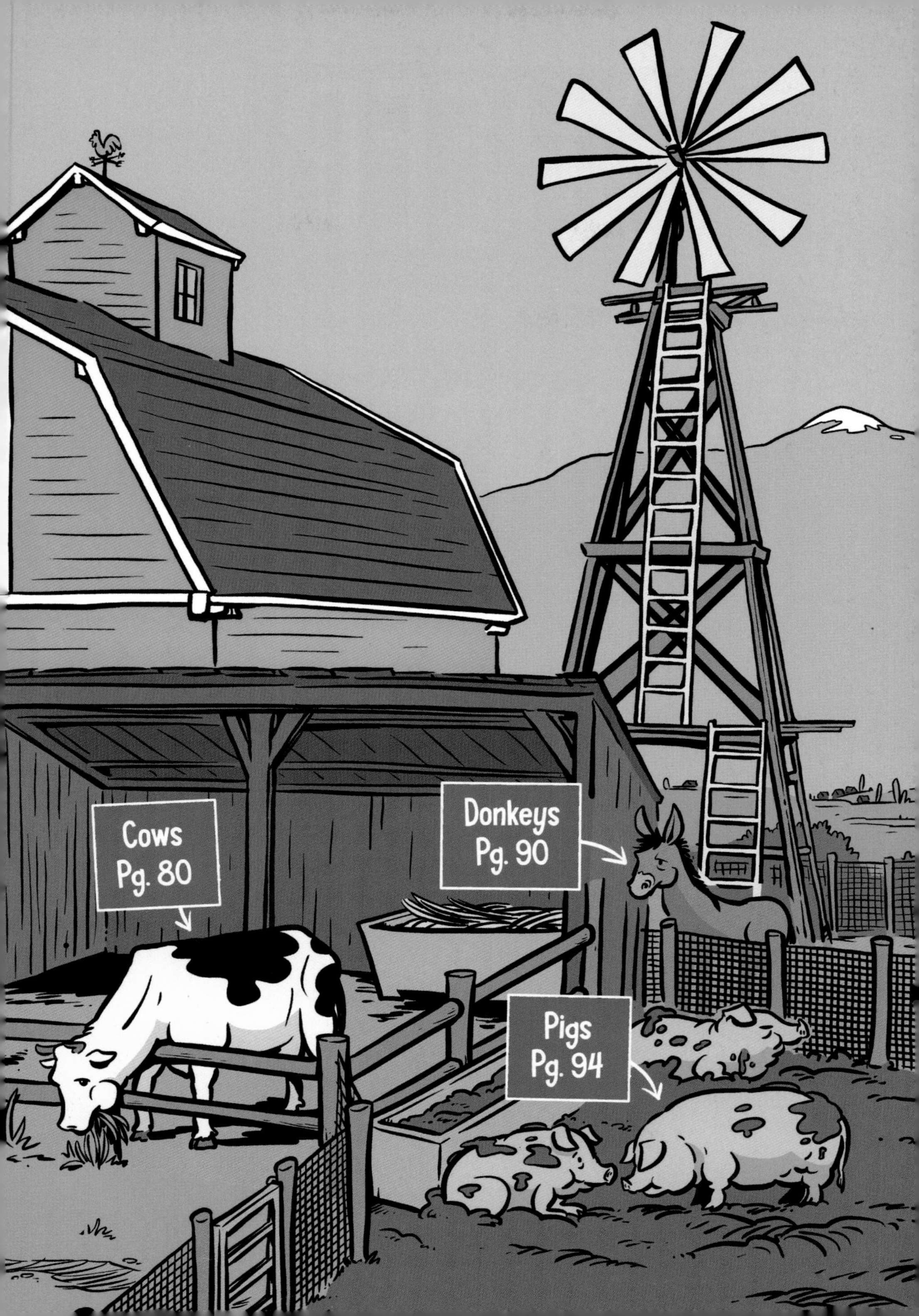
Cows
Pg. 80
Donkeys
Pg. 90
Pigs
Pg. 94

SHEEP

Ovis aries

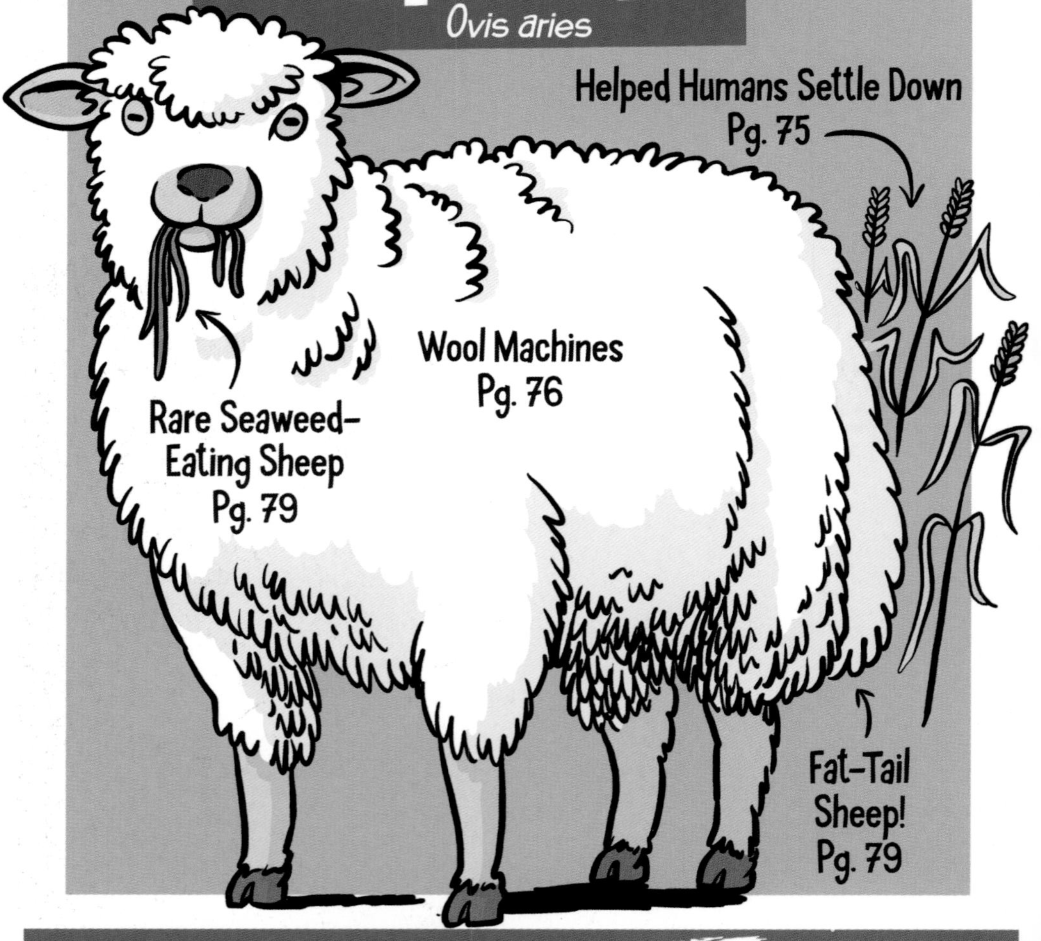

Helped Humans Settle Down
Pg. 75
Wool Machines
Pg. 76
Rare Seaweed-
Eating Sheep
Pg. 79
Fat-Tail
Sheep!
Pg. 79

Size: 2-4 ft (at the shoulder)

Weight: 90-350 lbs

Oddball Anatomy: Rectangular pupils for 320° vision

Diet: Grass, dandelions, clover

(Likely)
Wild Origin:

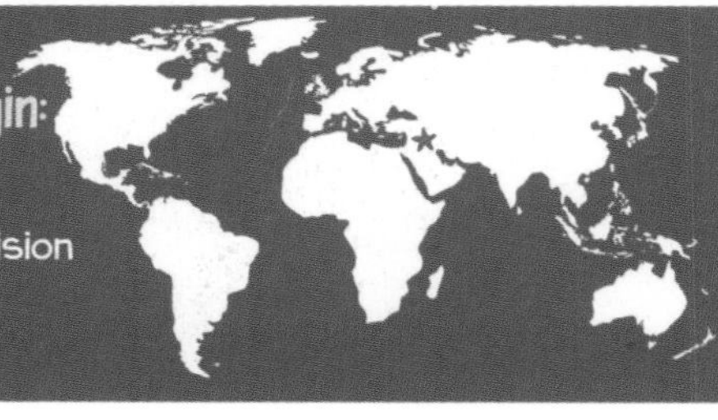

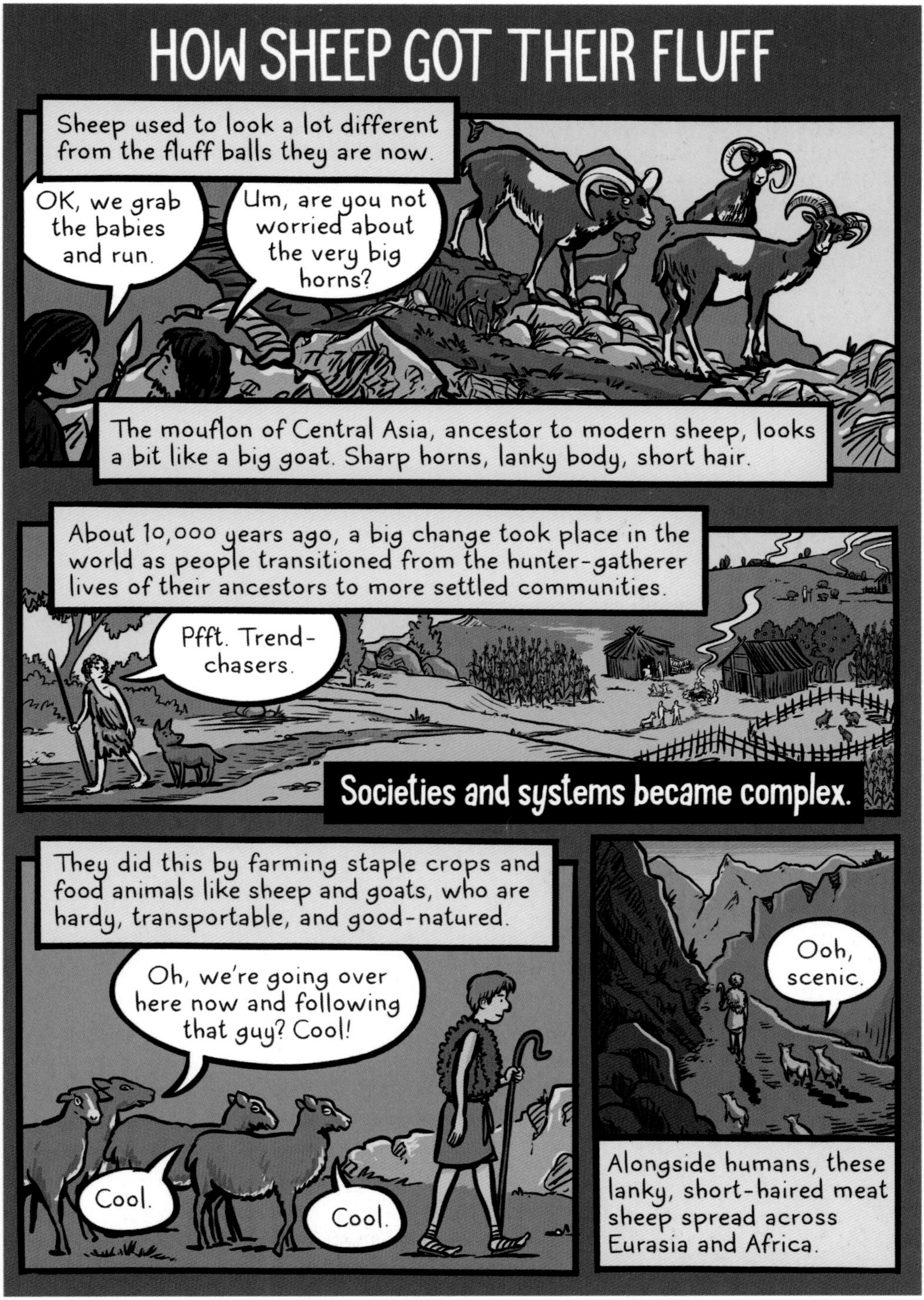
HOW SHEEP GOT THEIR FLUFF
Sheep used to look a lot different from the fluff balls they are now.
OK, we grab the babies and run.
Um, are you not worried about the very big horns?
The mouflon of Central Asia, ancestor to modern sheep, looks a bit like a big goat. Sharp horns, lanky body, short hair.
About 10,000 years ago, a big change took place in the world as people transitioned from the hunter-gatherer lives of their ancestors to more settled communities.
Pfft. Trend-chasers.
Societies and systems became complex.
They did this by farming staple crops and food animals like sheep and goats, who are hardy, transportable, and good-natured.
Oh, we're going over here now and following that guy? Cool!
Cool.
Cool.
Ooh, scenic.
Alongside humans, these lanky, short-haired meat sheep spread across Eurasia and Africa.

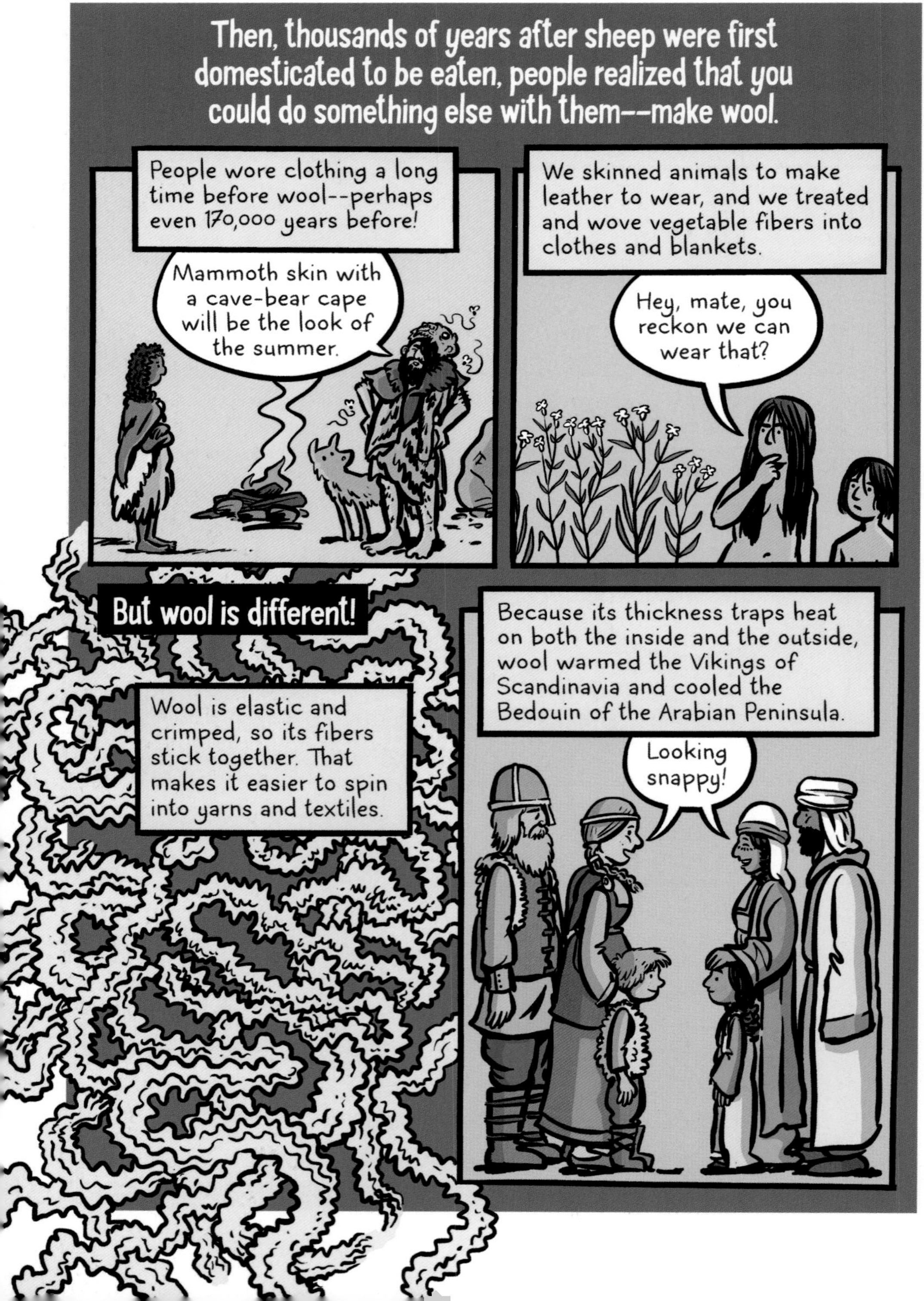
Then, thousands of years after sheep were first domesticated to be eaten, people realized that you could do something else with them--make wool.
People wore clothing a long time before wool--perhaps even 170,000 years before!
Mammoth skin with a cave-bear cape will be the look of the summer.
We skinned animals to make leather to wear, and we treated and wove vegetable fibers into clothes and blankets.
Hey, mate, you reckon we can wear that?
But wool is different!
Wool is elastic and crimped, so its fibers stick together. That makes it easier to spin into yarns and textiles.
Because its thickness traps heat on both the inside and the outside, wool warmed the Vikings of Scandinavia and cooled the Bedouin of the Arabian Peninsula.
Looking snappy!

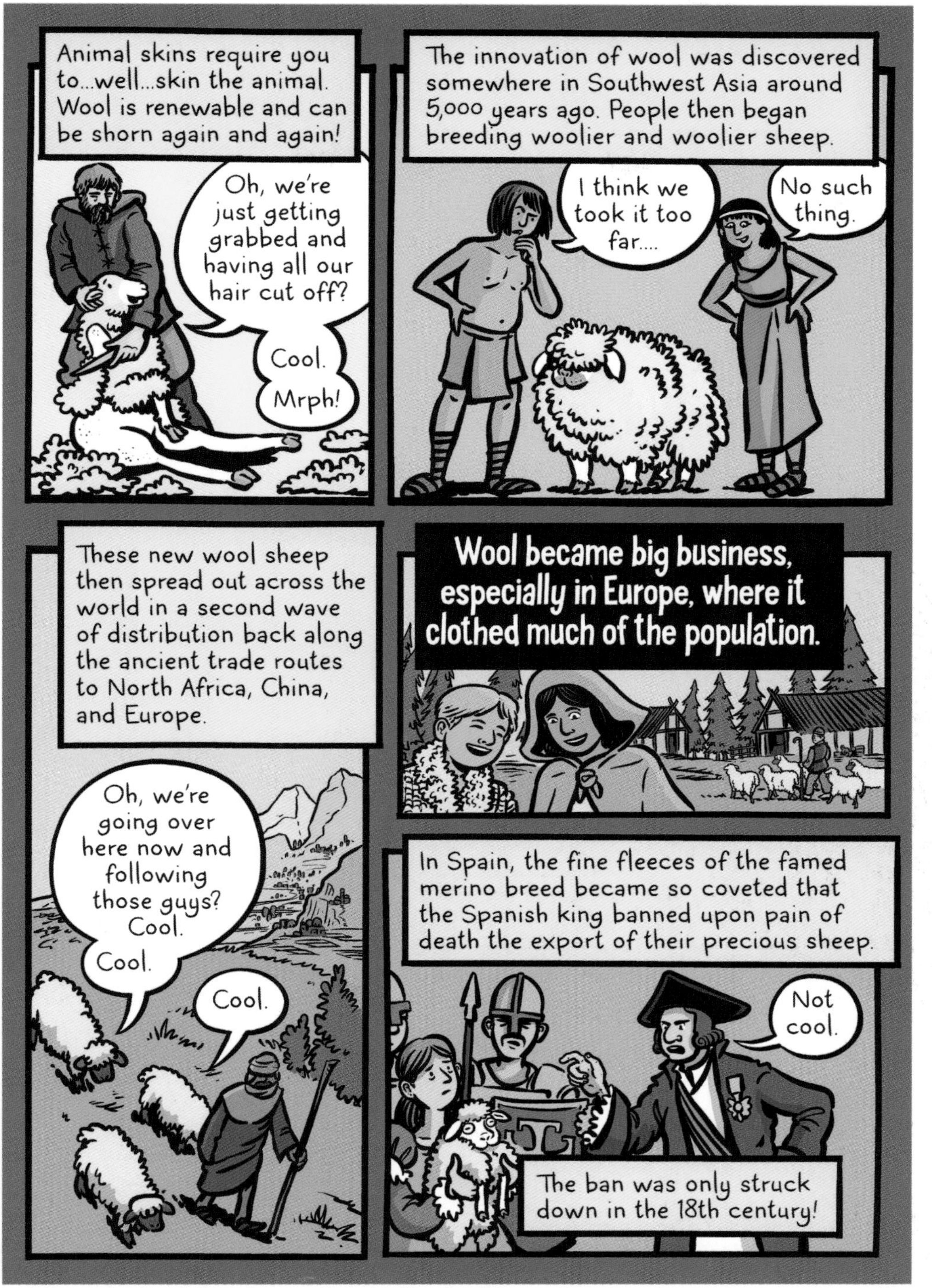
Animal skins require you to...well...skin the animal. Wool is renewable and can be shorn again and again!
Oh, we're just getting grabbed and having all our hair cut off?
Cool.
Mrph!
The innovation of wool was discovered somewhere in Southwest Asia around 5,000 years ago. People then began breeding woolier and woolier sheep.
I think we took it too far....
No such thing.
These new wool sheep then spread out across the world in a second wave of distribution back along the ancient trade routes to North Africa, China, and Europe.
Oh, we're going over here now and following those guys? Cool.
Cool.
Cool.
Wool became big business, especially in Europe, where it clothed much of the population.
In Spain, the fine fleeces of the famed merino breed became so coveted that the Spanish king banned upon pain of death the export of their precious sheep.
Not cool.
The ban was only struck down in the 18th century!

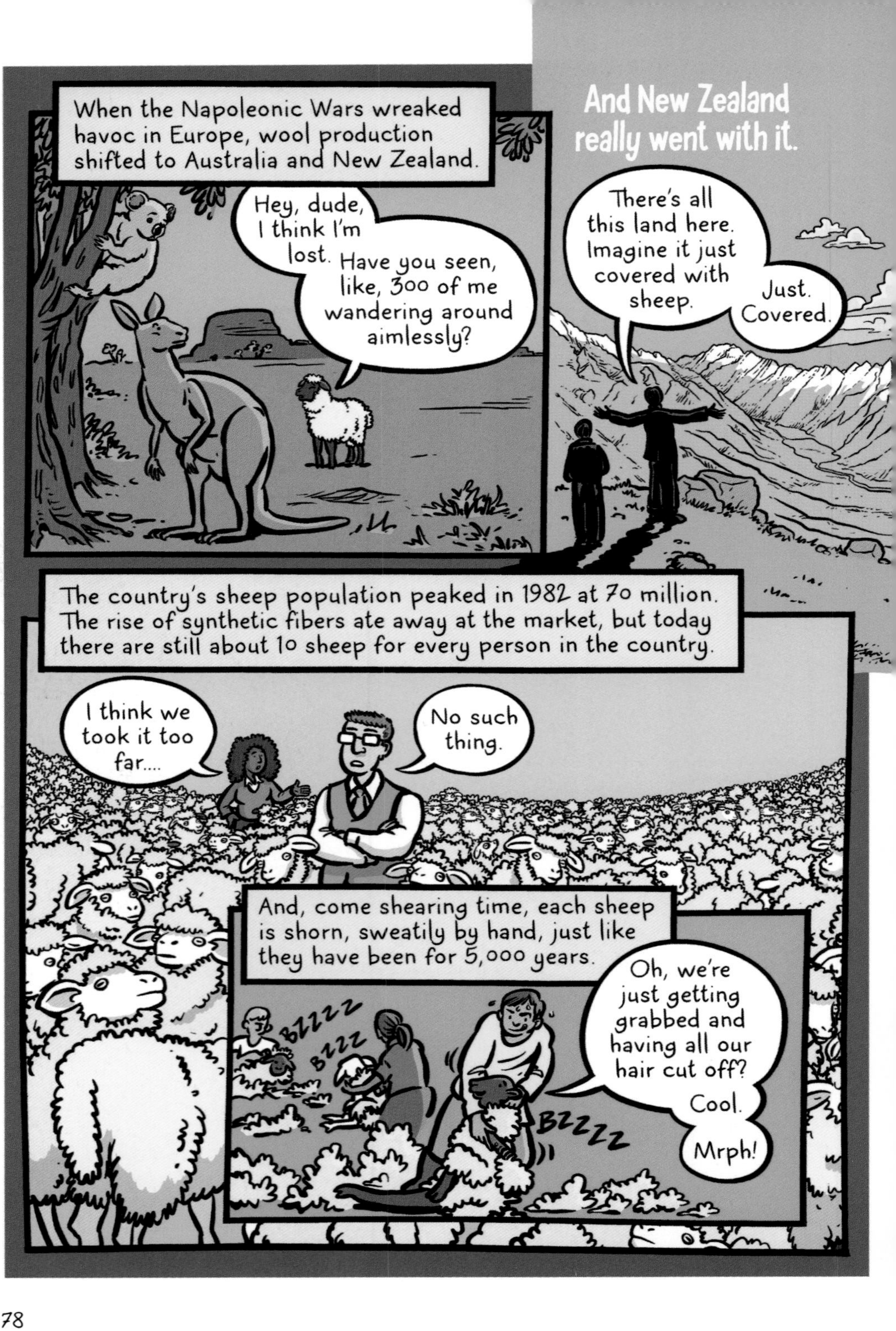

When the Napoleonic Wars wreaked havoc in Europe, wool production shifted to Australia and New Zealand.
Hey, dude, I think I'm lost.
Have you seen, like, 300 of me wandering around aimlessly?
And New Zealand really went with it.
There's all this land here. Imagine it just covered with sheep.
Just. Covered.
The country's sheep population peaked in 1982 at 70 million. The rise of synthetic fibers ate away at the market, but today there are still about 10 sheep for every person in the country.
I think we took it too far....
No such thing.
And, come shearing time, each sheep is shorn, sweatily by hand, just like they have been for 5,000 years.
BZZZZ
BZZZ
Oh, we're just getting grabbed and having all our hair cut off?
Cool.
BZZZZZ
Mrph!

SEAWEED-EATING SHEEP OF SCOTLAND

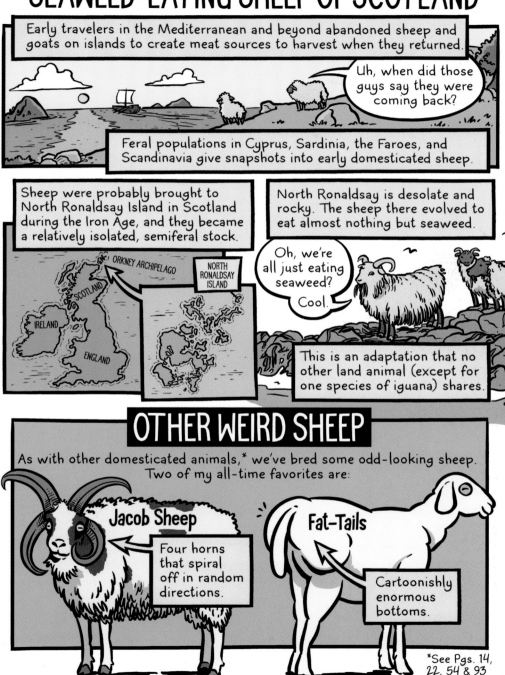

Early travelers in the Mediterranean and beyond abandoned sheep and goats on islands to create meat sources to harvest when they returned.

Uh, when did those guys say they were coming back?

Feral populations in Cyprus, Sardinia, the Faroes, and Scandinavia give snapshots into early domesticated sheep.

Sheep were probably brought to North Ronaldsay Island in Scotland during the Iron Age, and they became a relatively isolated, semiferal stock.

North Ronaldsay is desolate and rocky. The sheep there evolved to eat almost nothing but seaweed.

ORKNEY ARCHIPELAGO

NORTH RONALDSAY ISLAND

SCOTLAND

IRELAND

ENGLAND

Oh, we're all just eating seaweed? Cool.

This is an adaptation that no other land animal (except for one species of iguana) shares.

OTHER WEIRD SHEEP

As with other domesticated animals,* we've bred some odd-looking sheep. Two of my all-time favorites are:

Jacob Sheep

Four horns that spiral off in random directions.

Fat-Tails

Cartoonishly enormous bottoms.

*See Pgs. 14, 22, 54 & 93

79

COWS

Bos taurus

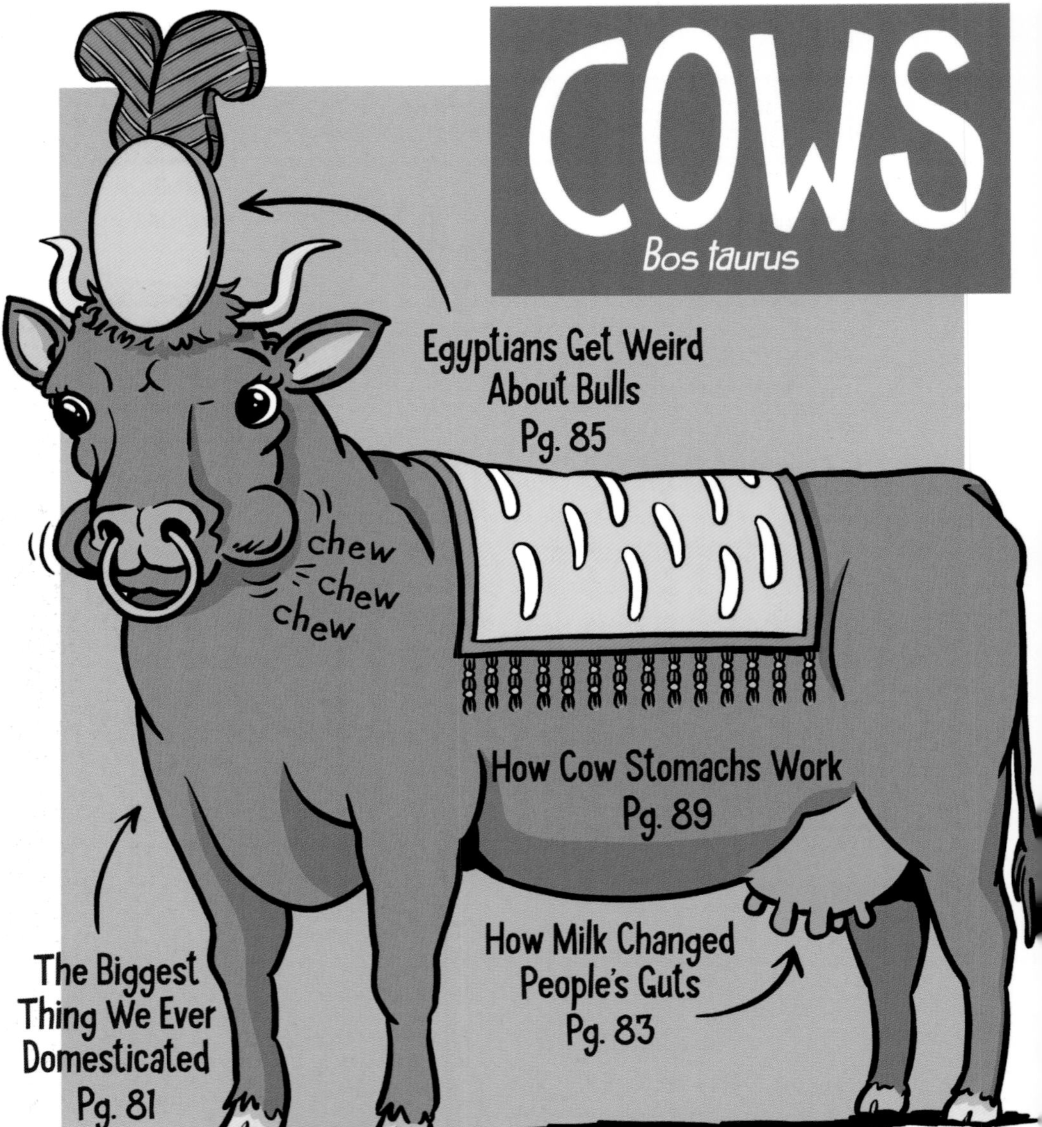

Egyptians Get Weird About Bulls Pg. 85
chew chew chew
How Cow Stomachs Work Pg. 89
How Milk Changed People's Guts Pg. 83
The Biggest Thing We Ever Domesticated Pg. 81

Size: 2-4 ft (at the shoulder)

Weight: 90-350 lbs

Majestic Mastication: Chews for 8 hours a day

Diet: Grass, dandelions, clover

(Likely) **Wild Origin:**

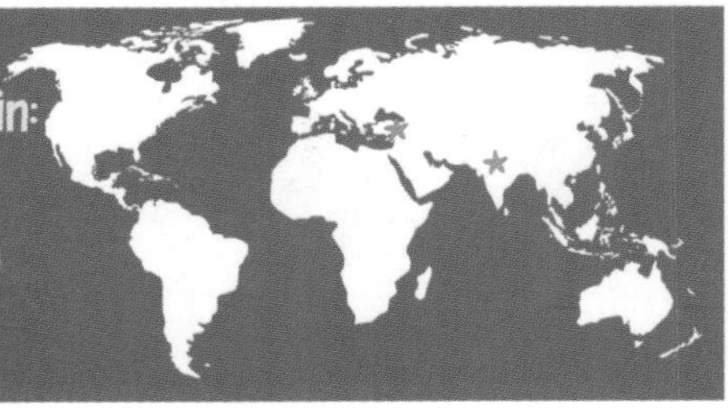

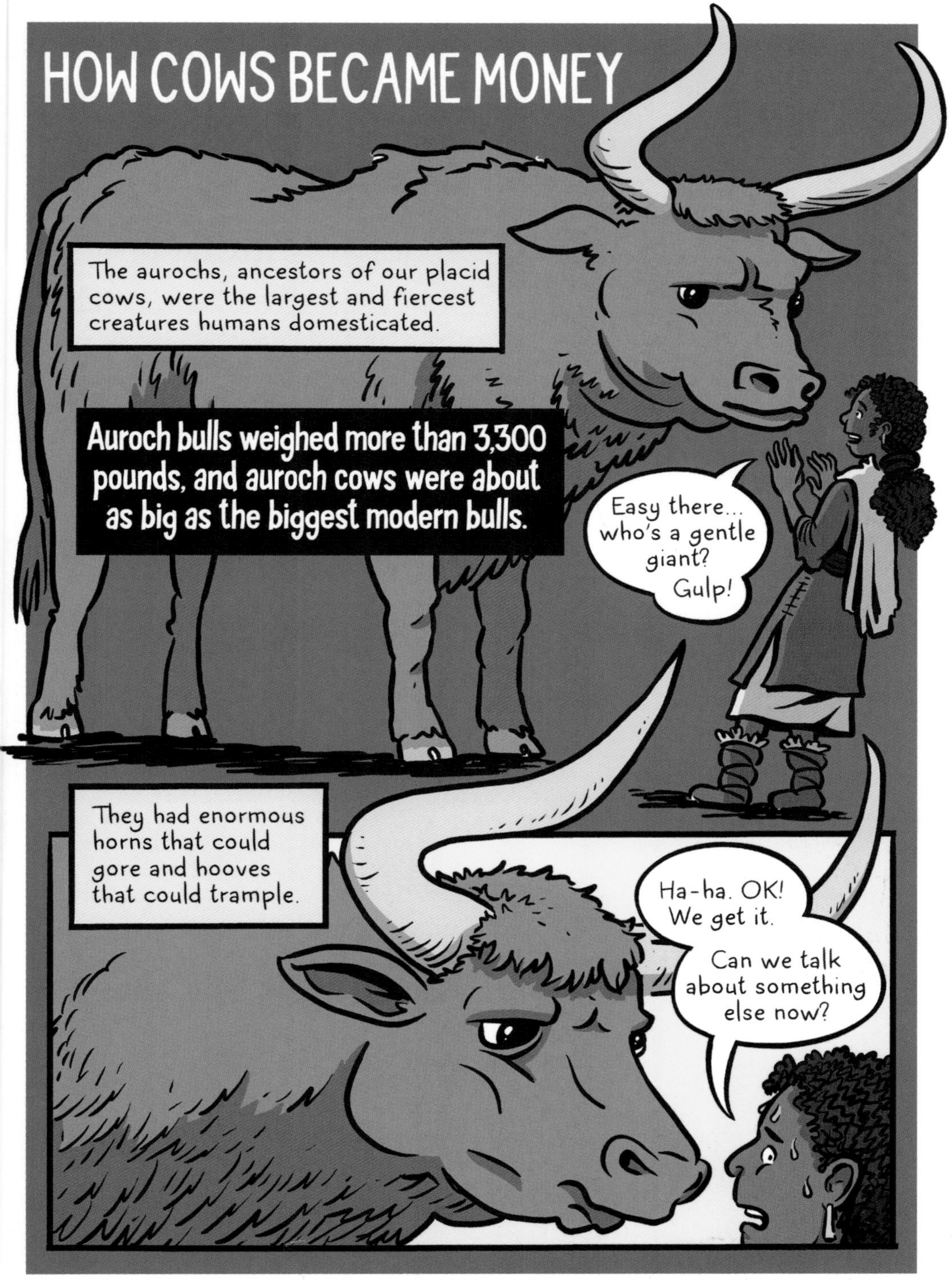
HOW COWS BECAME MONEY
The aurochs, ancestors of our placid cows, were the largest and fiercest creatures humans domesticated.
Auroch bulls weighed more than 3,300 pounds, and auroch cows were about as big as the biggest modern bulls.
Easy there... who's a gentle giant? Gulp!
They had enormous horns that could gore and hooves that could trample.
Ha-ha. OK! We get it.
Can we talk about something else now?

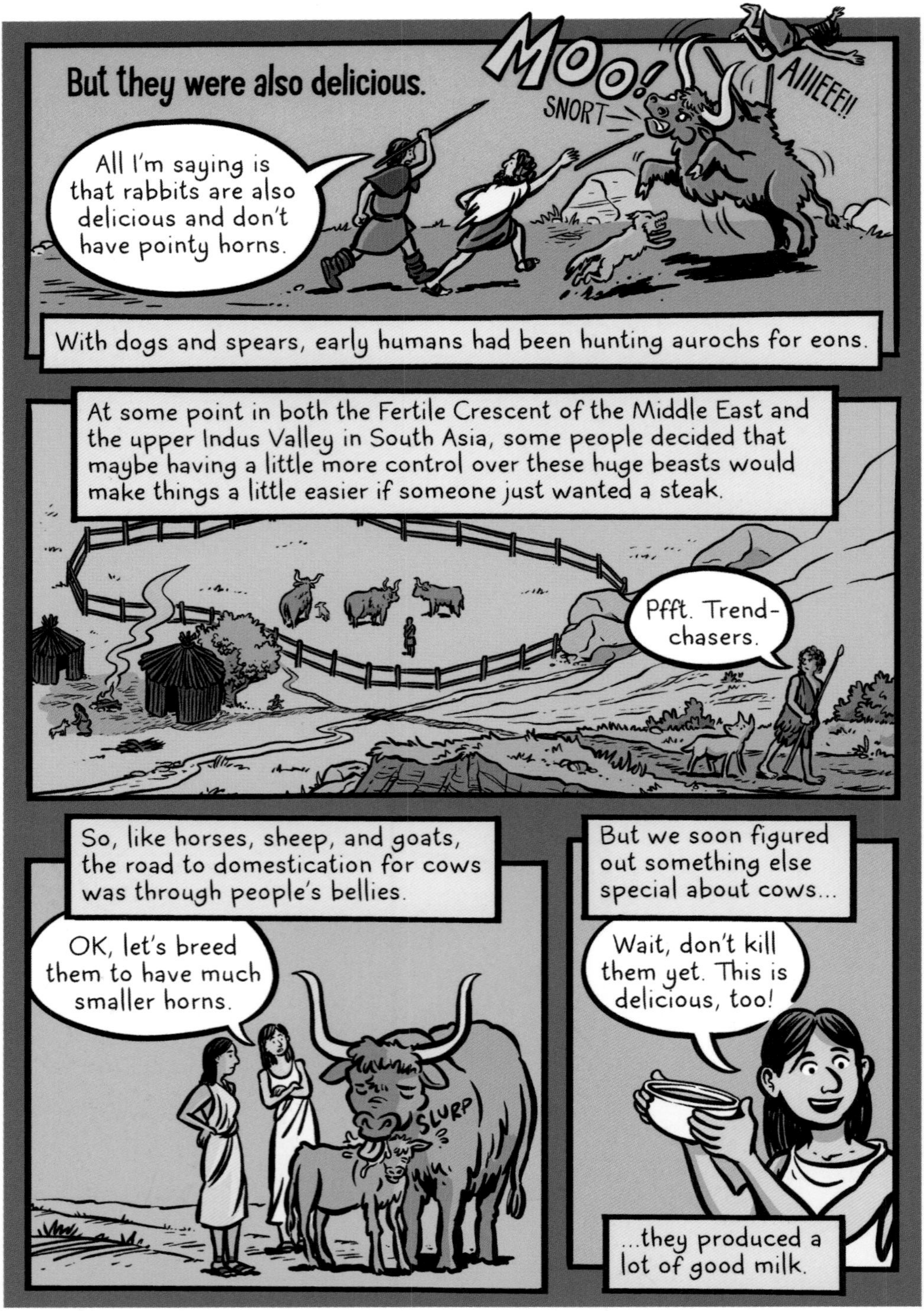
But they were also delicious.
MOO!
SNORT
AIIIEEE!!
All I'm saying is that rabbits are also delicious and don't have pointy horns.
With dogs and spears, early humans had been hunting aurochs for eons.
At some point in both the Fertile Crescent of the Middle East and the upper Indus Valley in South Asia, some people decided that maybe having a little more control over these huge beasts would make things a little easier if someone just wanted a steak.
Pfft. Trend-chasers.
So, like horses, sheep, and goats, the road to domestication for cows was through people's bellies.
OK, let's breed them to have much smaller horns.
SLURP
But we soon figured out something else special about cows...
Wait, don't kill them yet. This is delicious, too!
...they produced a lot of good milk.

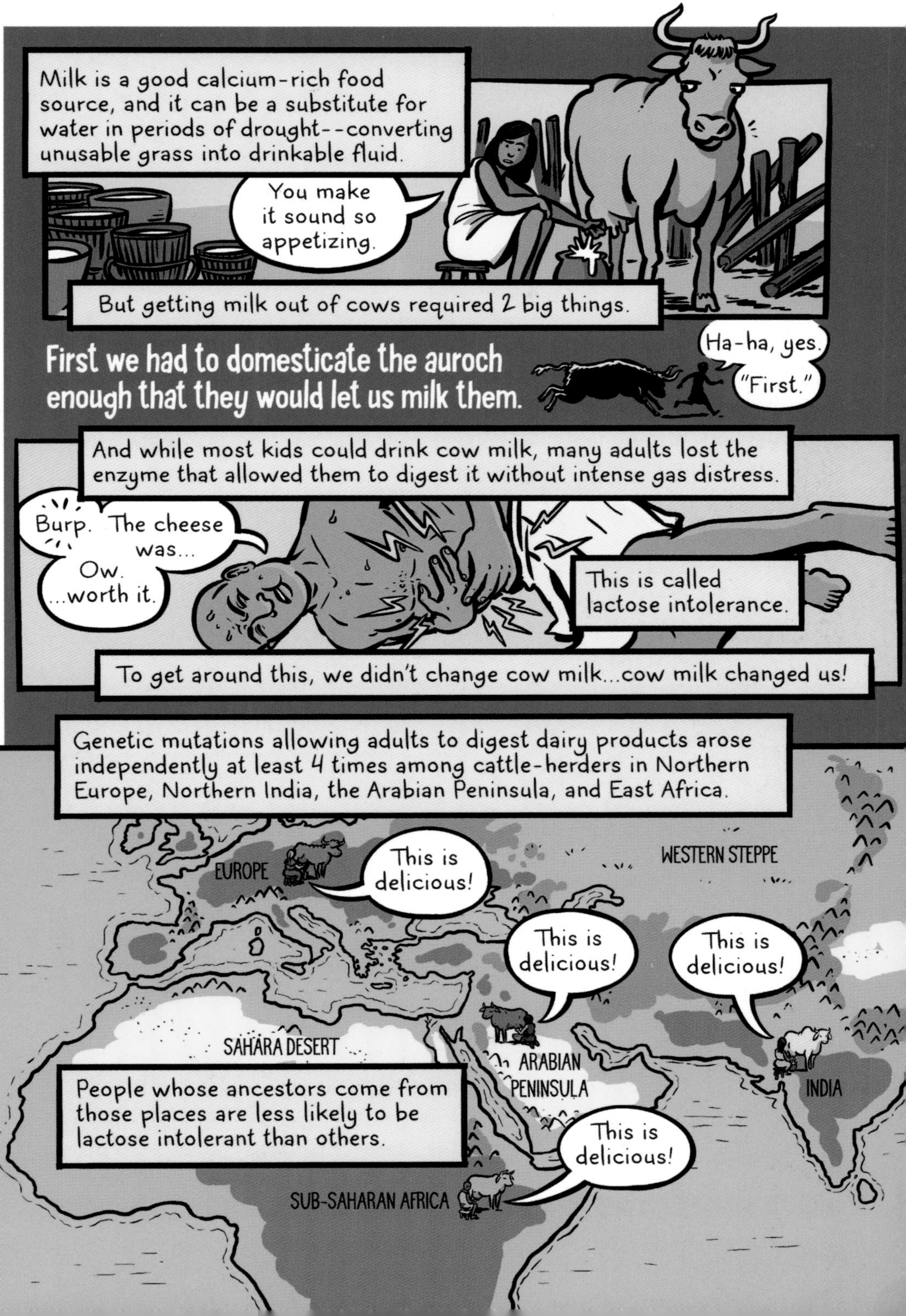
Milk is a good calcium-rich food source, and it can be a substitute for water in periods of drought--converting unusable grass into drinkable fluid.
You make it sound so appetizing.
But getting milk out of cows required 2 big things.
First we had to domesticate the auroch enough that they would let us milk them.
Ha-ha, yes. "First."
And while most kids could drink cow milk, many adults lost the enzyme that allowed them to digest it without intense gas distress.
Burp. The cheese was... Ow. ...worth it.
This is called lactose intolerance.
To get around this, we didn't change cow milk...cow milk changed us!
Genetic mutations allowing adults to digest dairy products arose independently at least 4 times among cattle-herders in Northern Europe, Northern India, the Arabian Peninsula, and East Africa.
EUROPE
This is delicious!
WESTERN STEPPE
This is delicious!
This is delicious!
SAHARA DESERT
ARABIAN PENINSULA
INDIA
People whose ancestors come from those places are less likely to be lactose intolerant than others.
This is delicious!
SUB-SAHARAN AFRICA

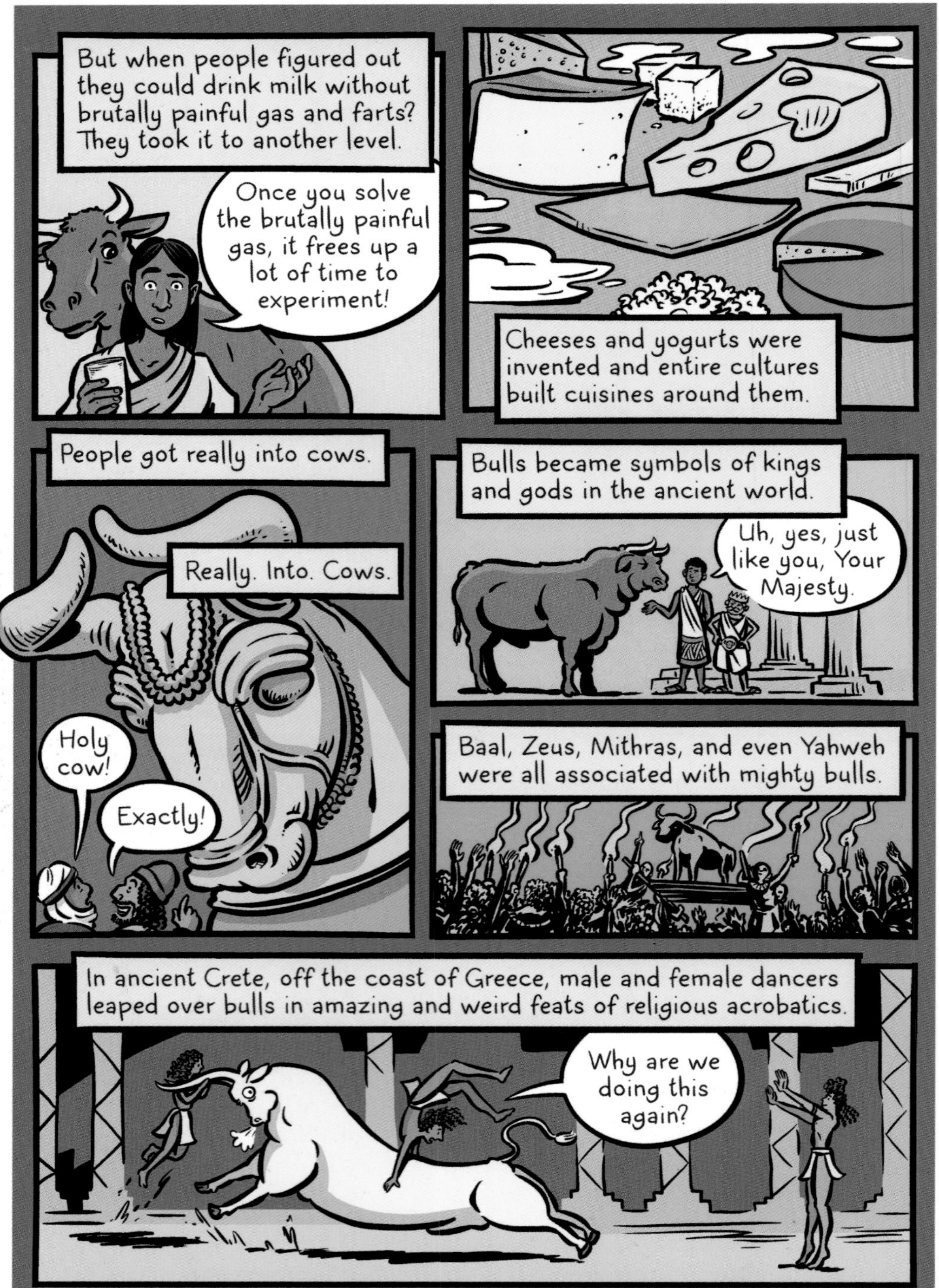
But when people figured out they could drink milk without brutally painful gas and farts? They took it to another level.
Once you solve the brutally painful gas, it frees up a lot of time to experiment!
Cheeses and yogurts were invented and entire cultures built cuisines around them.
People got really into cows.
Really. Into. Cows.
Holy cow!
Exactly!
Bulls became symbols of kings and gods in the ancient world.
Uh, yes, just like you, Your Majesty.
Baal, Zeus, Mithras, and even Yahweh were all associated with mighty bulls.
In ancient Crete, off the coast of Greece, male and female dancers leaped over bulls in amazing and weird feats of religious acrobatics.
Why are we doing this again?

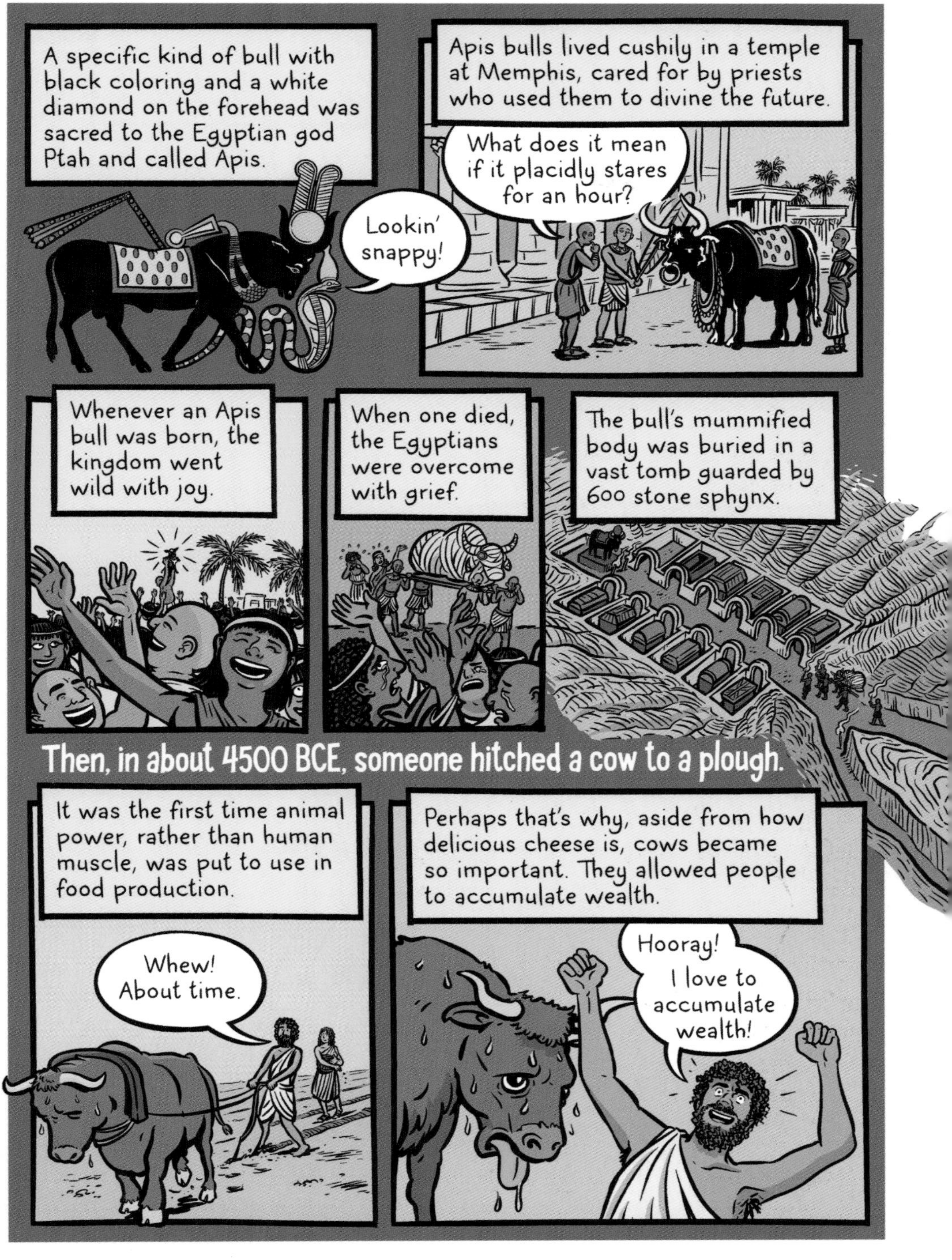
A specific kind of bull with black coloring and a white diamond on the forehead was sacred to the Egyptian god Ptah and called Apis.
Lookin' snappy!
Apis bulls lived cushily in a temple at Memphis, cared for by priests who used them to divine the future.
What does it mean if it placidly stares for an hour?
Whenever an Apis bull was born, the kingdom went wild with joy.
When one died, the Egyptians were overcome with grief.
The bull's mummified body was buried in a vast tomb guarded by 600 stone sphynx.
Then, in about 4500 BCE, someone hitched a cow to a plough.
It was the first time animal power, rather than human muscle, was put to use in food production.
Whew! About time.
Perhaps that's why, aside from how delicious cheese is, cows became so important. They allowed people to accumulate wealth.
Hooray!
I love to accumulate wealth!

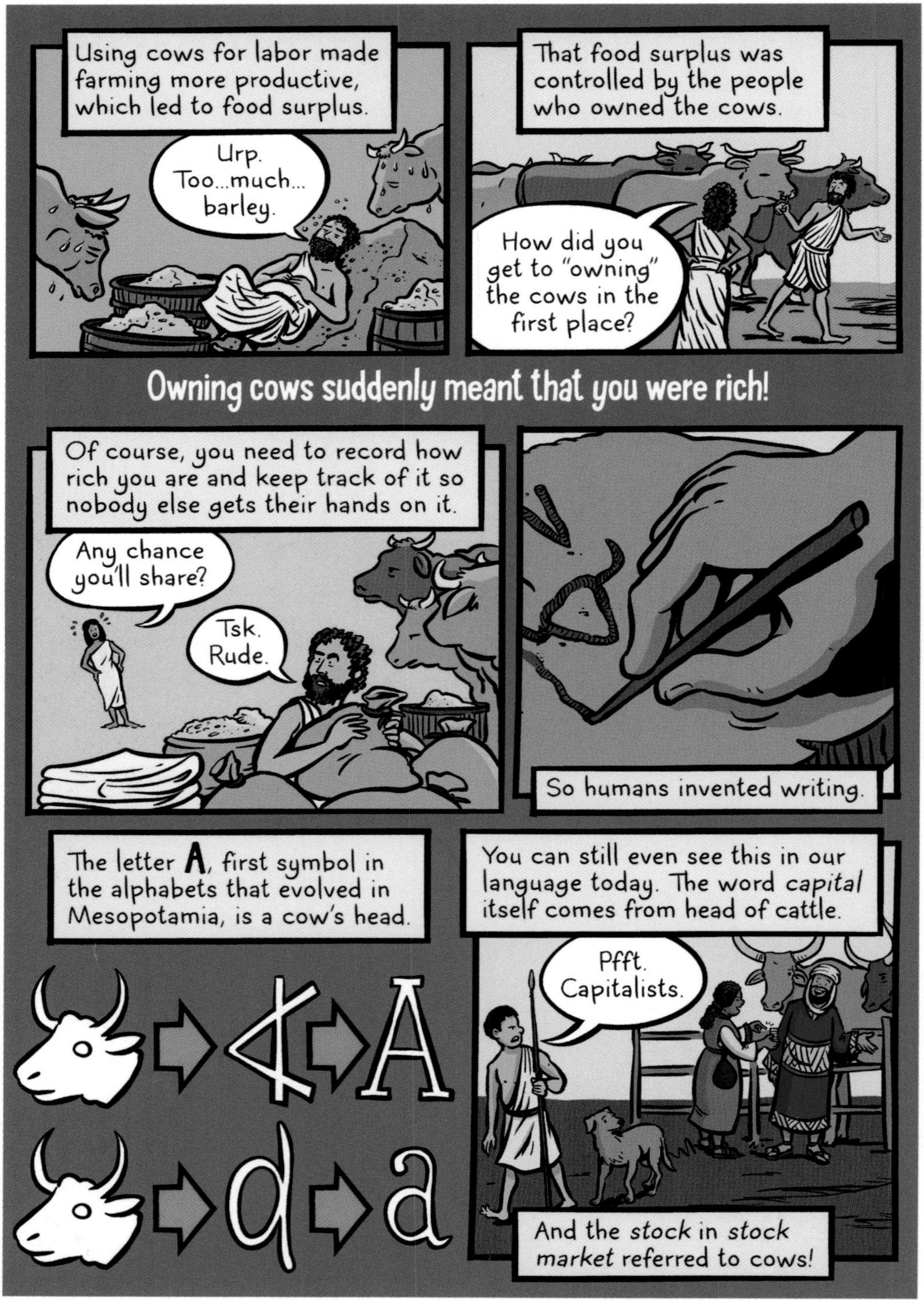
Using cows for labor made farming more productive, which led to food surplus.
Urp. Too...much... barley.
That food surplus was controlled by the people who owned the cows.
How did you get to "owning" the cows in the first place?
Owning cows suddenly meant that you were rich!
Of course, you need to record how rich you are and keep track of it so nobody else gets their hands on it.
Any chance you'll share?
Tsk. Rude.
So humans invented writing.
The letter A, first symbol in the alphabets that evolved in Mesopotamia, is a cow's head.
A
a
You can still even see this in our language today. The word *capital* itself comes from head of cattle.
Pfft. Capitalists.
And the *stock* in *stock market* referred to cows!

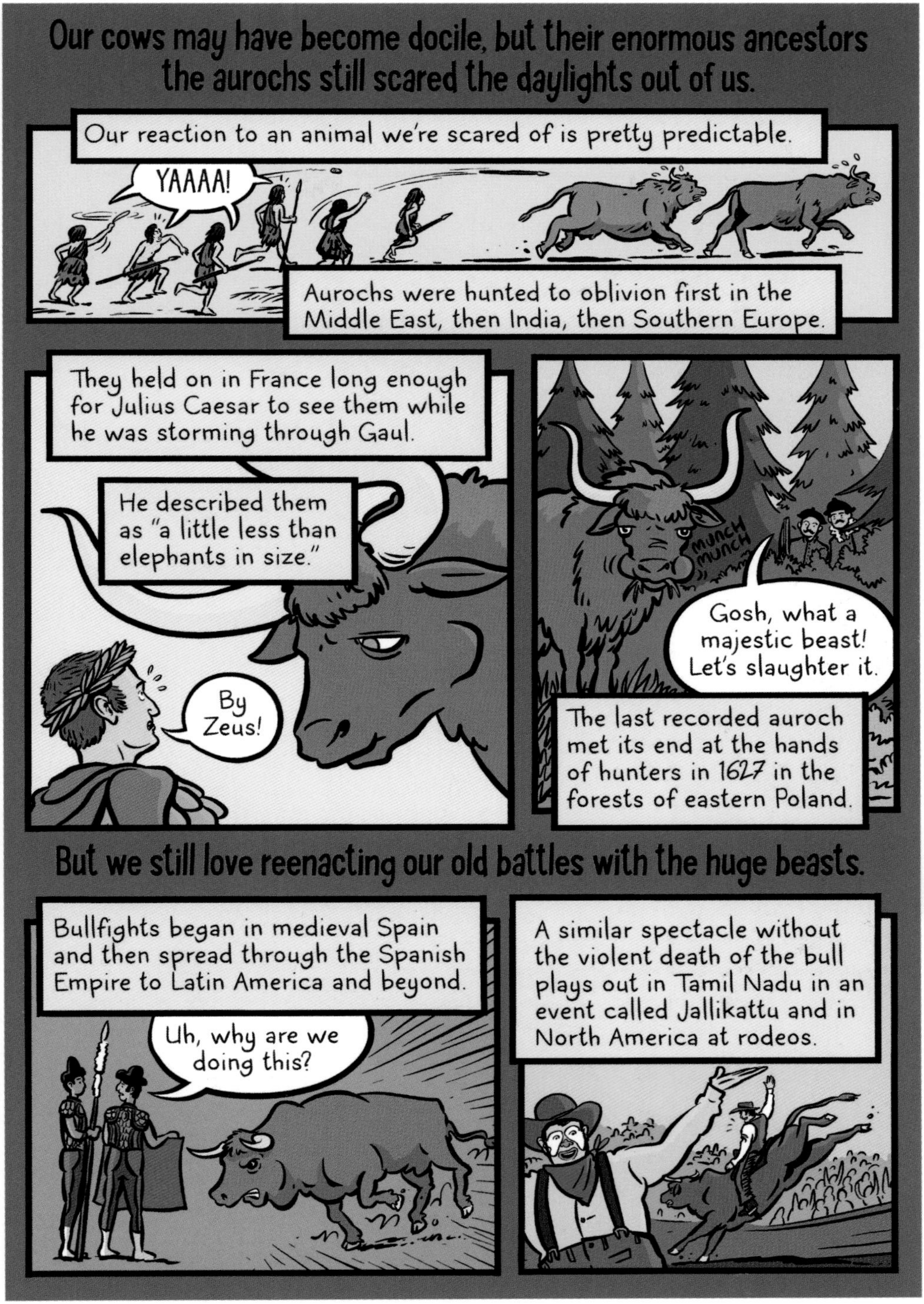
Our cows may have become docile, but their enormous ancestors the aurochs still scared the daylights out of us.
Our reaction to an animal we're scared of is pretty predictable.
YAAAA!
Aurochs were hunted to oblivion first in the Middle East, then India, then Southern Europe.
They held on in France long enough for Julius Caesar to see them while he was storming through Gaul.
He described them as "a little less than elephants in size."
By Zeus!
MUNCH MUNCH
Gosh, what a majestic beast! Let's slaughter it.
The last recorded auroch met its end at the hands of hunters in 1627 in the forests of eastern Poland.
But we still love reenacting our old battles with the huge beasts.
Bullfights began in medieval Spain and then spread through the Spanish Empire to Latin America and beyond.
Uh, why are we doing this?
A similar spectacle without the violent death of the bull plays out in Tamil Nadu in an event called Jallikattu and in North America at rodeos.

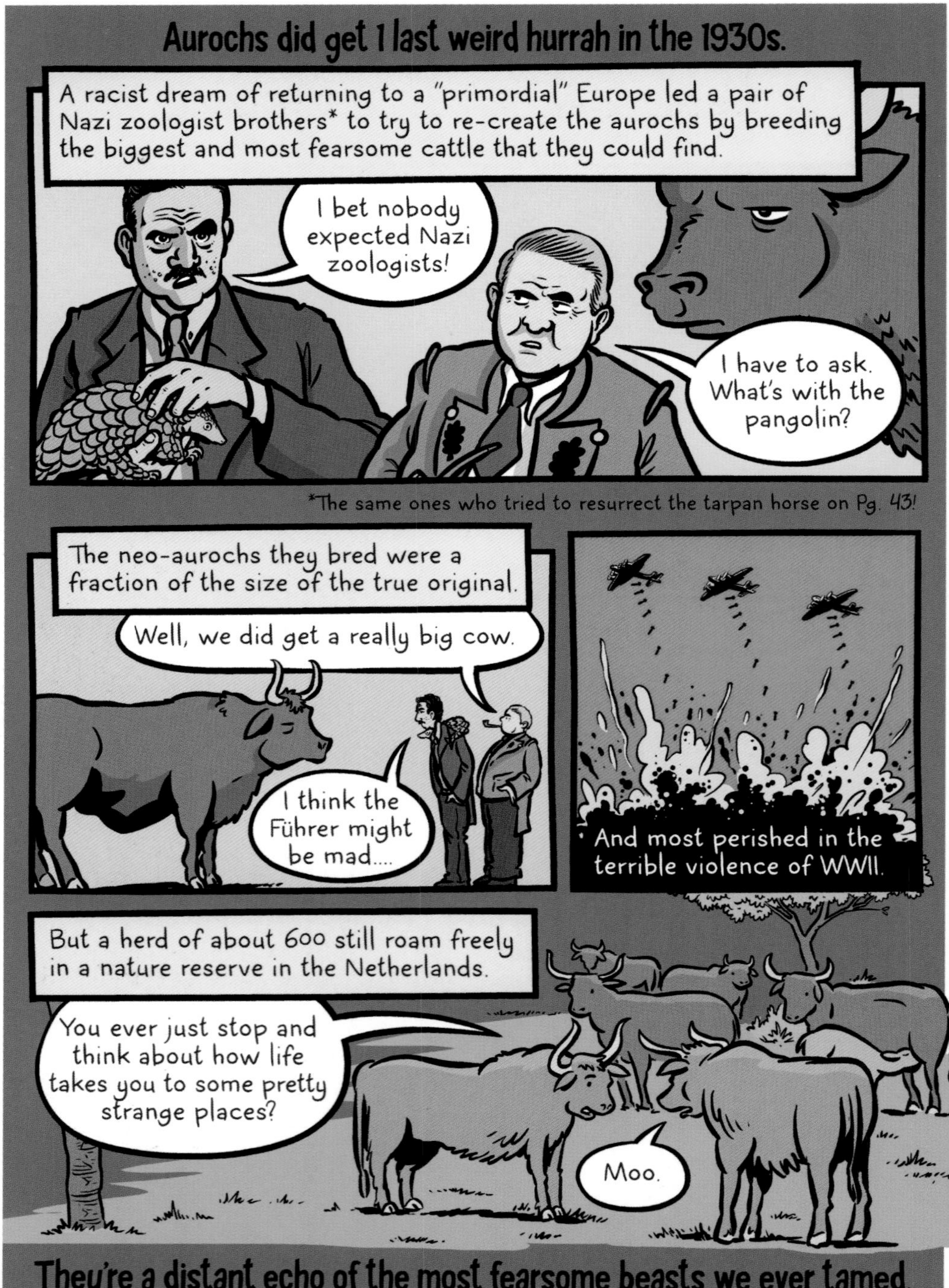

Aurochs did get 1 last weird hurrah in the 1930s.
A racist dream of returning to a "primordial" Europe led a pair of Nazi zoologist brothers* to try to re-create the aurochs by breeding the biggest and most fearsome cattle that they could find.
I bet nobody expected Nazi zoologists!
I have to ask. What's with the pangolin?
*The same ones who tried to resurrect the tarpan horse on Pg. 43!
The neo-aurochs they bred were a fraction of the size of the true original.
Well, we did get a really big cow.
I think the Führer might be mad....
And most perished in the terrible violence of WWII.
But a herd of about 600 still roam freely in a nature reserve in the Netherlands.
You ever just stop and think about how life takes you to some pretty strange places?
Moo.
They're a distant echo of the most fearsome beasts we ever tamed.

HOW COW STOMACHS WORK
Cows have 1 really big stomach with 4 parts.

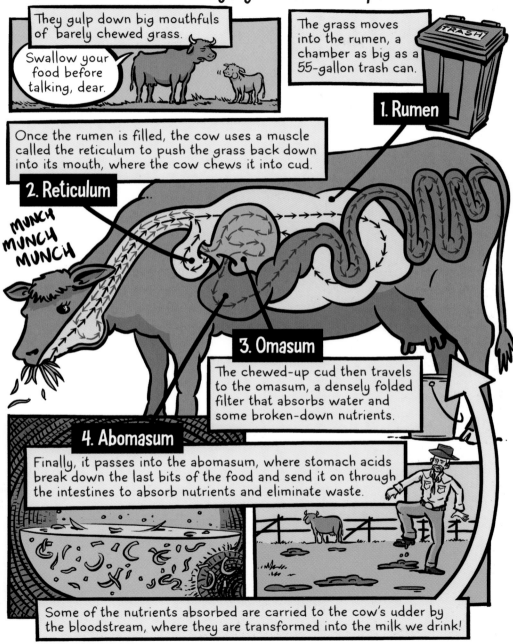

They gulp down big mouthfuls of barely chewed grass.

Swallow your food before talking, dear.

The grass moves into the rumen, a chamber as big as a 55-gallon trash can.

TRASH

1. Rumen

Once the rumen is filled, the cow uses a muscle called the reticulum to push the grass back down into its mouth, where the cow chews it into cud.

2. Reticulum

MUNCH MUNCH MUNCH

3. Omasum

The chewed-up cud then travels to the omasum, a densely folded filter that absorbs water and some broken-down nutrients.

4. Abomasum

Finally, it passes into the abomasum, where stomach acids break down the last bits of the food and send it on through the intestines to absorb nutrients and eliminate waste.

Some of the nutrients absorbed are carried to the cow's udder by the bloodstream, where they are transformed into the milk we drink!

DONKEYS

Equus africanus asinus

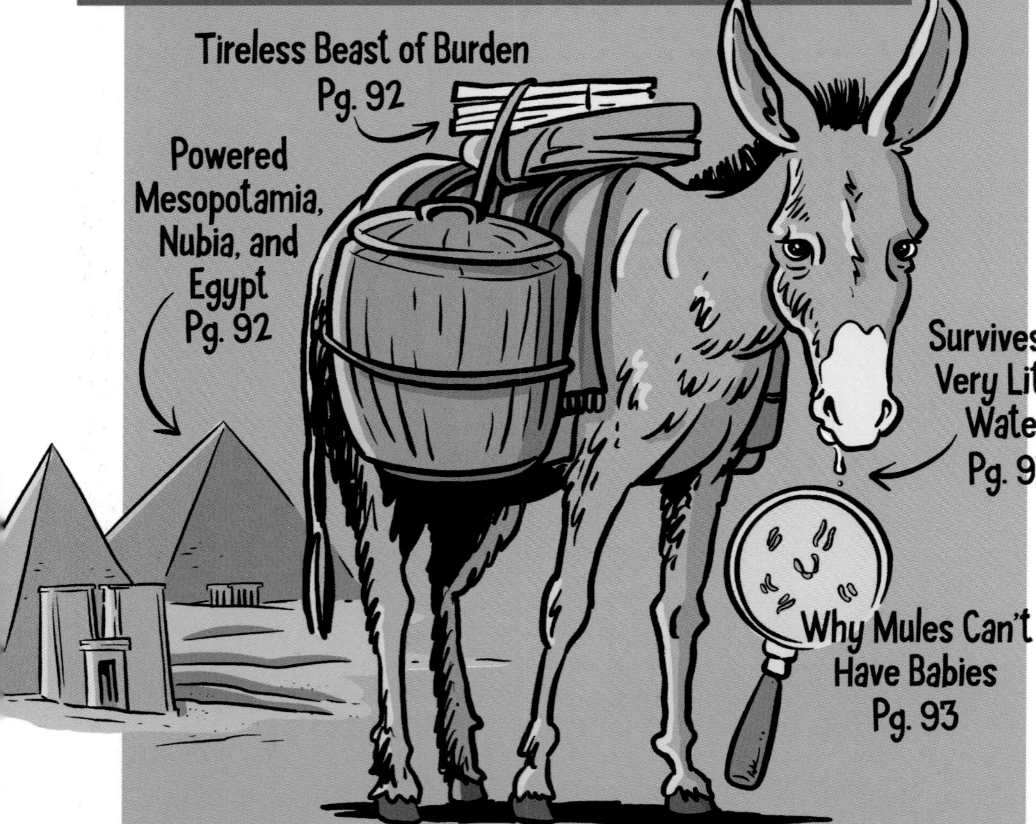

Tireless Beast of Burden
Pg. 92
Powered Mesopotamia, Nubia, and Egypt
Pg. 92
Survives on Very Little Water
Pg. 91
Why Mules Can't Have Babies
Pg. 93

Size: 2.6–5.2 ft (at the shoulder)

Weight: 180–1,100 lbs

Unusual Use: Guarding sheep from coyotes

Diet: Straw, hay, and grass

(Likely)
Wild Origin:

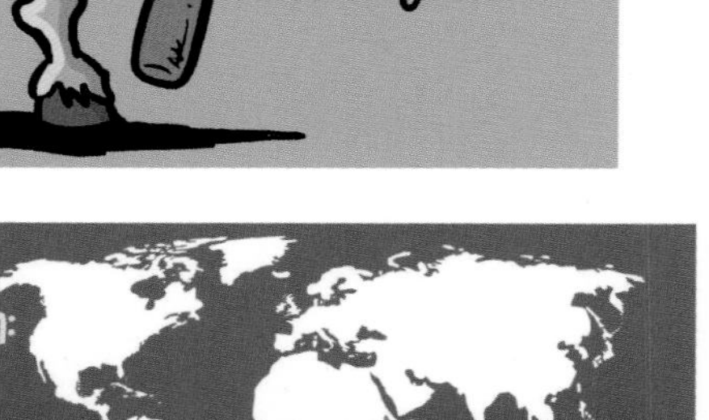

DONKEYS, MULES, AND CHROMOSOMES

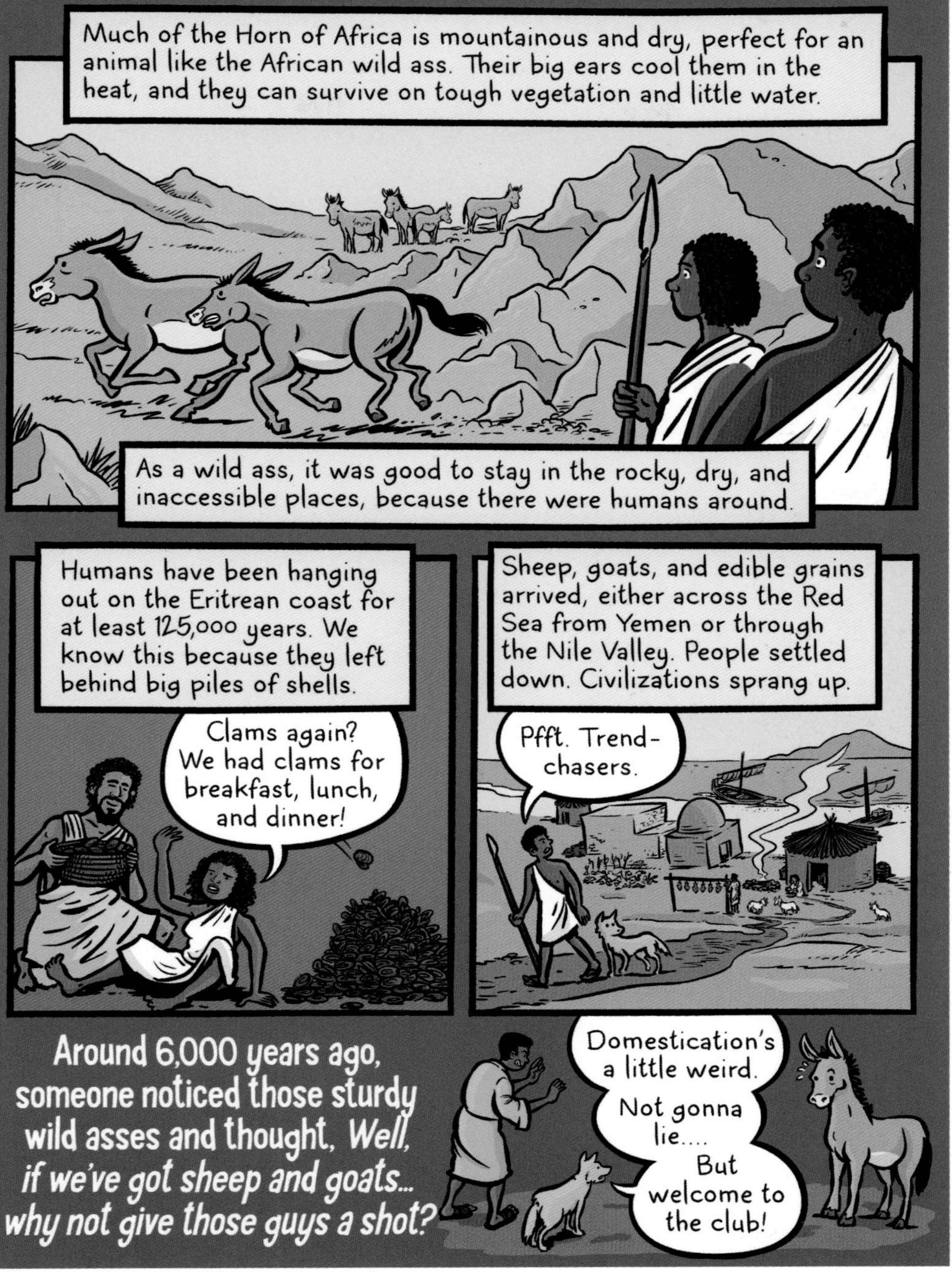
Much of the Horn of Africa is mountainous and dry, perfect for an animal like the African wild ass. Their big ears cool them in the heat, and they can survive on tough vegetation and little water.
As a wild ass, it was good to stay in the rocky, dry, and inaccessible places, because there were humans around.
Humans have been hanging out on the Eritrean coast for at least 125,000 years. We know this because they left behind big piles of shells.
Clams again? We had clams for breakfast, lunch, and dinner!
Sheep, goats, and edible grains arrived, either across the Red Sea from Yemen or through the Nile Valley. People settled down. Civilizations sprang up.
Pfft. Trend-chasers.
Around 6,000 years ago, someone noticed those sturdy wild asses and thought, *Well, if we've got sheep and goats... why not give those guys a shot?*
Domestication's a little weird.
Not gonna lie....
But welcome to the club!

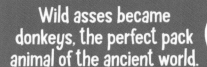 Wild asses became donkeys, the perfect pack animal of the ancient world.

 Believe me, I'd help. But my paw has been acting up.

Sigh.

Increasingly complex societies depended on the hardy animals, from the pyramid-building Egyptians and Nubians of the Nile Valley to the ziggurat-building Mesopotamians of the Fertile Crescent.

 Sigh.

 Sigh.

While they don't have as much raw power, donkeys don't need to stop to chew their cud like oxen, and so can keep pretty much plodding along at whatever they're doing.

 So, because I don't chew as much, I get to carry the heavy stuff. Great.

 Sigh.

So they plodded along with trading caravans to Europe and East Asia, then eventually to Australia and the Americas. Not as pretty as a horse, but useful.

Speaking of horses, at some point, someone got to thinking about how similar they are to donkeys...

...and tried breeding them together.

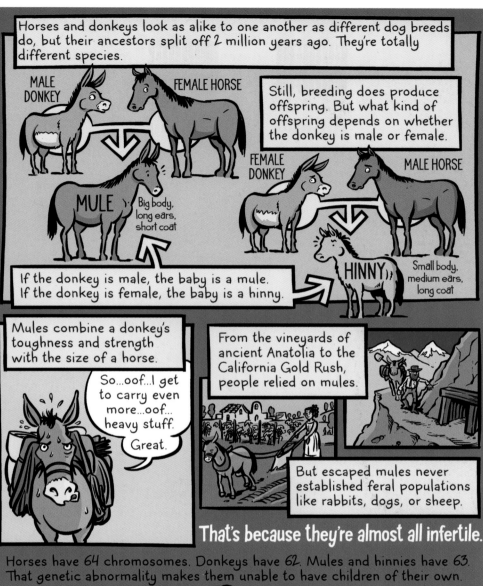

Horses and donkeys look as alike to one another as different dog breeds do, but their ancestors split off 2 million years ago. They're totally different species.

MALE DONKEY

FEMALE HORSE

MULE — Big body, long ears, short coat

Still, breeding does produce offspring. But what kind of offspring depends on whether the donkey is male or female.

FEMALE DONKEY

MALE HORSE

HINNY — Small body, medium ears, long coat

If the donkey is male, the baby is a mule. If the donkey is female, the baby is a hinny.

Mules combine a donkey's toughness and strength with the size of a horse.

So...oof...I get to carry even more...oof... heavy stuff. Great.

From the vineyards of ancient Anatolia to the California Gold Rush, people relied on mules.

But escaped mules never established feral populations like rabbits, dogs, or sheep.

That's because they're almost all infertile.

Horses have 64 chromosomes. Donkeys have 62. Mules and hinnies have 63. That genetic abnormality makes them unable to have children of their own.

Another way to look at it is that humans have been doing some weird science experiments for a lot longer than you'd think!

93

PIGS

Sus domesticus

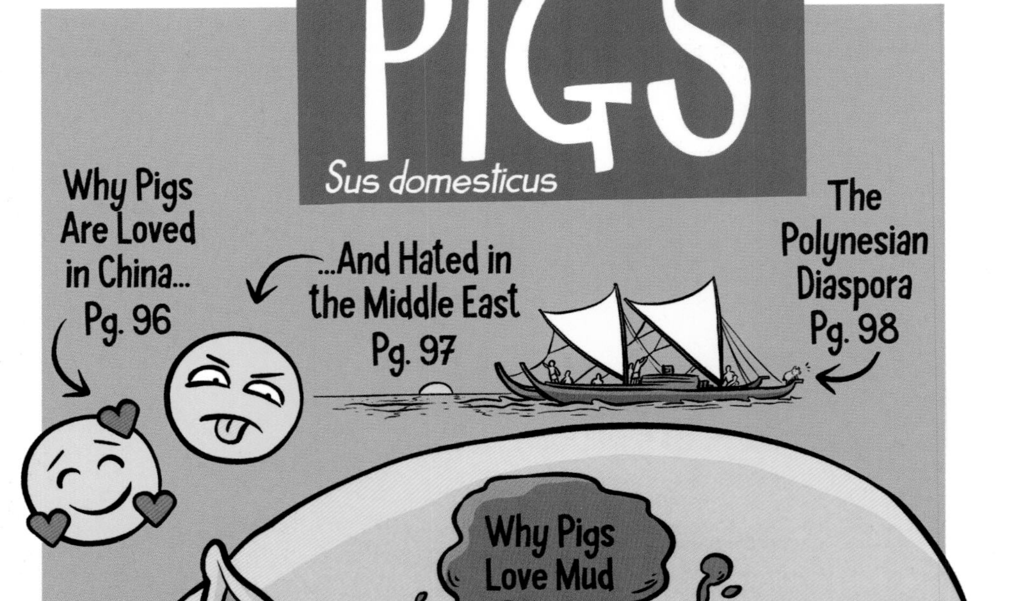

Why Pigs Are Loved in China... Pg. 96
...And Hated in the Middle East Pg. 97
The Polynesian Diaspora Pg. 98
Why Pigs Love Mud Pg. 100
Garbage Disposals Pg. 96

Size: 3-4.7 ft (at the shoulder)

Weight: 110–1,000 lbs

Unusual Use: Hunting for truffles (an edible wild fungus)

Diet: Almost anything they get ahold of

(Likely) **Wild Origin:**

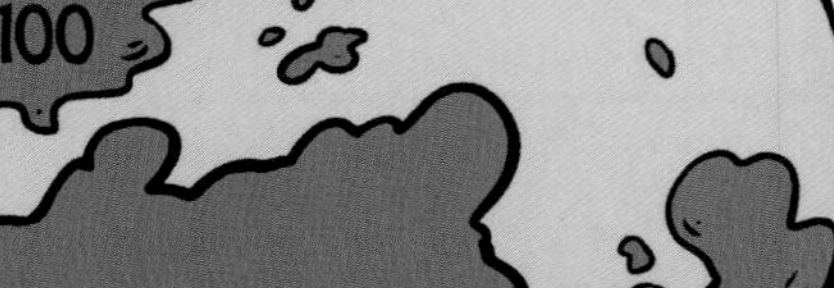

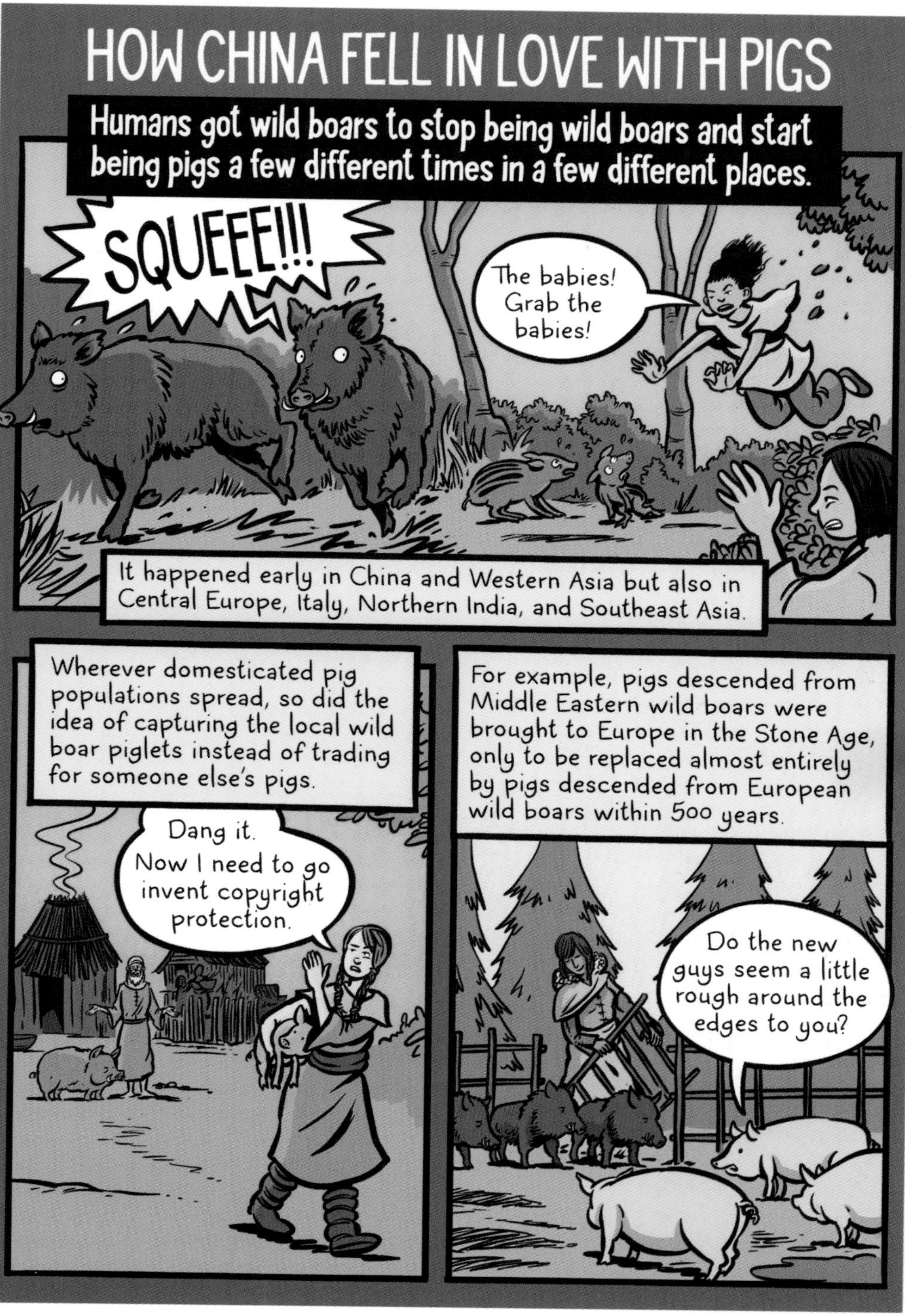
HOW CHINA FELL IN LOVE WITH PIGS
Humans got wild boars to stop being wild boars and start being pigs a few different times in a few different places.
SQUEEEE!!!
The babies! Grab the babies!
It happened early in China and Western Asia but also in Central Europe, Italy, Northern India, and Southeast Asia.
Wherever domesticated pig populations spread, so did the idea of capturing the local wild boar piglets instead of trading for someone else's pigs.
Dang it. Now I need to go invent copyright protection.
For example, pigs descended from Middle Eastern wild boars were brought to Europe in the Stone Age, only to be replaced almost entirely by pigs descended from European wild boars within 500 years.
Do the new guys seem a little rough around the edges to you?

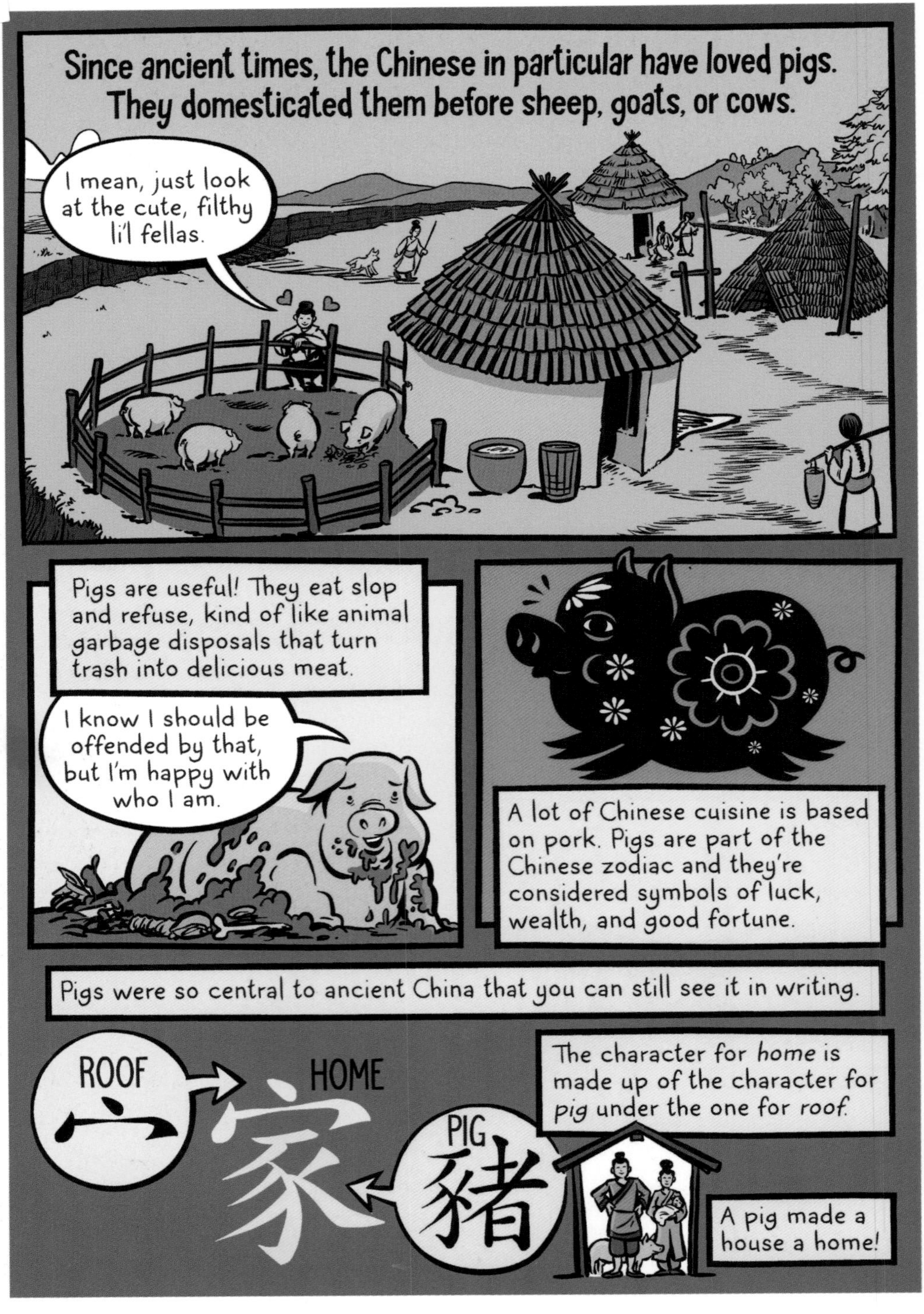
Since ancient times, the Chinese in particular have loved pigs. They domesticated them before sheep, goats, or cows.
I mean, just look at the cute, filthy li'l fellas.
Pigs are useful! They eat slop and refuse, kind of like animal garbage disposals that turn trash into delicious meat.
I know I should be offended by that, but I'm happy with who I am.
A lot of Chinese cuisine is based on pork. Pigs are part of the Chinese zodiac and they're considered symbols of luck, wealth, and good fortune.
Pigs were so central to ancient China that you can still see it in writing.
ROOF
宀
HOME
家
PIG
豬
The character for *home* is made up of the character for *pig* under the one for *roof*.
A pig made a house a home!

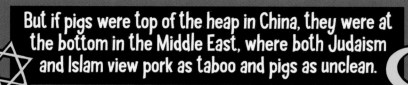

But if pigs were top of the heap in China, they were at the bottom in the Middle East, where both Judaism and Islam view pork as taboo and pigs as unclean.

According to the biblical book of Leviticus, the ban is rooted in how difficult it was for people to categorize pigs.

Hmmm....

Pigs have cleft feet, like cattle, sheep, and goats. Unlike those other animals, pigs do not chew their cud.*

Gasp!

As such, they were deemed weird and even sinister.

*See Pg. 89

But Middle Eastern distaste for pigs goes back further than any records of the Israelites. It goes all the way back to ancient Egypt!

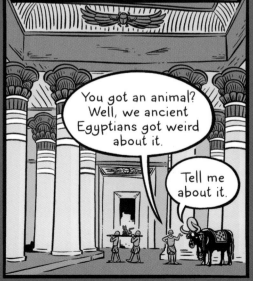

You got an animal? Well, we ancient Egyptians got weird about it.

Tell me about it.

Predynastic Egyptians ate plenty of pigs, especially in the fertile Nile Delta.

It was there that they became associated with the mysterious storm god Set, who was held in high regard.

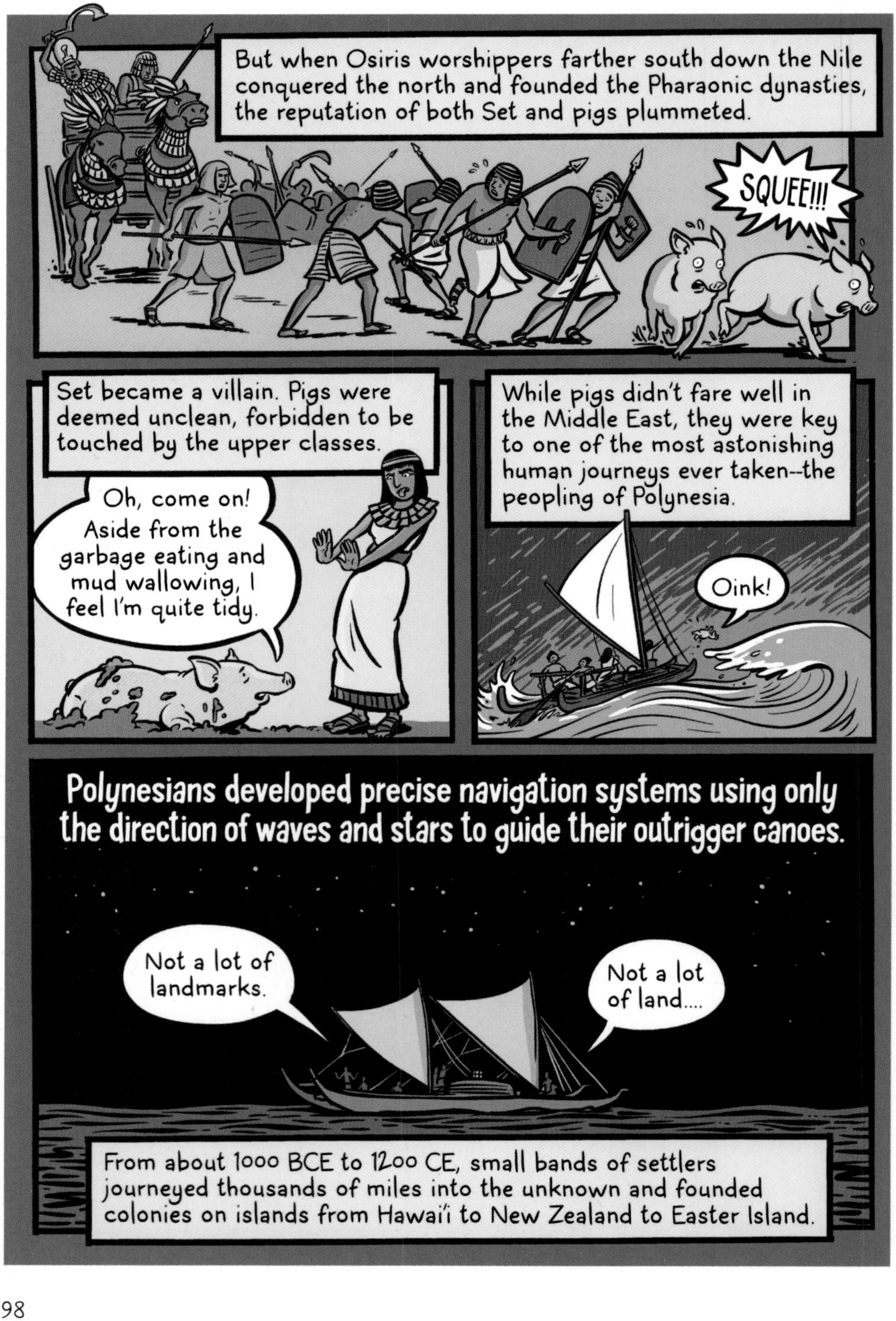
But when Osiris worshippers farther south down the Nile conquered the north and founded the Pharaonic dynasties, the reputation of both Set and pigs plummeted.
SQUEE!!!
Set became a villain. Pigs were deemed unclean, forbidden to be touched by the upper classes.
Oh, come on! Aside from the garbage eating and mud wallowing, I feel I'm quite tidy.
While pigs didn't fare well in the Middle East, they were key to one of the most astonishing human journeys ever taken--the peopling of Polynesia.
Oink!
Polynesians developed precise navigation systems using only the direction of waves and stars to guide their outrigger canoes.
Not a lot of landmarks.
Not a lot of land....
From about 1000 BCE to 1200 CE, small bands of settlers journeyed thousands of miles into the unknown and founded colonies on islands from Hawai'i to New Zealand to Easter Island.

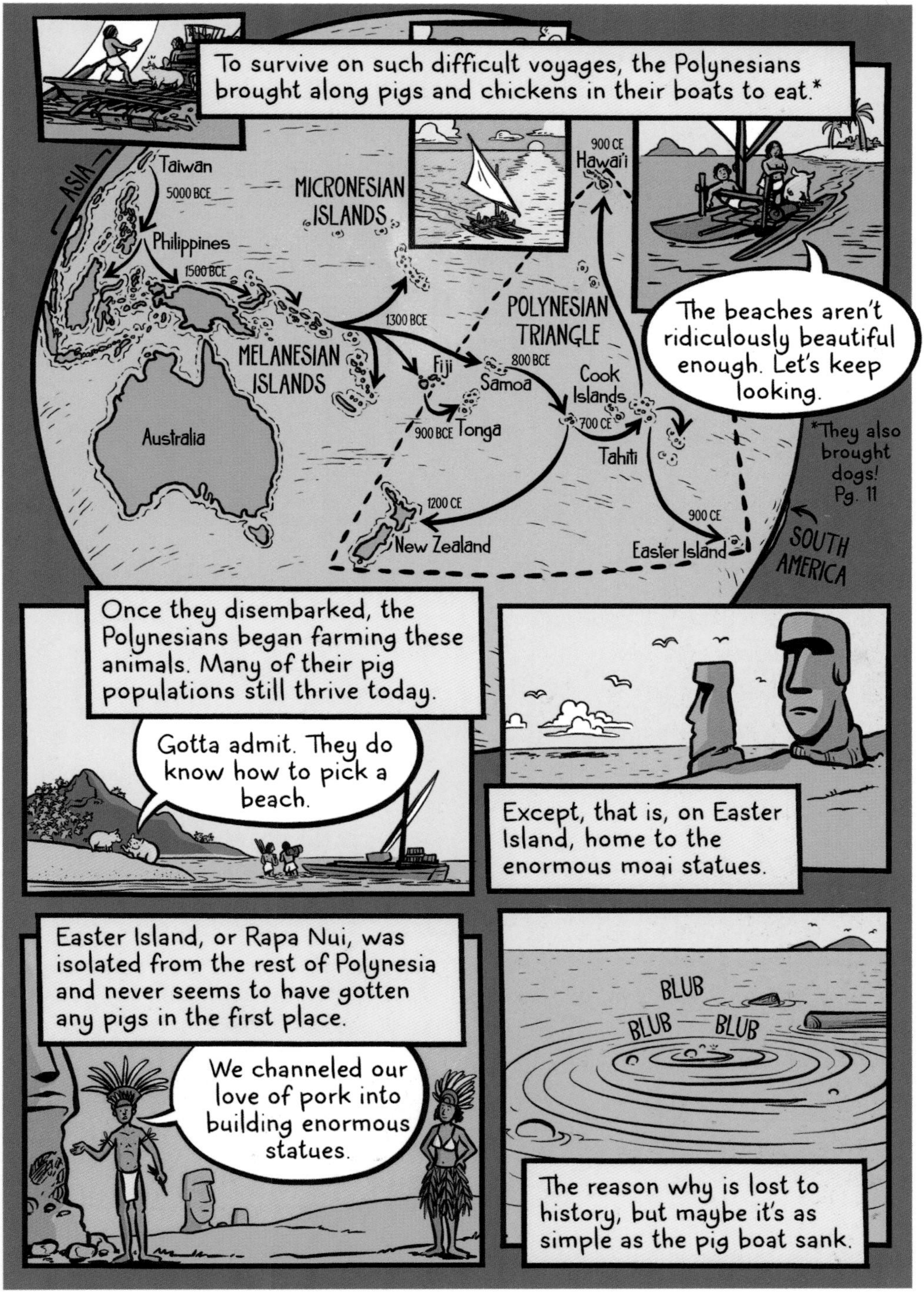
To survive on such difficult voyages, the Polynesians brought along pigs and chickens in their boats to eat.*
ASIA
Taiwan
5000 BCE
Philippines
1500 BCE
MICRONESIAN ISLANDS
1300 BCE
MELANESIAN ISLANDS
Australia
Fiji
Samoa
800 BCE
Tonga
900 BCE
POLYNESIAN TRIANGLE
900 CE
Hawai'i
Cook Islands
700 CE
Tahiti
1200 CE
New Zealand
900 CE
Easter Island
SOUTH AMERICA
The beaches aren't ridiculously beautiful enough. Let's keep looking.
*They also brought dogs! Pg. 11
Once they disembarked, the Polynesians began farming these animals. Many of their pig populations still thrive today.
Gotta admit. They do know how to pick a beach.
Except, that is, on Easter Island, home to the enormous moai statues.
Easter Island, or Rapa Nui, was isolated from the rest of Polynesia and never seems to have gotten any pigs in the first place.
We channeled our love of pork into building enormous statues.
BLUB
BLUB
BLUB
The reason why is lost to history, but maybe it's as simple as the pig boat sank.

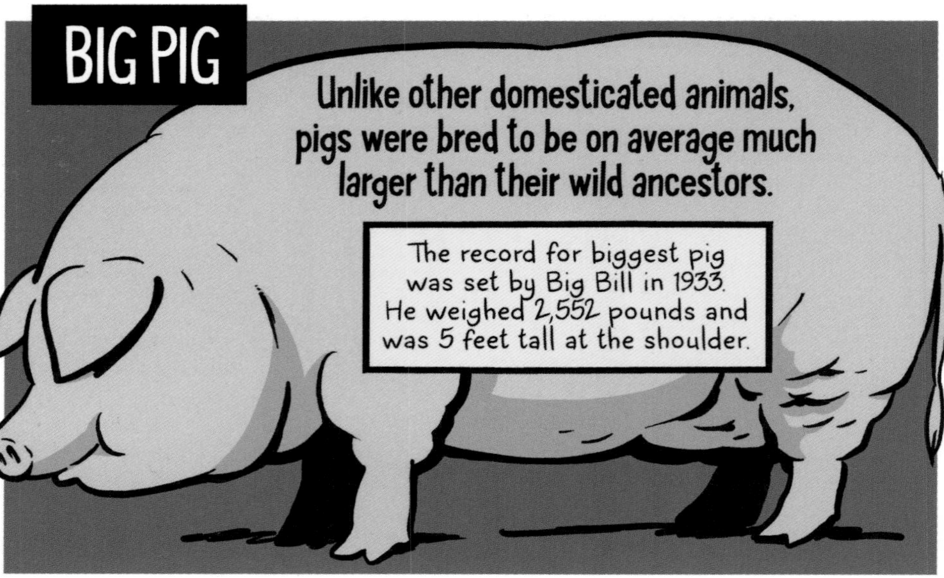
BIG PIG
Unlike other domesticated animals, pigs were bred to be on average much larger than their wild ancestors.
The record for biggest pig was set by Big Bill in 1933. He weighed 2,552 pounds and was 5 feet tall at the shoulder.

PIGSKINS, SUNBURNS, AND MUD

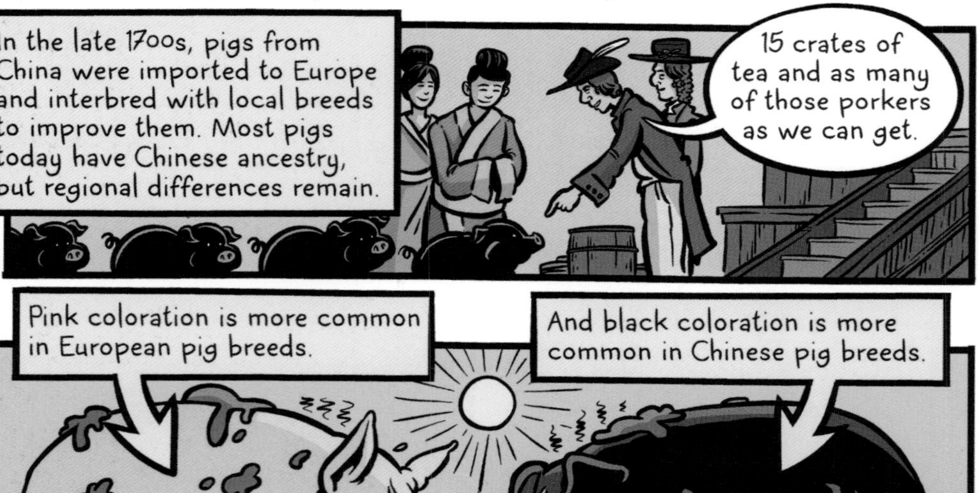
In the late 1700s, pigs from China were imported to Europe and interbred with local breeds to improve them. Most pigs today have Chinese ancestry, but regional differences remain.
15 crates of tea and as many of those porkers as we can get.

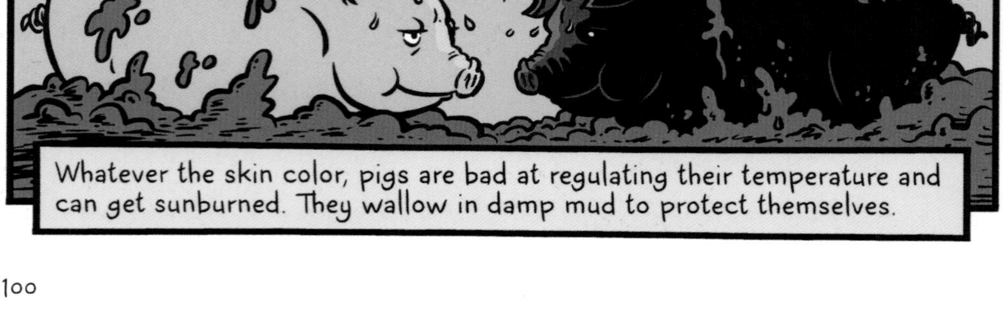
Pink coloration is more common in European pig breeds.
And black coloration is more common in Chinese pig breeds.
Whatever the skin color, pigs are bad at regulating their temperature and can get sunburned. They wallow in damp mud to protect themselves.

THE DICTATOR AND SWINE FLU

Because of our close proximity to animals, many of our diseases originally come from them.* *See Pg. 134 & 135

Swine flu, which originates in pigs, made the jump to humans a few times, often with deadly results.

COUGH! HACK!

COUGH! HACK!

One of the bigger recent outbreaks was in 2009, when the world feared a pandemic.

In a blundering response, Egypt's dictator ordered all 300,000 pigs in his country be killed.

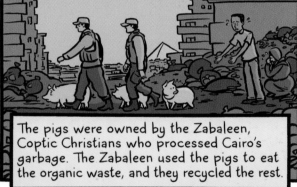

The pigs were owned by the Zabaleen, Coptic Christians who processed Cairo's garbage. The Zabaleen used the pigs to eat the organic waste, and they recycled the rest.

With the pigs gone, the streets of Cairo soon filled with garbage and rats.

munch munch munch

The ban was lifted after the dictatorship was toppled in 2011 by an uprising. Egyptian politics are still messy, but at least the pigs of Cairo have returned.

BEES

Apis mellifera

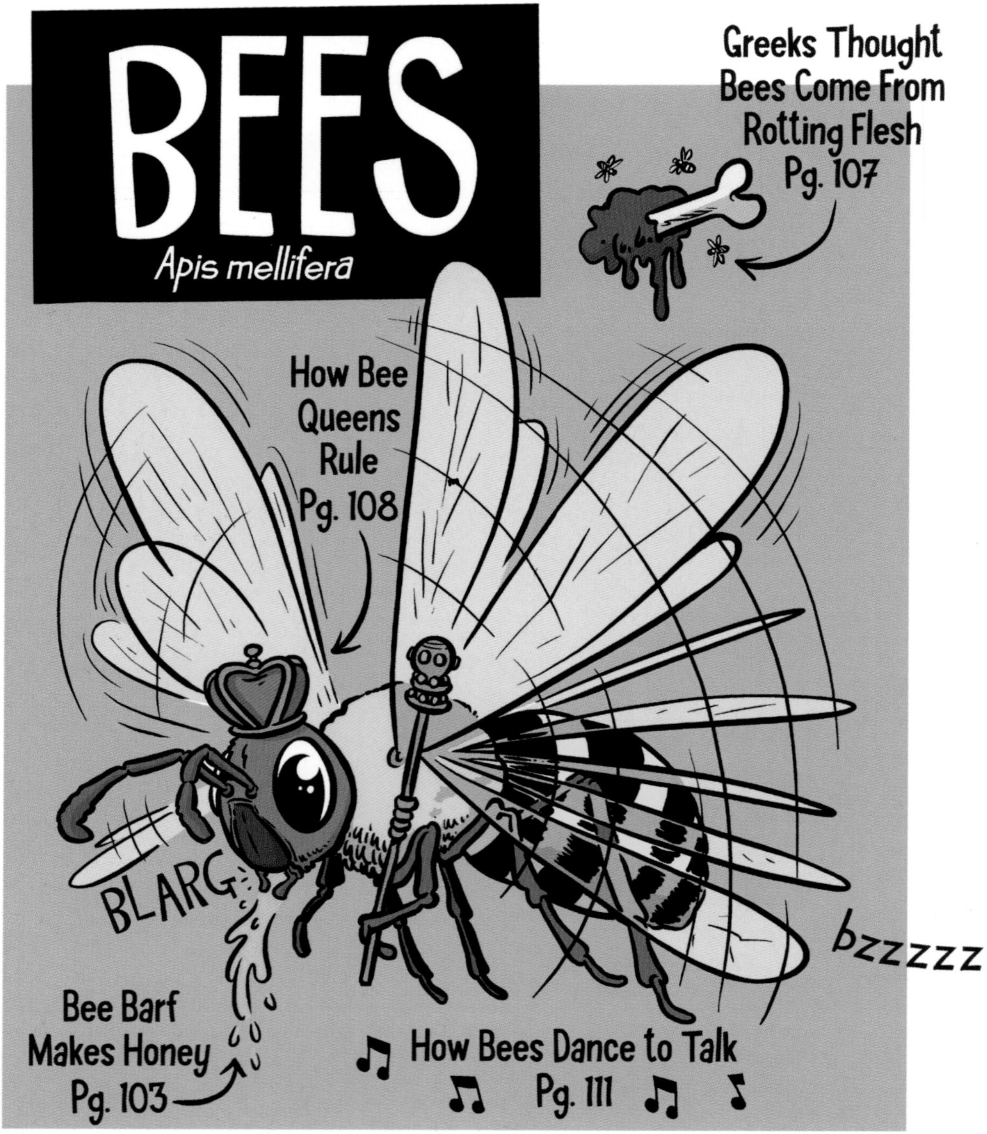

Greeks Thought Bees Come From Rotting Flesh Pg. 107
How Bee Queens Rule Pg. 108
BLARG
bzzzzz
Bee Barf Makes Honey Pg. 103
How Bees Dance to Talk Pg. 111

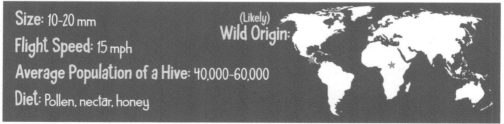

Size: 10-20 mm

Flight Speed: 15 mph

Average Population of a Hive: 40,000-60,000

Diet: Pollen, nectar, honey

(Likely) **Wild Origin:**

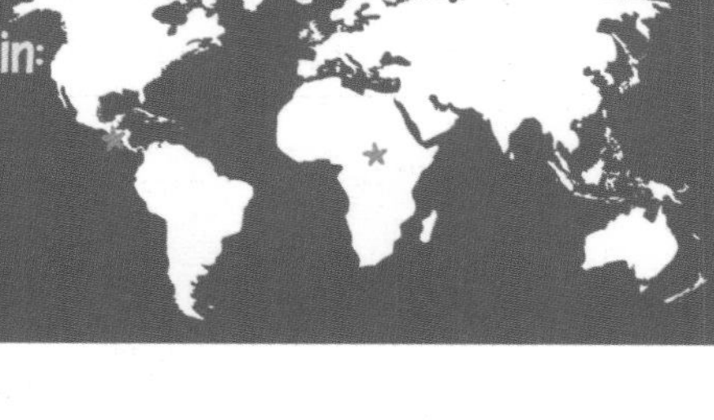

BEES ARE WEIRDER THAN YOU THINK

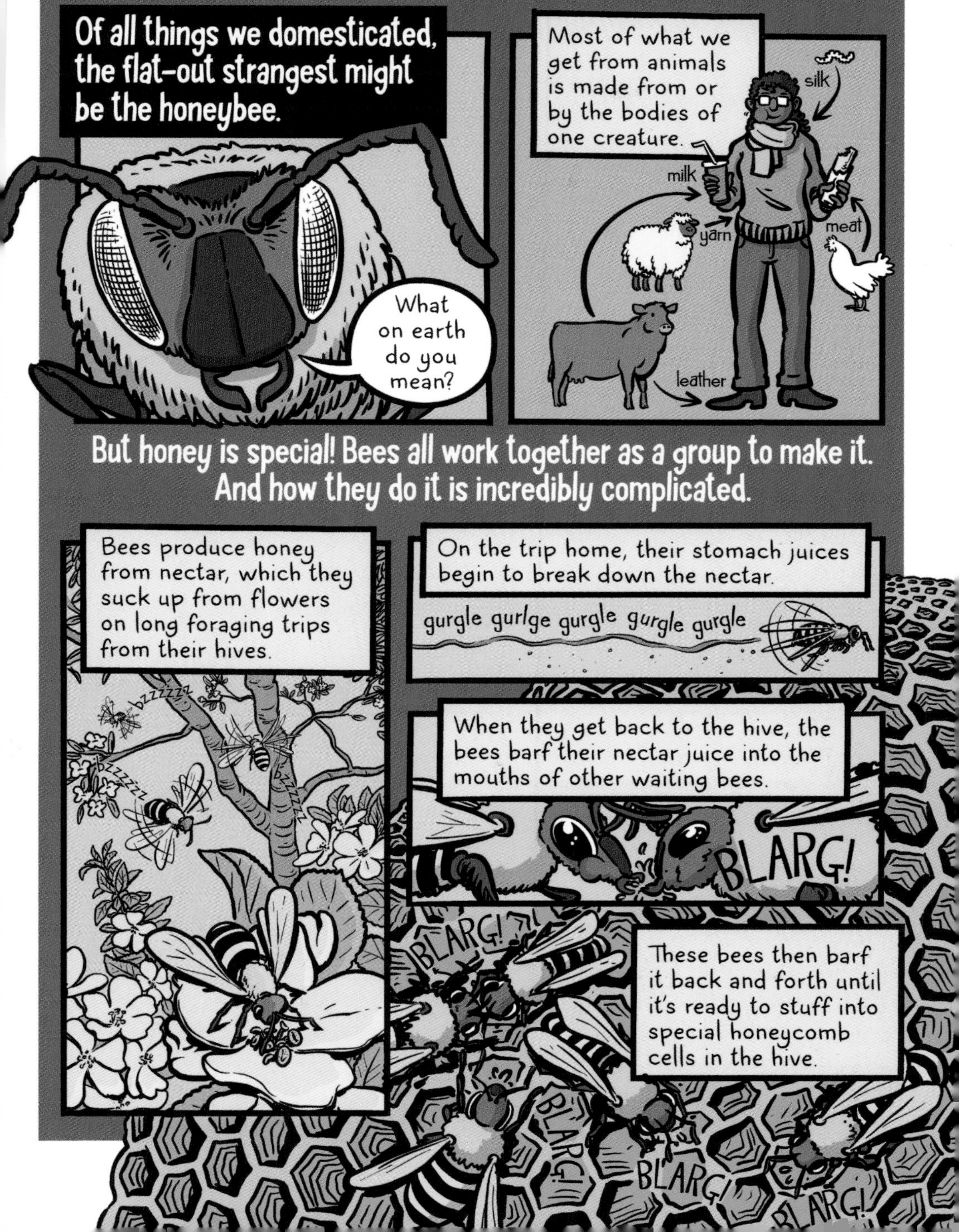

Of all things we domesticated, the flat-out strangest might be the honeybee.
What on earth do you mean?
Most of what we get from animals is made from or by the bodies of one creature.
silk
milk
yarn
meat
leather
But honey is special! Bees all work together as a group to make it. And how they do it is incredibly complicated.
Bees produce honey from nectar, which they suck up from flowers on long foraging trips from their hives.
bzzzzzz
bzzzzzz
bzzzzz
On the trip home, their stomach juices begin to break down the nectar.
gurgle gurlge gurgle gurgle gurgle
When they get back to the hive, the bees barf their nectar juice into the mouths of other waiting bees.
BLARG!
These bees then barf it back and forth until it's ready to stuff into special honeycomb cells in the hive.
BLARG!
BLARG!
BLARG!
BLARG!
BLARG!

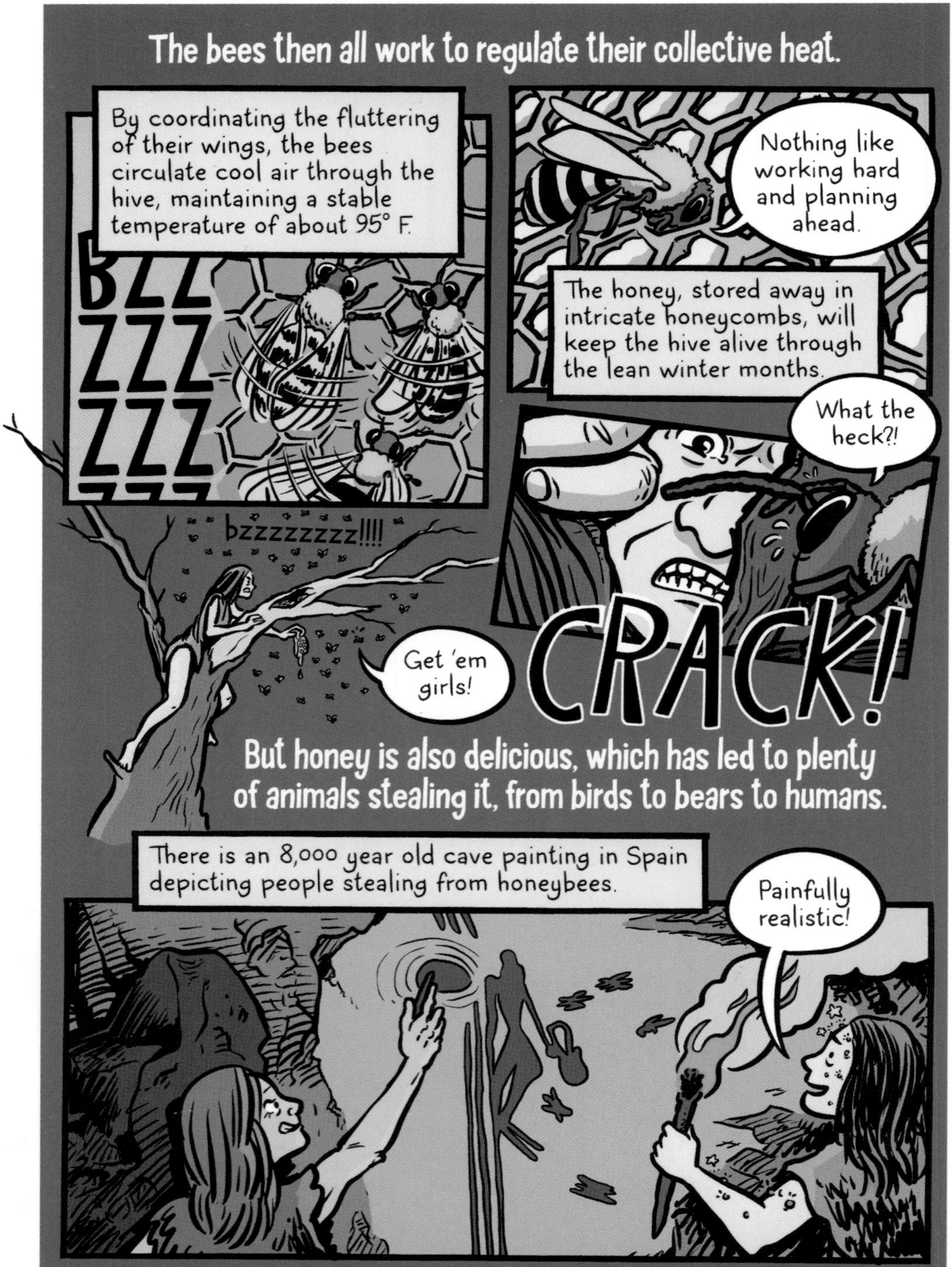
The bees then all work to regulate their collective heat.
By coordinating the fluttering of their wings, the bees circulate cool air through the hive, maintaining a stable temperature of about 95° F.
BZZ ZZZ ZZZ
Nothing like working hard and planning ahead.
The honey, stored away in intricate honeycombs, will keep the hive alive through the lean winter months.
What the heck?!
bzzzzzzzz!!!!
CRACK!
Get 'em girls!
But honey is also delicious, which has led to plenty of animals stealing it, from birds to bears to humans.
There is an 8,000 year old cave painting in Spain depicting people stealing from honeybees.
Painfully realistic!

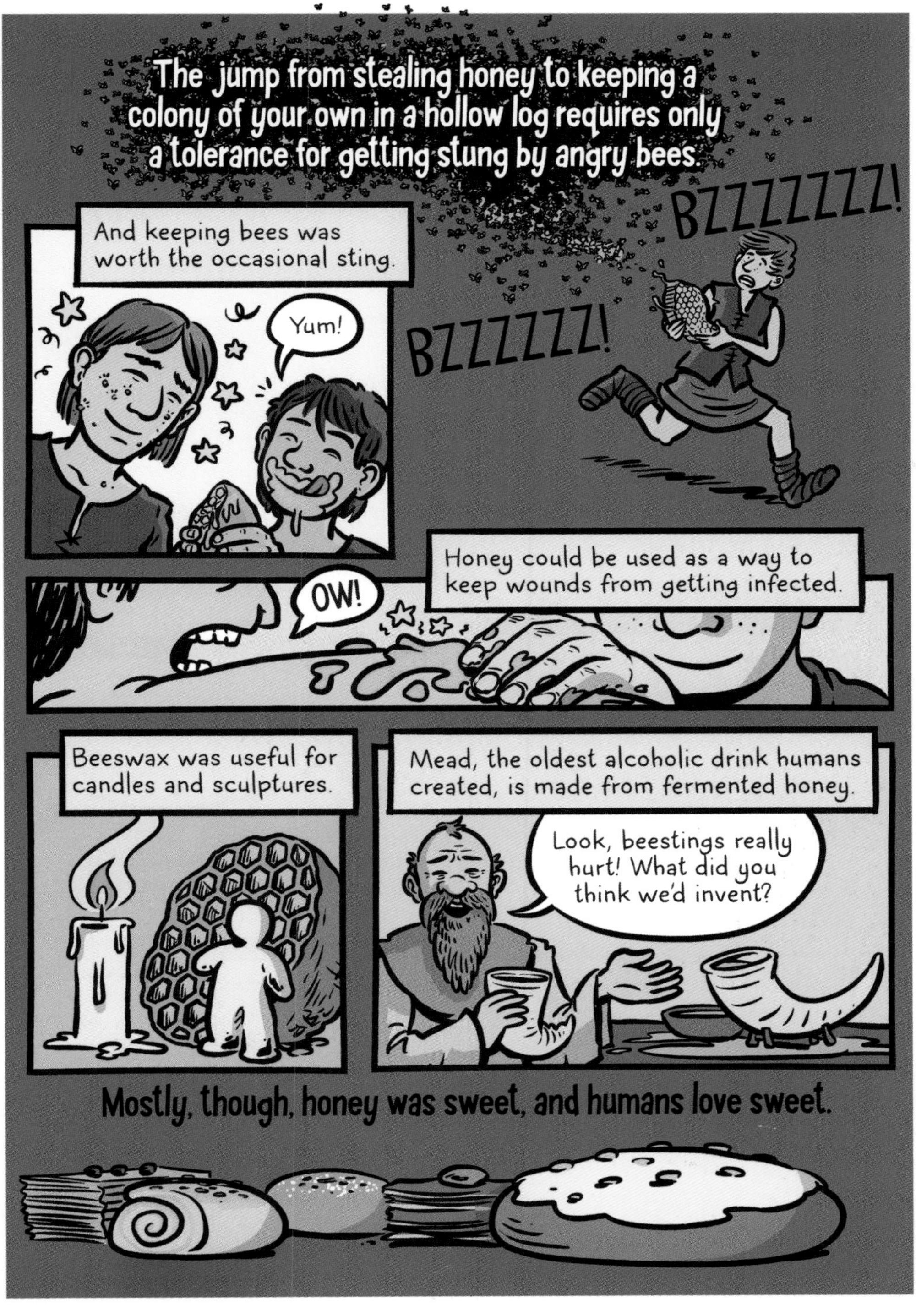

The jump from stealing honey to keeping a colony of your own in a hollow log requires only a tolerance for getting stung by angry bees.
BZZZZZZZZ!
BZZZZZZZ!
And keeping bees was worth the occasional sting.
Yum!
Honey could be used as a way to keep wounds from getting infected.
OW!
Beeswax was useful for candles and sculptures.
Mead, the oldest alcoholic drink humans created, is made from fermented honey.
Look, beestings really hurt! What did you think we'd invent?
Mostly, though, honey was sweet, and humans love sweet.

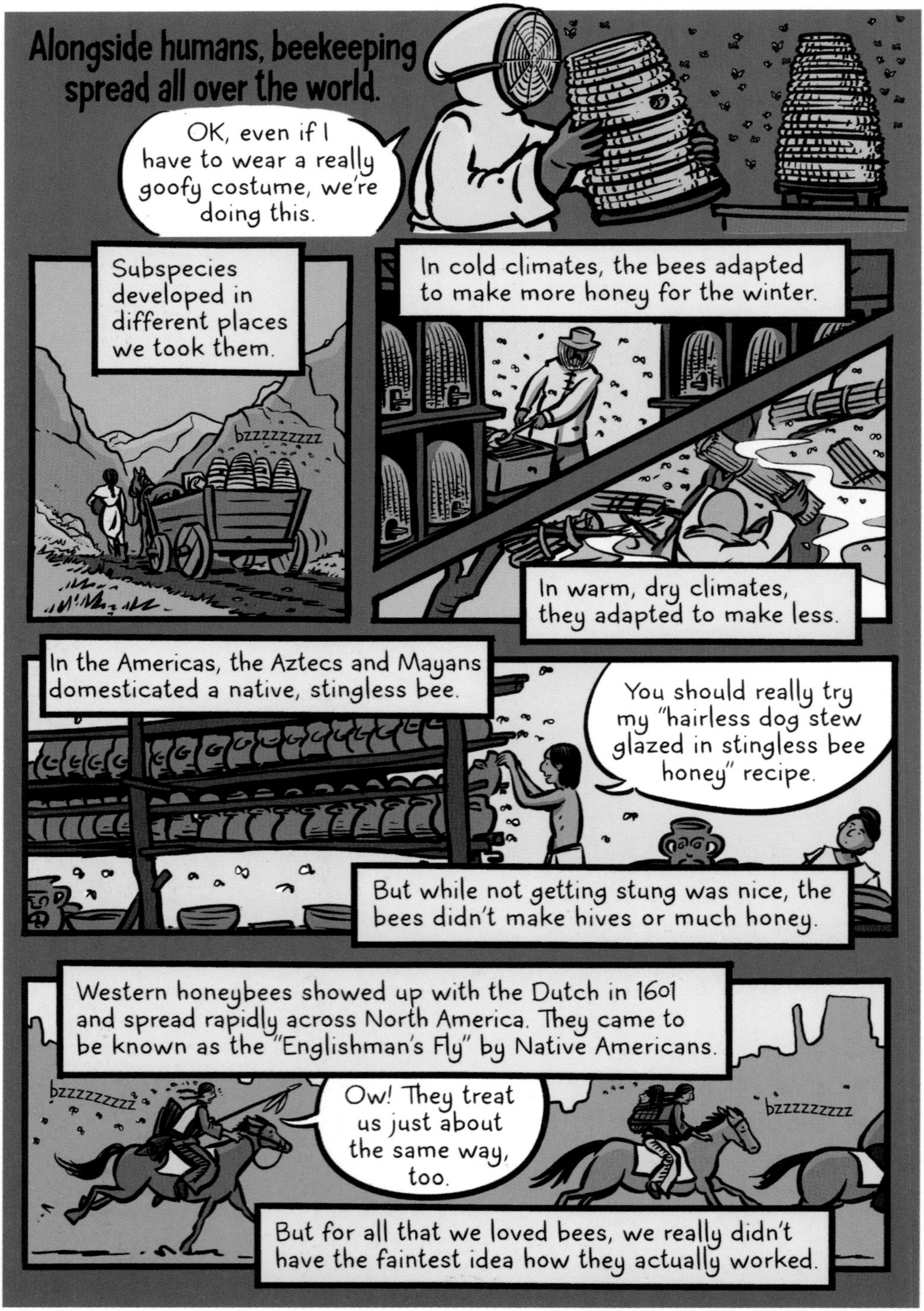
Alongside humans, beekeeping spread all over the world.
OK, even if I have to wear a really goofy costume, we're doing this.
Subspecies developed in different places we took them.
bzzzzzzzzz
In cold climates, the bees adapted to make more honey for the winter.
In warm, dry climates, they adapted to make less.
In the Americas, the Aztecs and Mayans domesticated a native, stingless bee.
You should really try my "hairless dog stew glazed in stingless bee honey" recipe.
But while not getting stung was nice, the bees didn't make hives or much honey.
Western honeybees showed up with the Dutch in 1601 and spread rapidly across North America. They came to be known as the "Englishman's Fly" by Native Americans.
bzzzzzzzzz
Ow! They treat us just about the same way, too.
bzzzzzzzzz
But for all that we loved bees, we really didn't have the faintest idea how they actually worked.

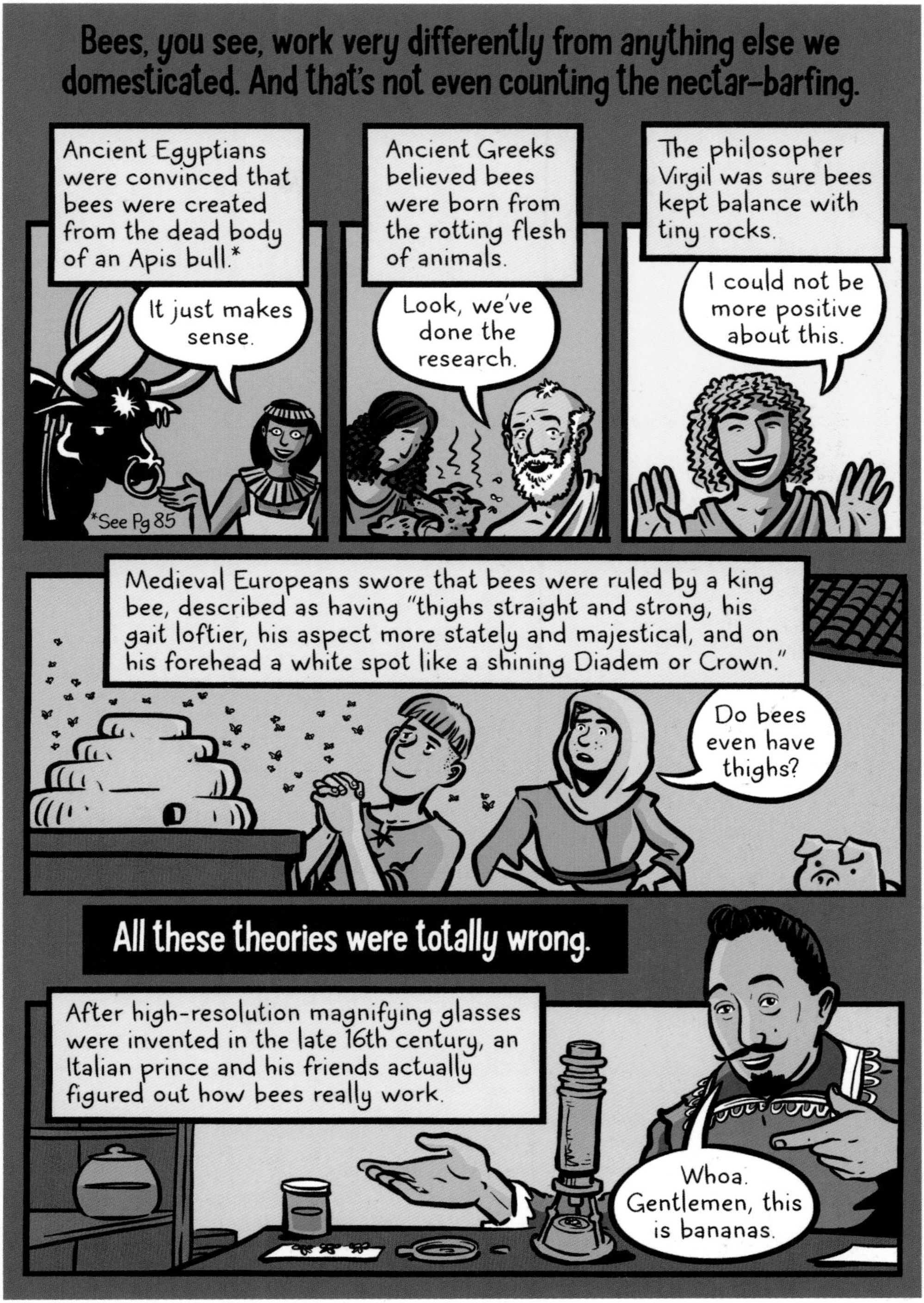
Bees, you see, work very differently from anything else we domesticated. And that's not even counting the nectar-barfing.
Ancient Egyptians were convinced that bees were created from the dead body of an Apis bull.*
It just makes sense.
*See Pg 85
Ancient Greeks believed bees were born from the rotting flesh of animals.
Look, we've done the research.
The philosopher Virgil was sure bees kept balance with tiny rocks.
I could not be more positive about this.
Medieval Europeans swore that bees were ruled by a king bee, described as having "thighs straight and strong, his gait loftier, his aspect more stately and majestical, and on his forehead a white spot like a shining Diadem or Crown."
Do bees even have thighs?
All these theories were totally wrong.
After high-resolution magnifying glasses were invented in the late 16th century, an Italian prince and his friends actually figured out how bees really work.
Whoa. Gentlemen, this is bananas.

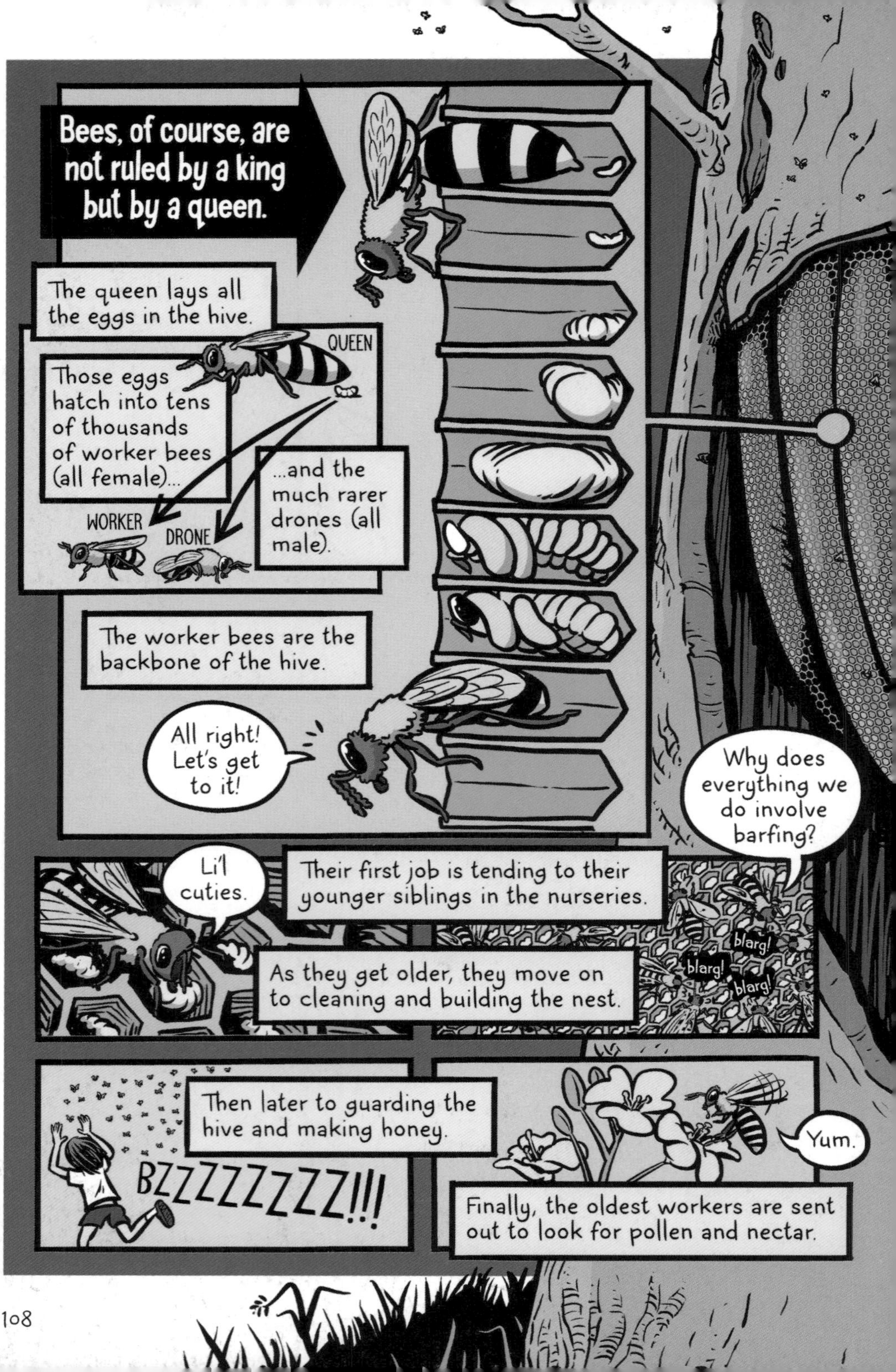

Bees, of course, are not ruled by a king but by a queen.
The queen lays all the eggs in the hive.
QUEEN
Those eggs hatch into tens of thousands of worker bees (all female)...
...and the much rarer drones (all male).
WORKER
DRONE
The worker bees are the backbone of the hive.
All right! Let's get to it!
Why does everything we do involve barfing?
Li'l cuties.
Their first job is tending to their younger siblings in the nurseries.
blarg!
blarg!
blarg!
As they get older, they move on to cleaning and building the nest.
Then later to guarding the hive and making honey.
BZZZZZZZZZ!!!
Yum.
Finally, the oldest workers are sent out to look for pollen and nectar.

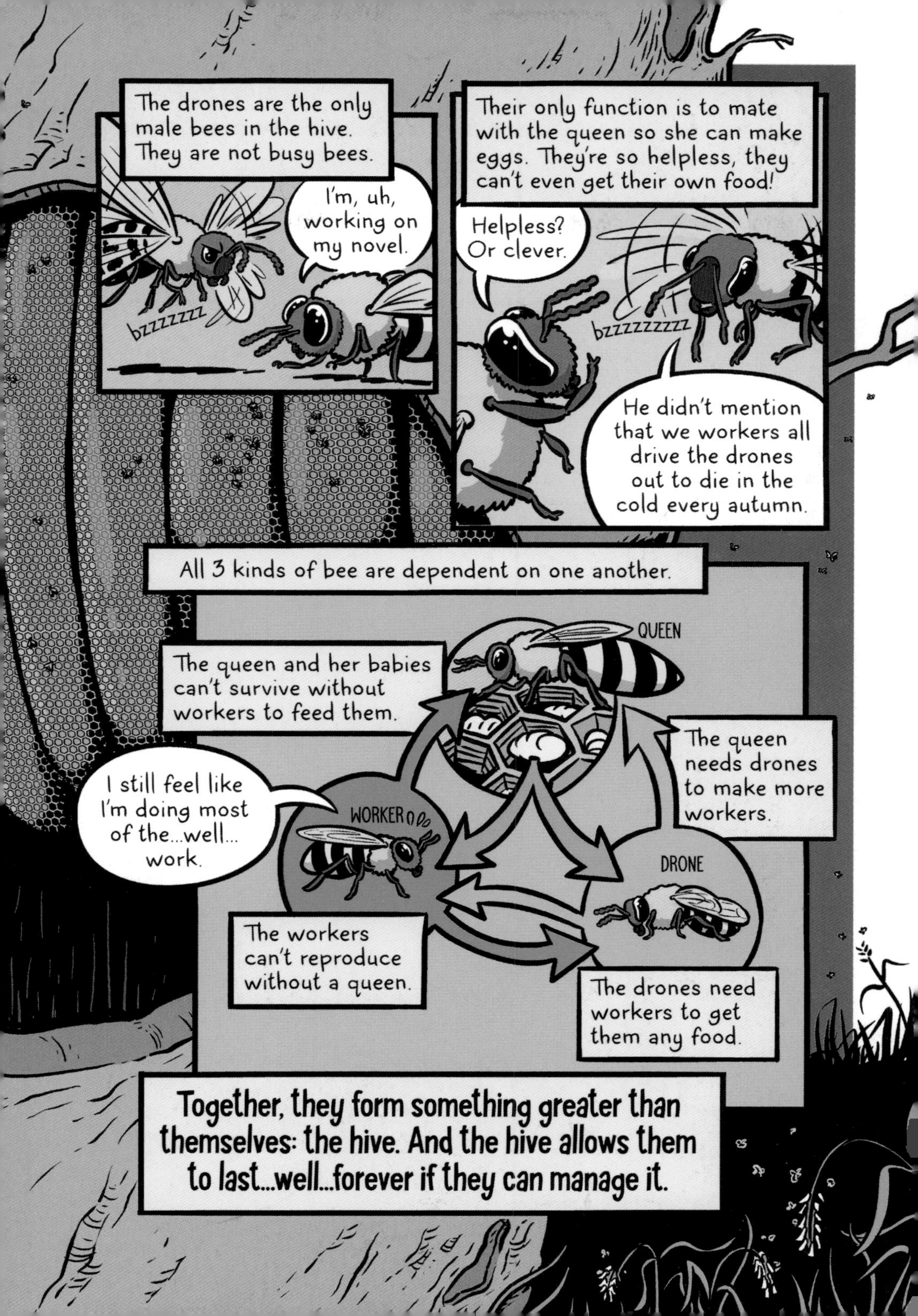

The drones are the only male bees in the hive. They are not busy bees.
bzzzzzzz
I'm, uh, working on my novel.
Their only function is to mate with the queen so she can make eggs. They're so helpless, they can't even get their own food!
Helpless? Or clever.
bzzzzzzzzz
He didn't mention that we workers all drive the drones out to die in the cold every autumn.
All 3 kinds of bee are dependent on one another.
QUEEN
The queen and her babies can't survive without workers to feed them.
The queen needs drones to make more workers.
I still feel like I'm doing most of the...well... work.
WORKER
DRONE
The workers can't reproduce without a queen.
The drones need workers to get them any food.
Together, they form something greater than themselves: the hive. And the hive allows them to last...well...forever if they can manage it.

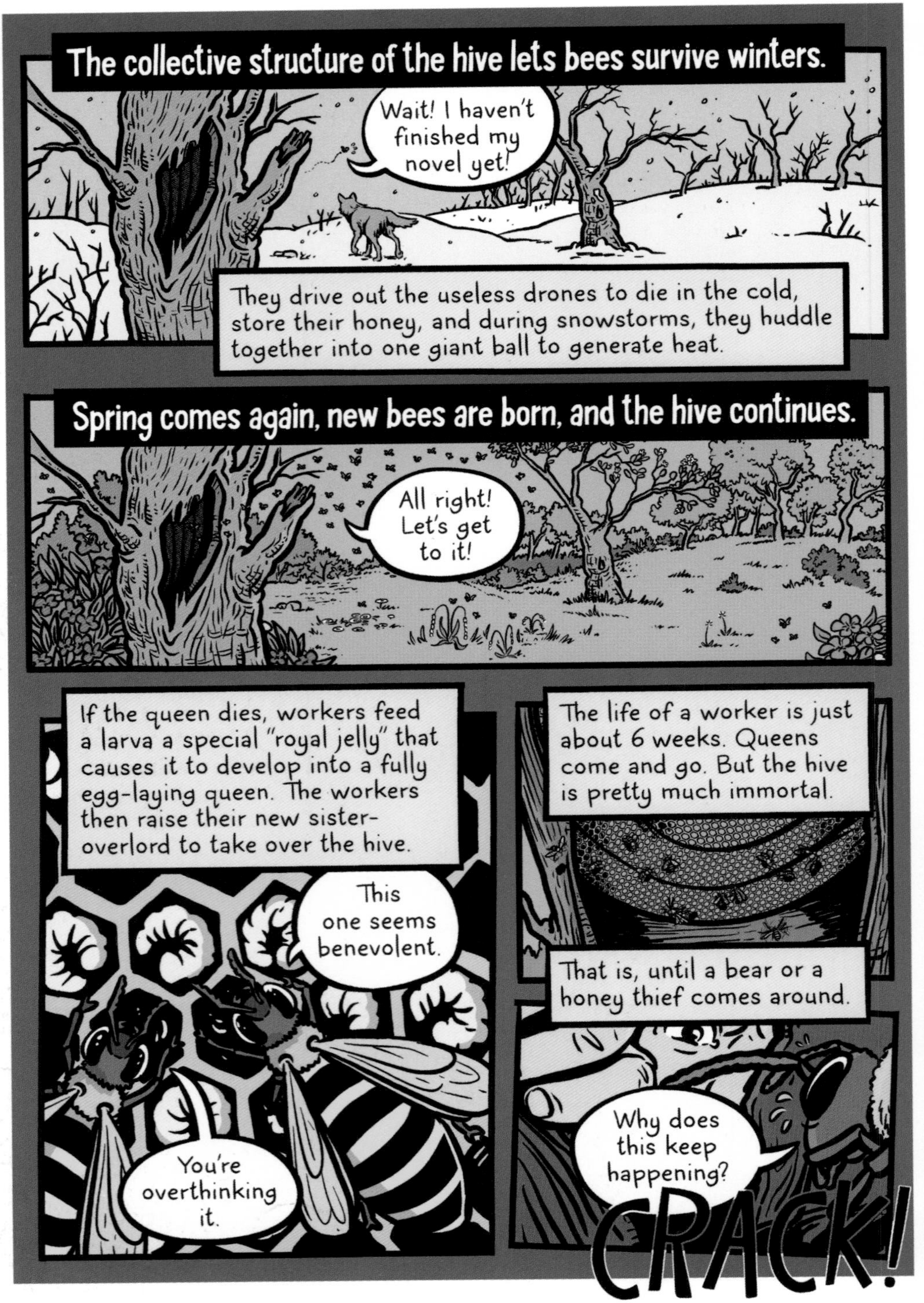
The collective structure of the hive lets bees survive winters.
Wait! I haven't finished my novel yet!
They drive out the useless drones to die in the cold, store their honey, and during snowstorms, they huddle together into one giant ball to generate heat.
Spring comes again, new bees are born, and the hive continues.
All right! Let's get to it!
If the queen dies, workers feed a larva a special "royal jelly" that causes it to develop into a fully egg-laying queen. The workers then raise their new sister-overlord to take over the hive.
This one seems benevolent.
You're overthinking it.
The life of a worker is just about 6 weeks. Queens come and go. But the hive is pretty much immortal.
That is, until a bear or a honey thief comes around.
Why does this keep happening?
CRACK!

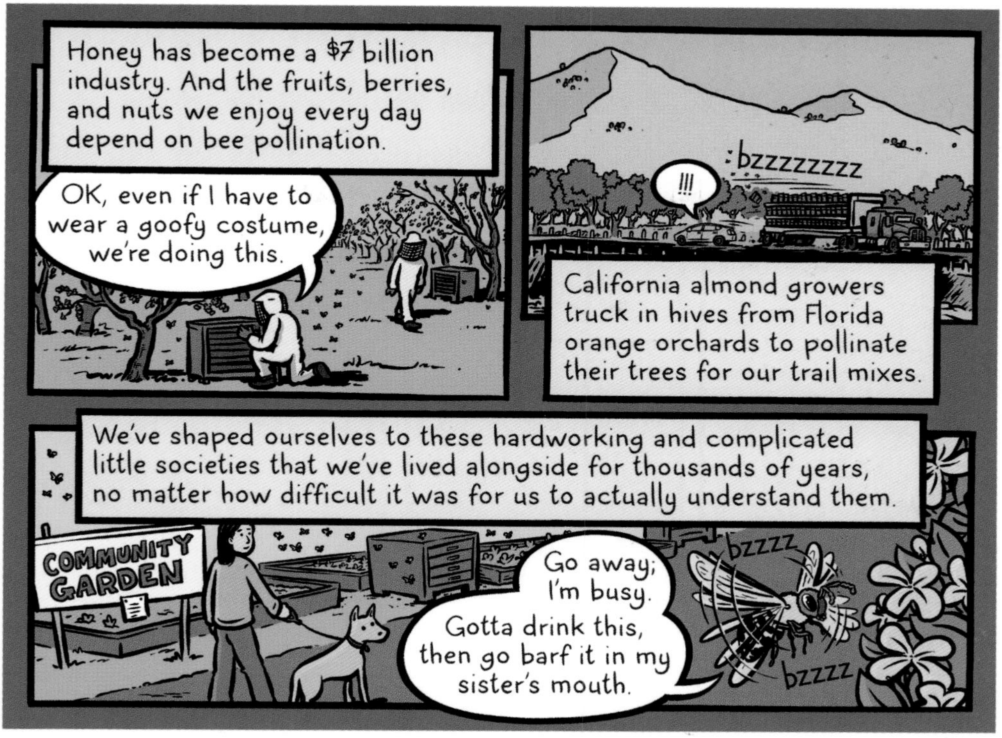
Honey has become a $7 billion industry. And the fruits, berries, and nuts we enjoy every day depend on bee pollination.
OK, even if I have to wear a goofy costume, we're doing this.
bzzzzzzzz
!!!
California almond growers truck in hives from Florida orange orchards to pollinate their trees for our trail mixes.
We've shaped ourselves to these hardworking and complicated little societies that we've lived alongside for thousands of years, no matter how difficult it was for us to actually understand them.
COMMUNITY GARDEN
Go away; I'm busy.
Gotta drink this, then go barf it in my sister's mouth.
bzzzz
bzzzz

HOW BEES TALK TO ONE ANOTHER

Bees can't hear, so they use 2 different dances to communicate.

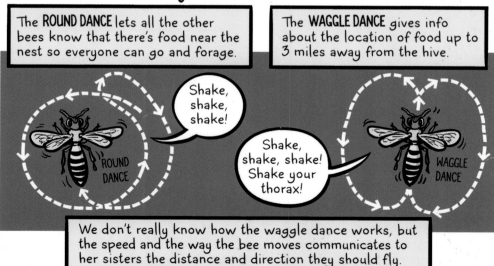

The **ROUND DANCE** lets all the other bees know that there's food near the nest so everyone can go and forage.

The **WAGGLE DANCE** gives info about the location of food up to 3 miles away from the hive.

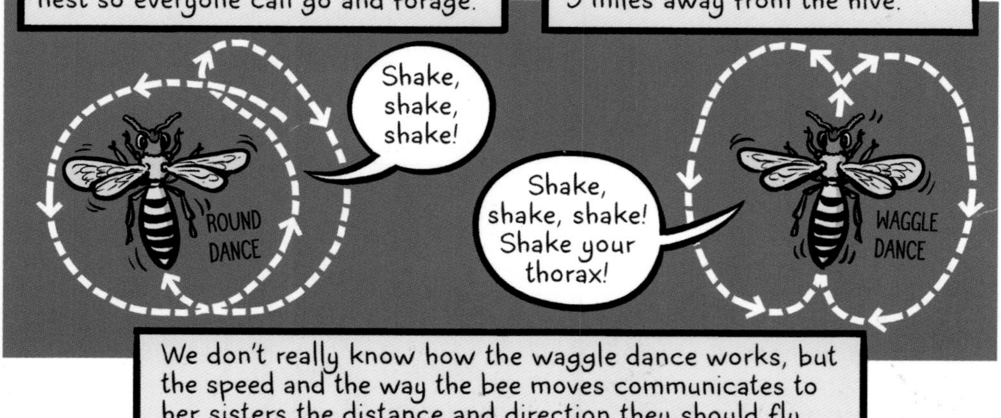
Shake, shake, shake!
ROUND DANCE
Shake, shake, shake! Shake your thorax!
WAGGLE DANCE

We don't really know how the waggle dance works, but the speed and the way the bee moves communicates to her sisters the distance and direction they should fly.

THE BEES OF DESERET

Because of the long and close relationship humans have had with bees, we've incorporated them into religions and myths and legends over the years.

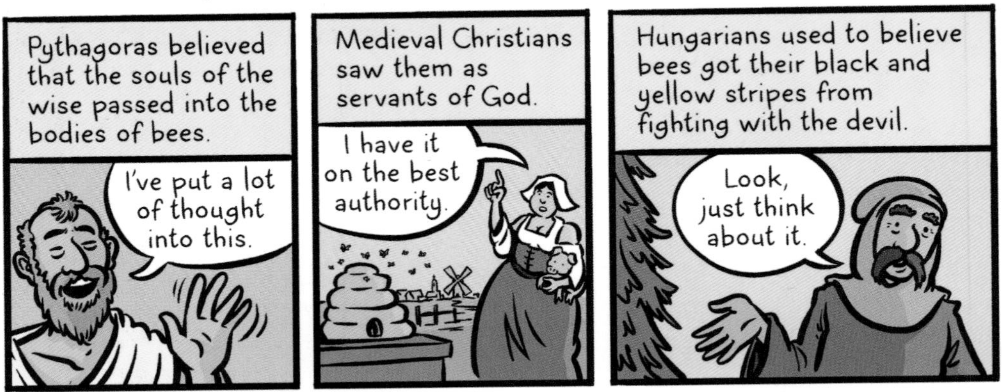
Pythagoras believed that the souls of the wise passed into the bodies of bees.
I've put a lot of thought into this.
Medieval Christians saw them as servants of God.
I have it on the best authority.
Hungarians used to believe bees got their black and yellow stripes from fighting with the devil.
Look, just think about it.

Mormons, following in this long tradition, really, really love bees.

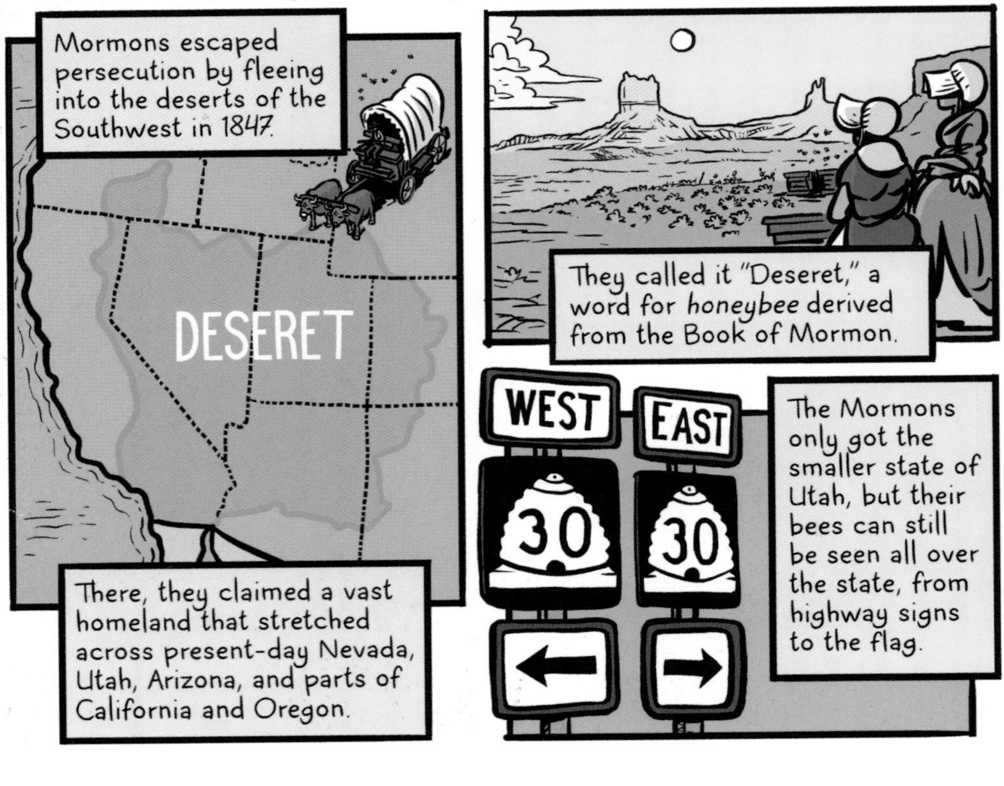
Mormons escaped persecution by fleeing into the deserts of the Southwest in 1847.
DESERET
There, they claimed a vast homeland that stretched across present-day Nevada, Utah, Arizona, and parts of California and Oregon.
They called it "Deseret," a word for *honeybee* derived from the Book of Mormon.
WEST
30
EAST
30
The Mormons only got the smaller state of Utah, but their bees can still be seen all over the state, from highway signs to the flag.

"KILLER" BEES

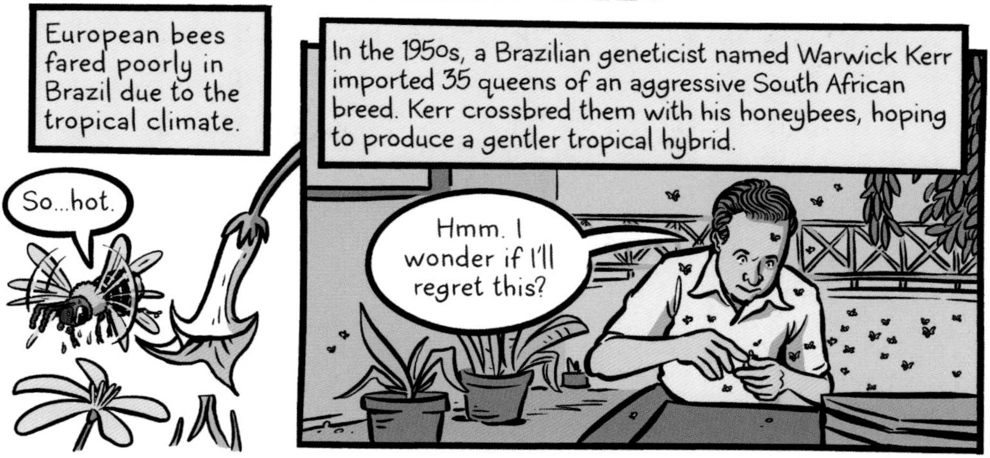
European bees fared poorly in Brazil due to the tropical climate.
So...hot.
In the 1950s, a Brazilian geneticist named Warwick Kerr imported 35 queens of an aggressive South African breed. Kerr crossbred them with his honeybees, hoping to produce a gentler tropical hybrid.
Hmm. I wonder if I'll regret this?

Then one of his beekeepers accidentally left open an entrance guard that kept the queens confined and 26 queens swarmed.

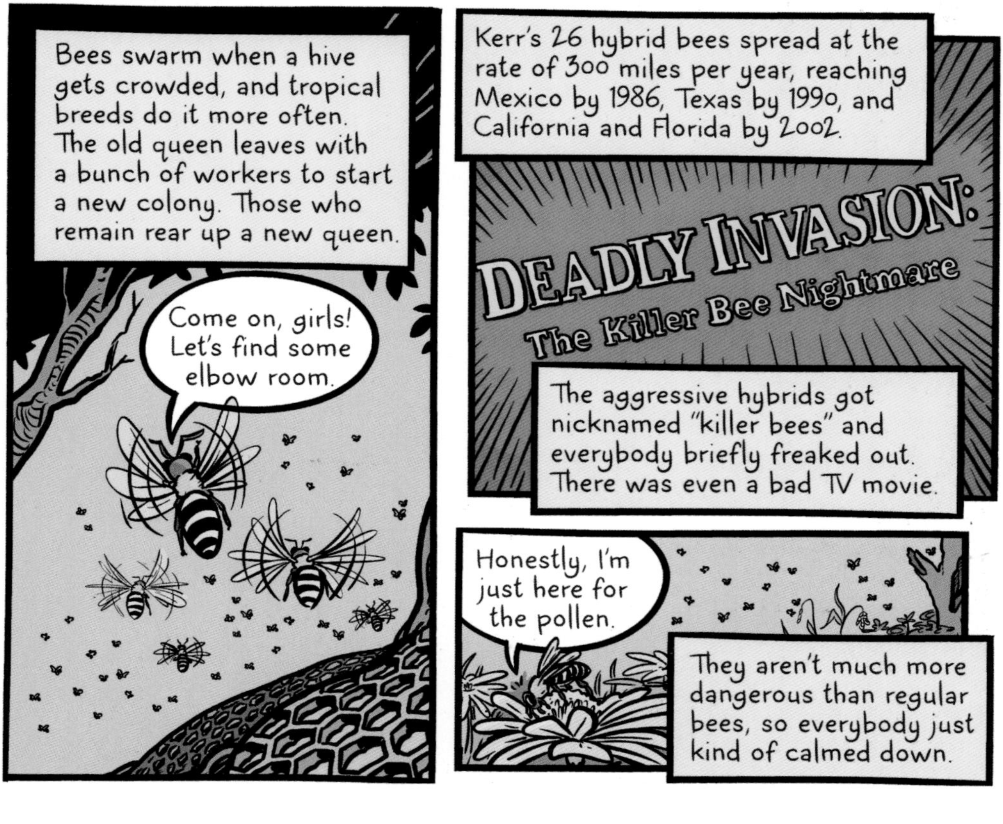
Bees swarm when a hive gets crowded, and tropical breeds do it more often. The old queen leaves with a bunch of workers to start a new colony. Those who remain rear up a new queen.
Come on, girls! Let's find some elbow room.
Kerr's 26 hybrid bees spread at the rate of 300 miles per year, reaching Mexico by 1986, Texas by 1990, and California and Florida by 2002.
DEADLY INVASION:
The Killer Bee Nightmare
The aggressive hybrids got nicknamed "killer bees" and everybody briefly freaked out. There was even a bad TV movie.
Honestly, I'm just here for the pollen.
They aren't much more dangerous than regular bees, so everybody just kind of calmed down.

CHICKENS

Gallus gallus domesticus

The Chicken of Tomorrow
Pg. 119

How Chickens Became Enormous Pg. 119

Royal Birds of the Indus Valley Pg. 116

Romans Loved Omelets Pg. 117

Size: 5–26 in

Weight: 15 oz–11 lbs

Population: 23.7 billion

Diet: Hay, vegetables, occasional fruits

(Likely) **Wild Origin:**

FROM THE JUNGLE TO THE NUGGET

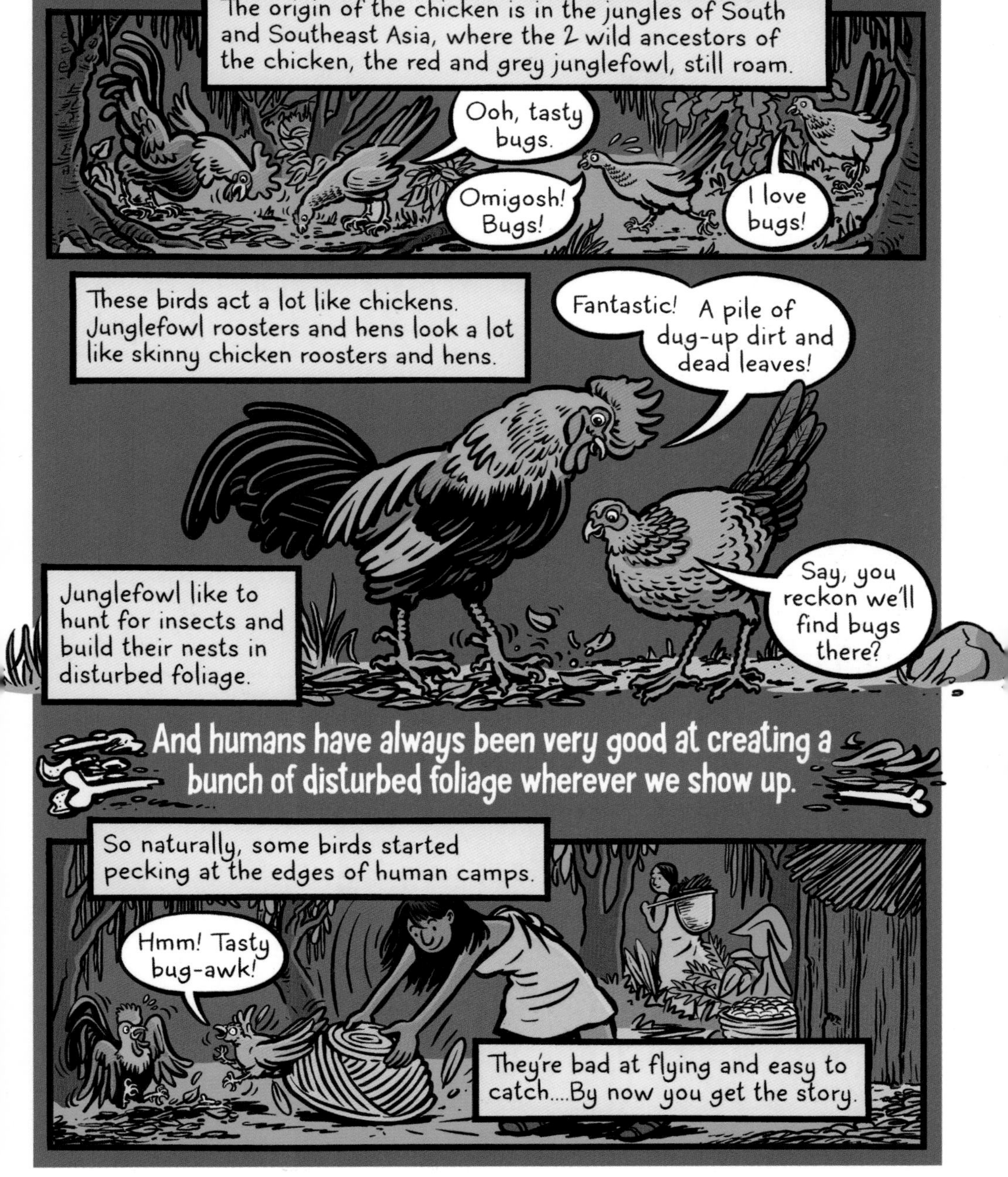
The origin of the chicken is in the jungles of South and Southeast Asia, where the 2 wild ancestors of the chicken, the red and grey junglefowl, still roam.
Ooh, tasty bugs.
Omigosh! Bugs!
I love bugs!
These birds act a lot like chickens. Junglefowl roosters and hens look a lot like skinny chicken roosters and hens.
Fantastic! A pile of dug-up dirt and dead leaves!
Junglefowl like to hunt for insects and build their nests in disturbed foliage.
Say, you reckon we'll find bugs there?
And humans have always been very good at creating a bunch of disturbed foliage wherever we show up.
So naturally, some birds started pecking at the edges of human camps.
Hmm! Tasty bug-awk!
They're bad at flying and easy to catch....By now you get the story.

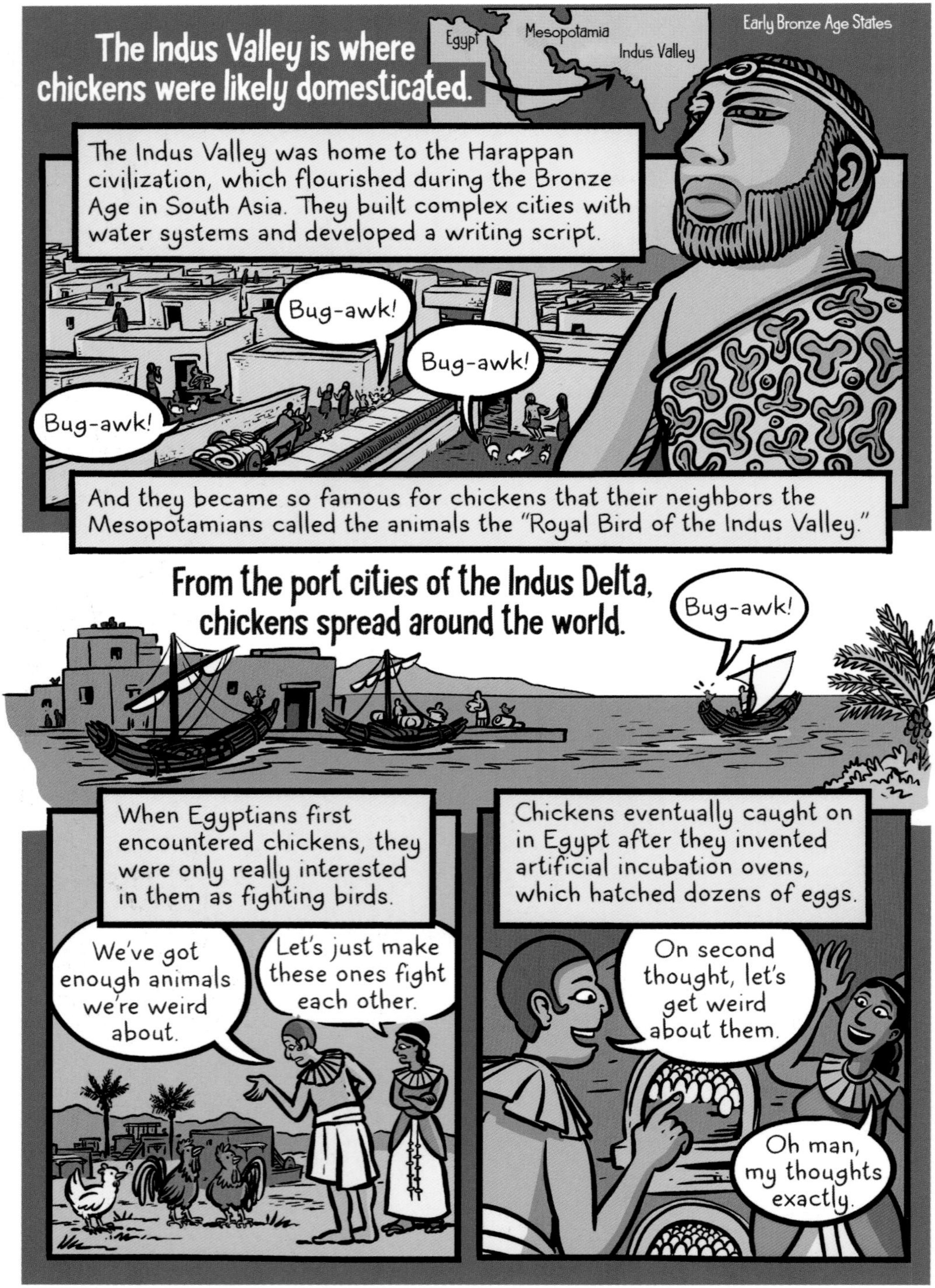
The Indus Valley is where chickens were likely domesticated.
Early Bronze Age States
Egypt
Mesopotamia
Indus Valley
The Indus Valley was home to the Harappan civilization, which flourished during the Bronze Age in South Asia. They built complex cities with water systems and developed a writing script.
Bug-awk!
Bug-awk!
Bug-awk!
And they became so famous for chickens that their neighbors the Mesopotamians called the animals the "Royal Bird of the Indus Valley."
From the port cities of the Indus Delta, chickens spread around the world.
Bug-awk!
When Egyptians first encountered chickens, they were only really interested in them as fighting birds.
We've got enough animals we're weird about.
Let's just make these ones fight each other.
Chickens eventually caught on in Egypt after they invented artificial incubation ovens, which hatched dozens of eggs.
On second thought, let's get weird about them.
Oh man, my thoughts exactly.

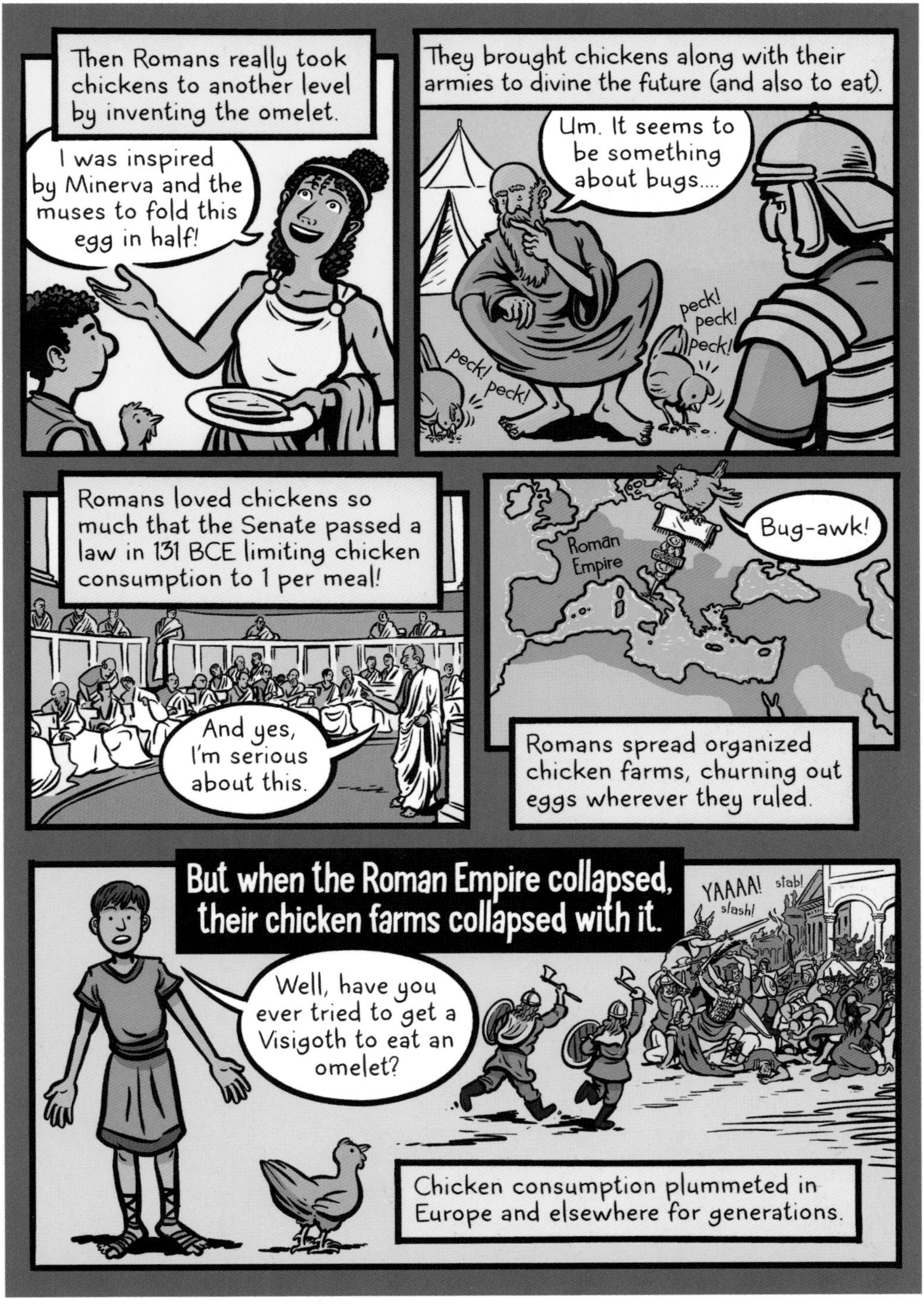
Then Romans really took chickens to another level by inventing the omelet.
I was inspired by Minerva and the muses to fold this egg in half!
They brought chickens along with their armies to divine the future (and also to eat).
Um. It seems to be something about bugs....
peck! peck! peck!
peck! peck!
Romans loved chickens so much that the Senate passed a law in 131 BCE limiting chicken consumption to 1 per meal!
And yes, I'm serious about this.
Roman Empire
Bug-awk!
Romans spread organized chicken farms, churning out eggs wherever they ruled.
But when the Roman Empire collapsed, their chicken farms collapsed with it.
YAAAA!
stab!
slash!
Well, have you ever tried to get a Visigoth to eat an omelet?
Chicken consumption plummeted in Europe and elsewhere for generations.

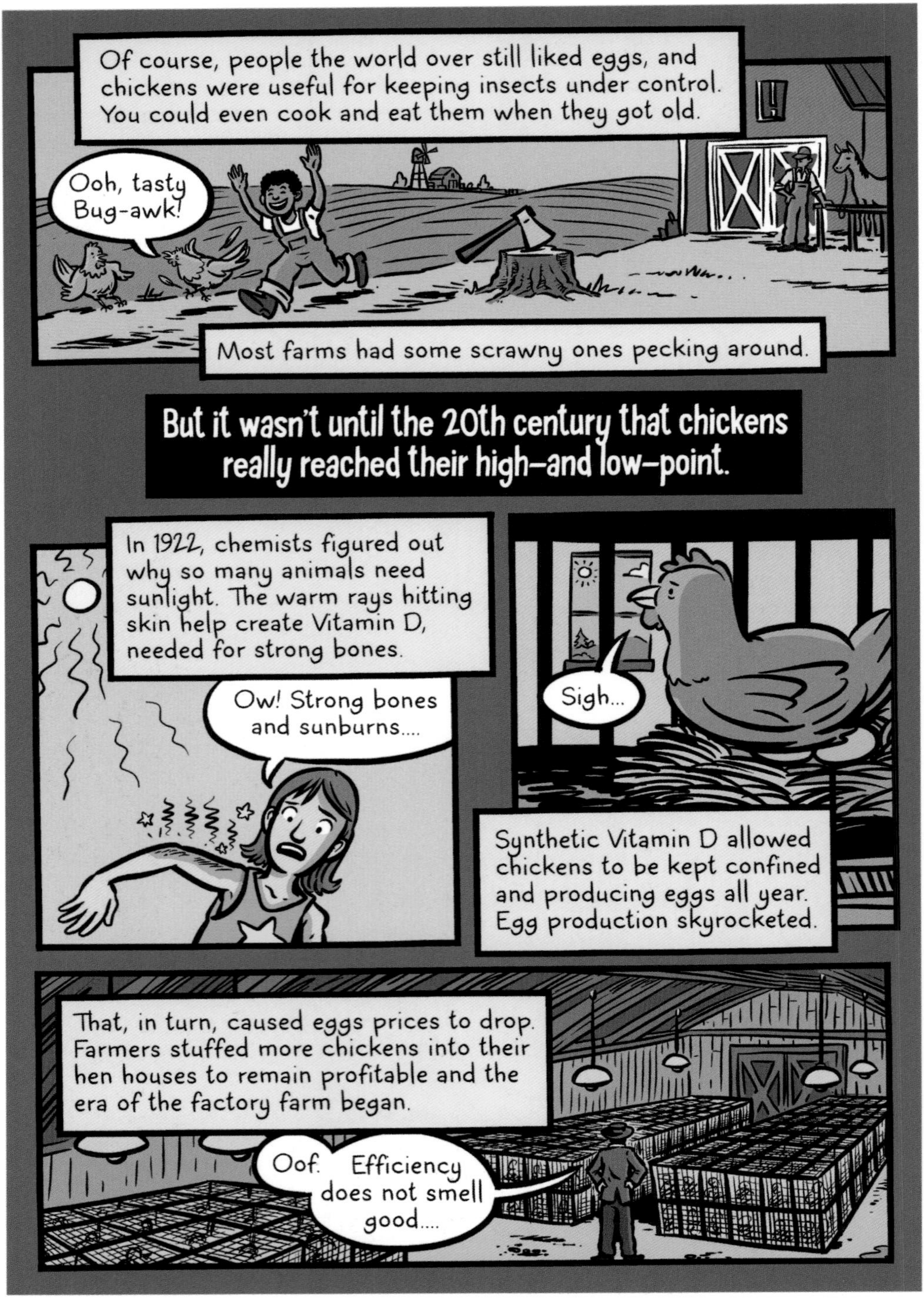

Of course, people the world over still liked eggs, and chickens were useful for keeping insects under control. You could even cook and eat them when they got old.
Ooh, tasty Bug-awk!
Most farms had some scrawny ones pecking around.
But it wasn't until the 20th century that chickens really reached their high–and low–point.
In 1922, chemists figured out why so many animals need sunlight. The warm rays hitting skin help create Vitamin D, needed for strong bones.
Ow! Strong bones and sunburns....
Sigh...
Synthetic Vitamin D allowed chickens to be kept confined and producing eggs all year. Egg production skyrocketed.
That, in turn, caused eggs prices to drop. Farmers stuffed more chickens into their hen houses to remain profitable and the era of the factory farm began.
Oof. Efficiency does not smell good....

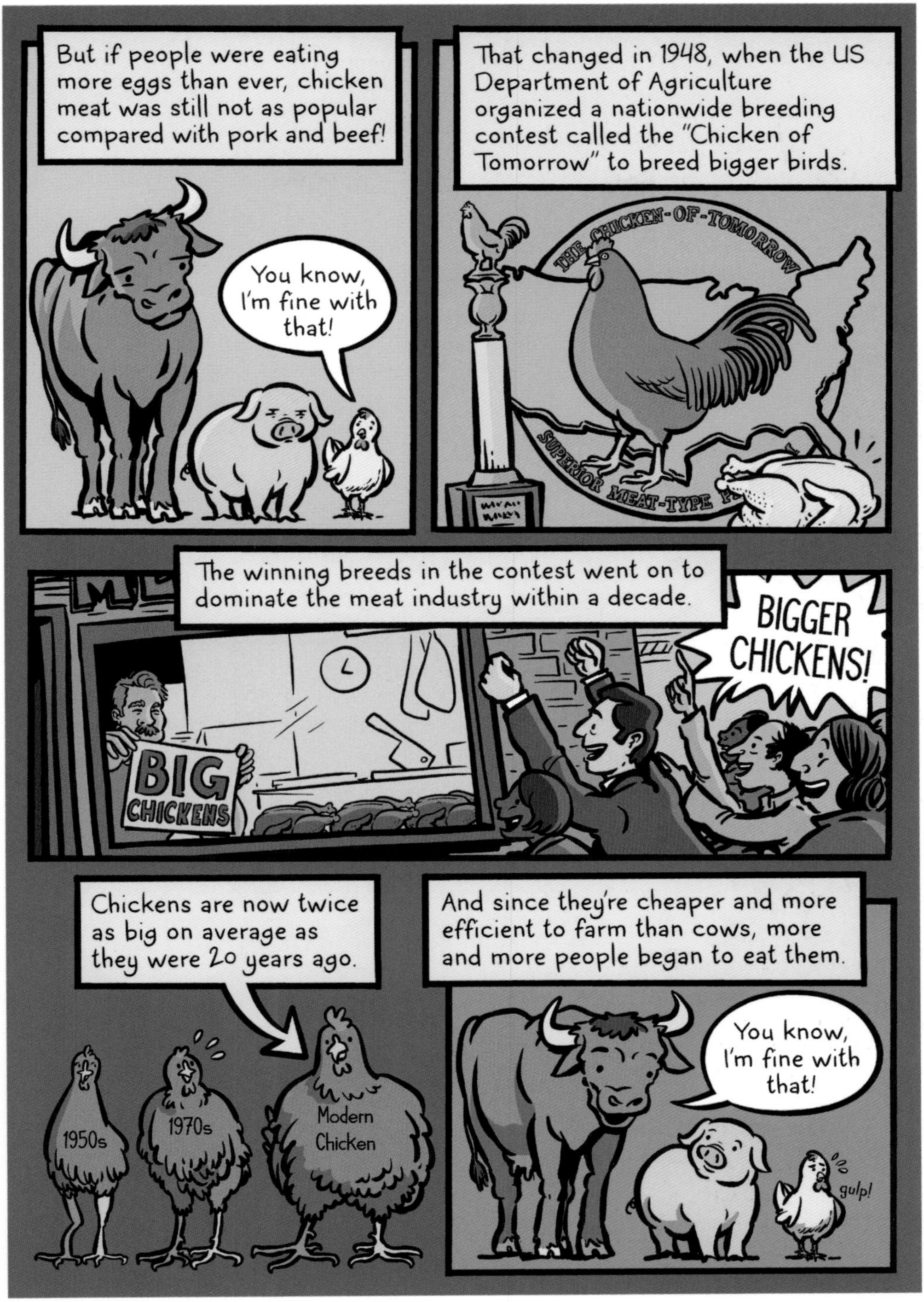

But if people were eating more eggs than ever, chicken meat was still not as popular compared with pork and beef!
You know, I'm fine with that!
That changed in 1948, when the US Department of Agriculture organized a nationwide breeding contest called the "Chicken of Tomorrow" to breed bigger birds.
THE CHICKEN-OF-TOMORROW
SUPERIOR MEAT-TYPE
The winning breeds in the contest went on to dominate the meat industry within a decade.
BIGGER CHICKENS!
BIG CHICKENS
Chickens are now twice as big on average as they were 20 years ago.
1950s
1970s
Modern Chicken
And since they're cheaper and more efficient to farm than cows, more and more people began to eat them.
You know, I'm fine with that!
gulp!

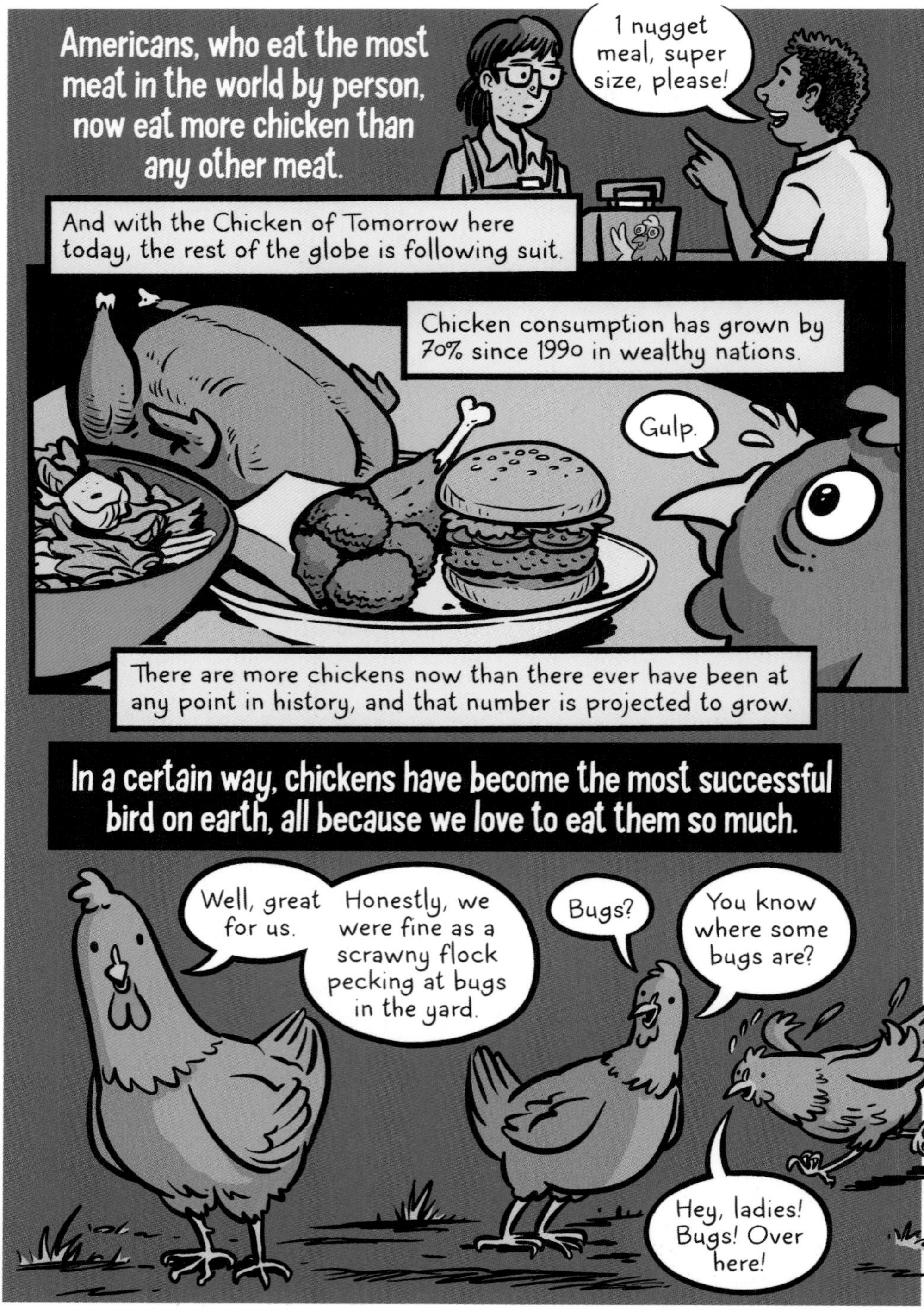

Americans, who eat the most meat in the world by person, now eat more chicken than any other meat.
1 nugget meal, super size, please!
And with the Chicken of Tomorrow here today, the rest of the globe is following suit.
Chicken consumption has grown by 70% since 1990 in wealthy nations.
Gulp.
There are more chickens now than there ever have been at any point in history, and that number is projected to grow.
In a certain way, chickens have become the most successful bird on earth, all because we love to eat them so much.
Well, great for us.
Honestly, we were fine as a scrawny flock pecking at bugs in the yard.
Bugs?
You know where some bugs are?
Hey, ladies! Bugs! Over here!

COCKFIGHTING

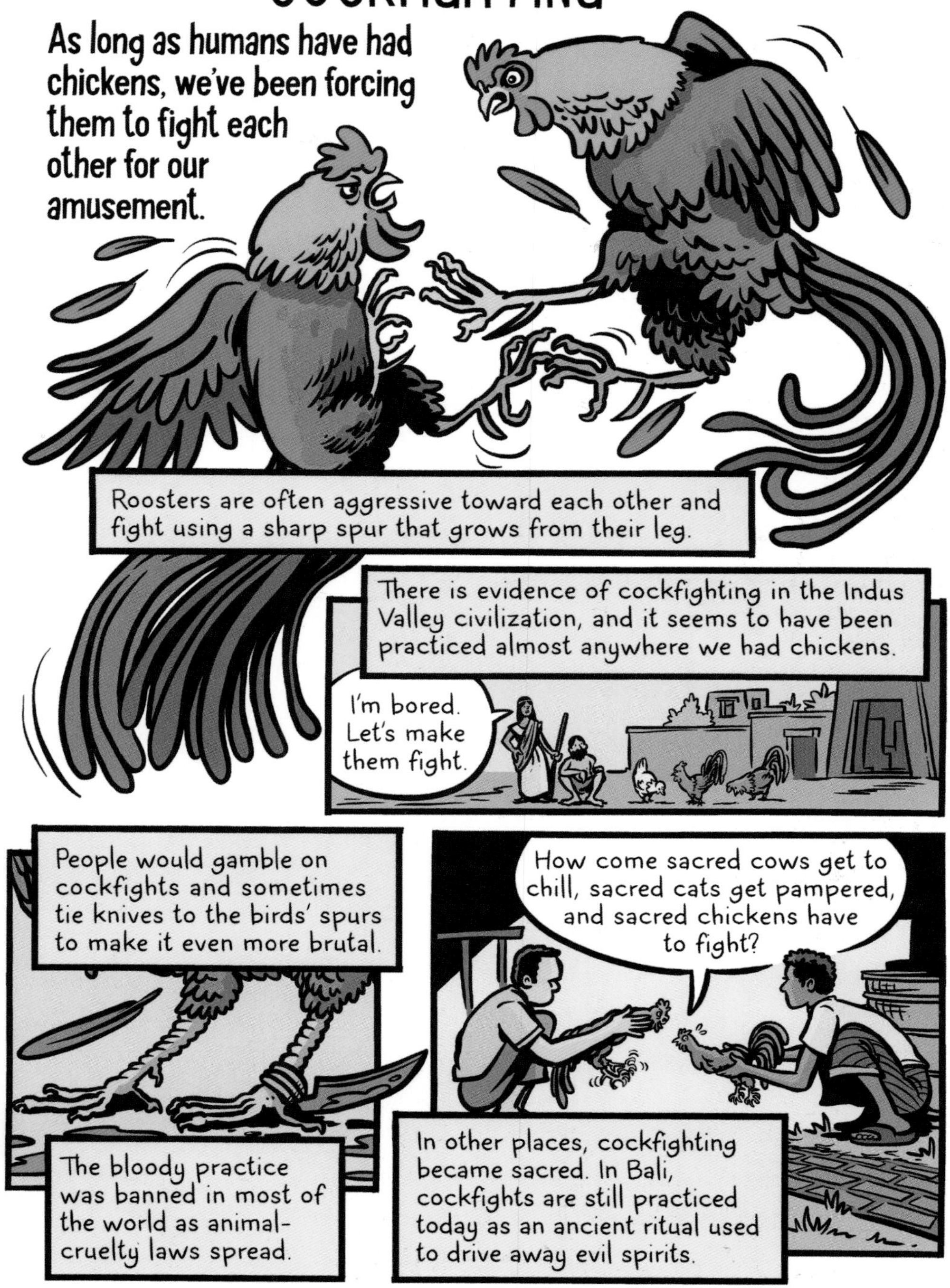

As long as humans have had chickens, we've been forcing them to fight each other for our amusement.
Roosters are often aggressive toward each other and fight using a sharp spur that grows from their leg.
There is evidence of cockfighting in the Indus Valley civilization, and it seems to have been practiced almost anywhere we had chickens.
I'm bored. Let's make them fight.
People would gamble on cockfights and sometimes tie knives to the birds' spurs to make it even more brutal.
The bloody practice was banned in most of the world as animal-cruelty laws spread.
How come sacred cows get to chill, sacred cats get pampered, and sacred chickens have to fight?
In other places, cockfighting became sacred. In Bali, cockfights are still practiced today as an ancient ritual used to drive away evil spirits.

CREATURES THAT FIND US USEFUL

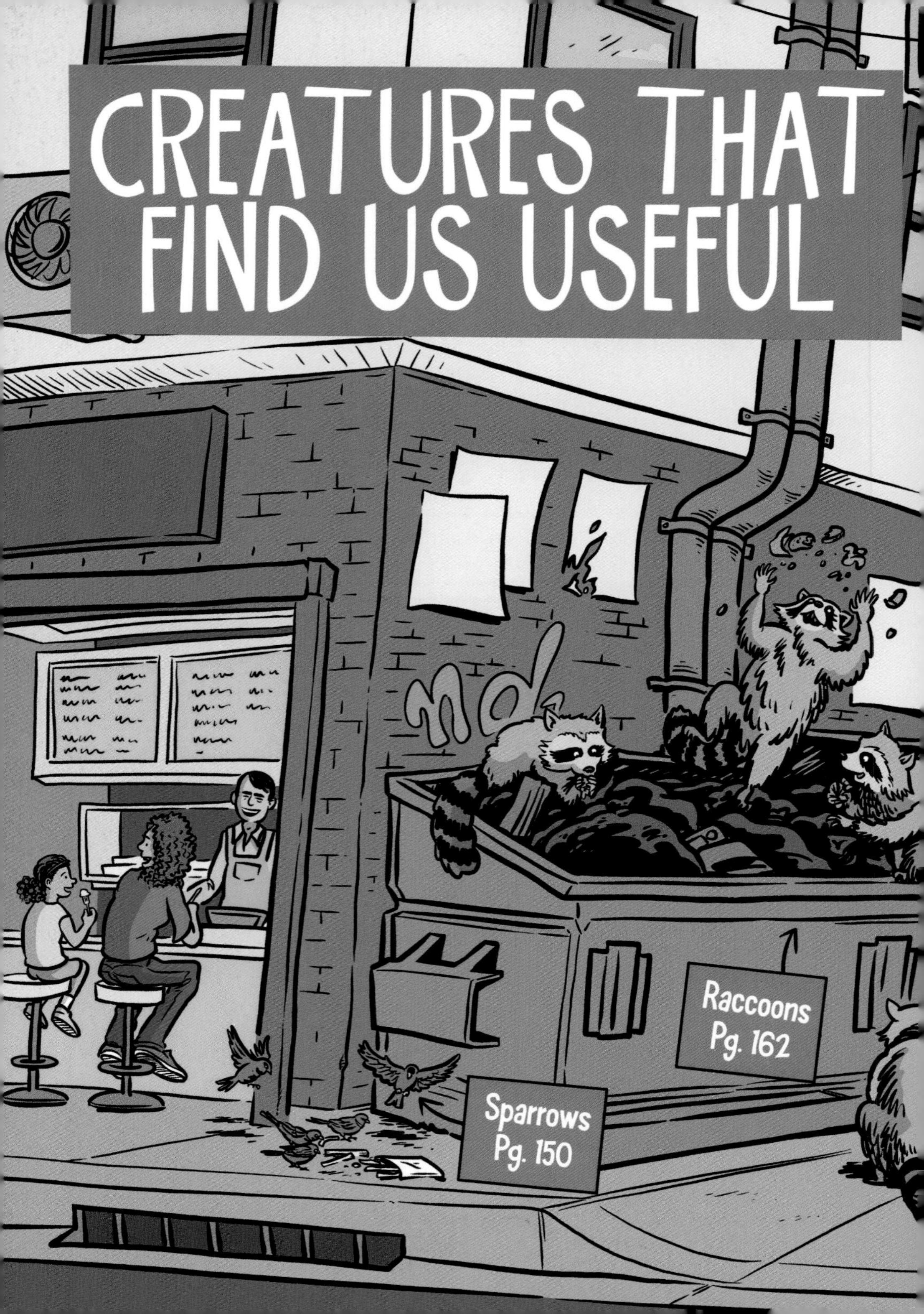
Raccoons
Pg. 162
Sparrows
Pg. 150

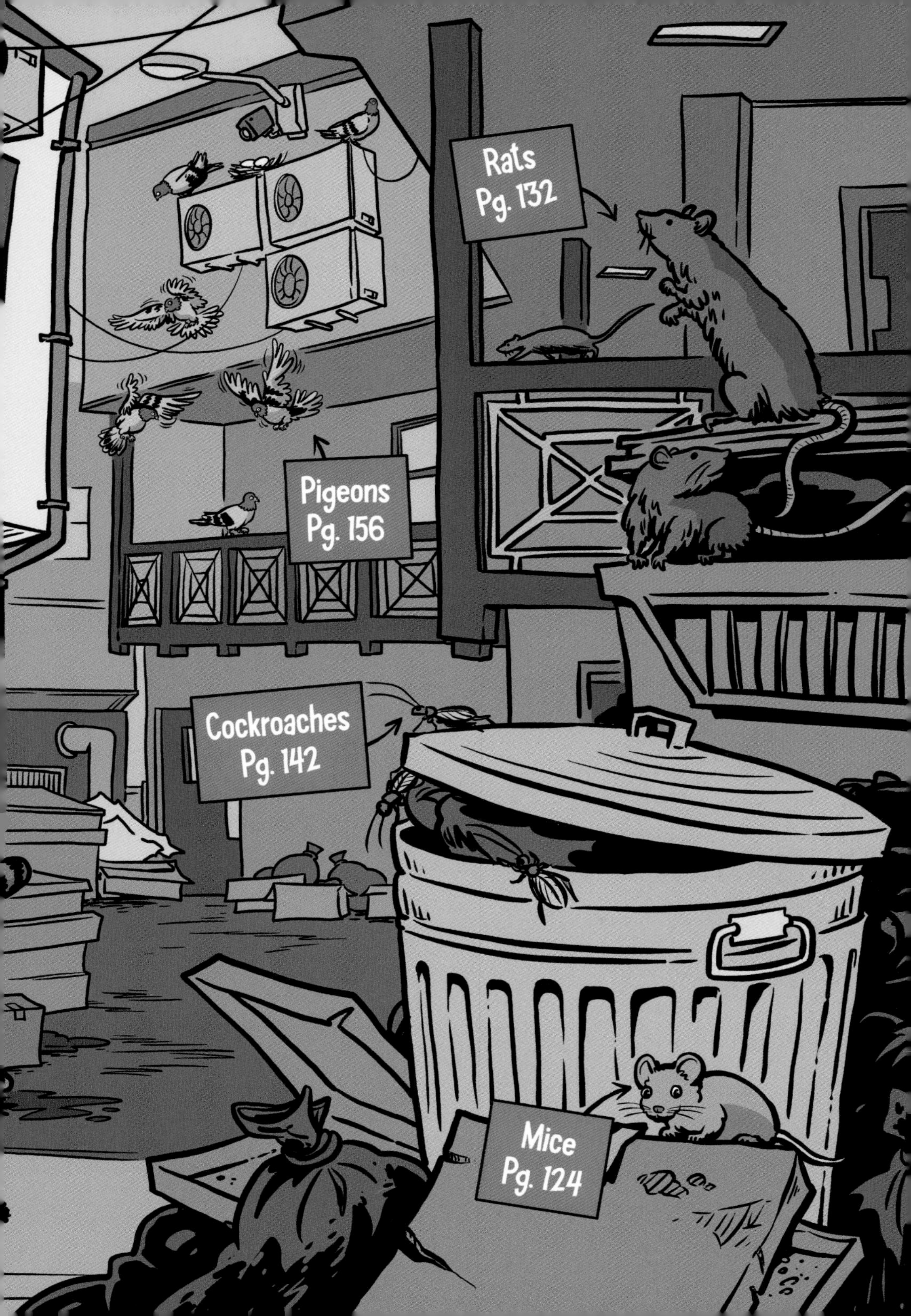
Rats
Pg. 132
Pigeons
Pg. 156
Cockroaches
Pg. 142
Mice
Pg. 124

MICE

Mus musculus

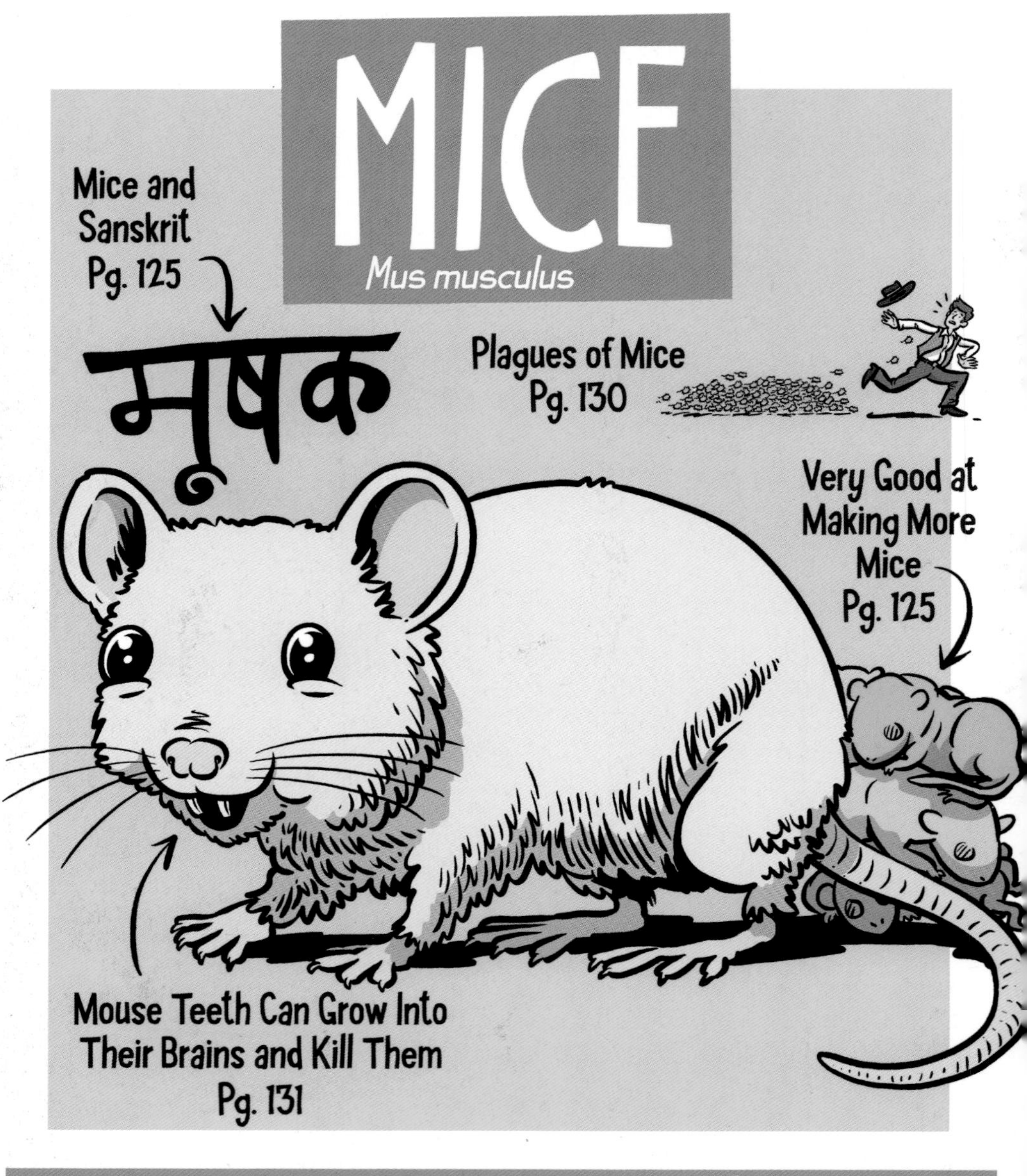

Mice and Sanskrit Pg. 125
मूषक
Plagues of Mice Pg. 130
Very Good at Making More Mice Pg. 125
Mouse Teeth Can Grow Into Their Brains and Kill Them Pg. 131

Size: 5–7 in

Weight: 0.68 oz

Athletic Ability: Can squeeze through a 6-mm opening

Diet: Grains, seeds, fruits, trash, sometimes each other

(Likely) **Wild Origin:**

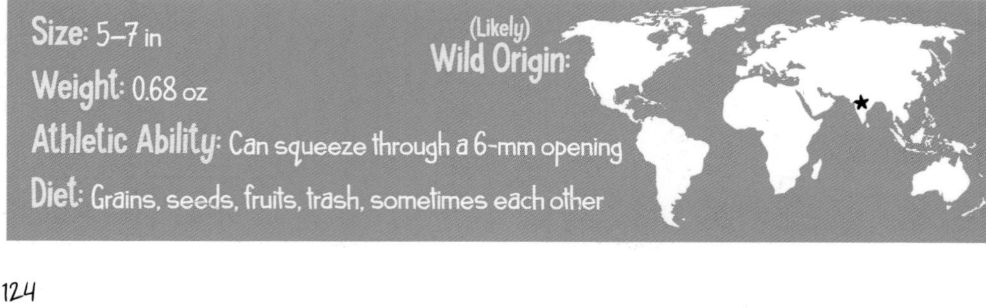

HOW MICE CONQUERED BOTH THE STOREROOM AND THE LABORATORY

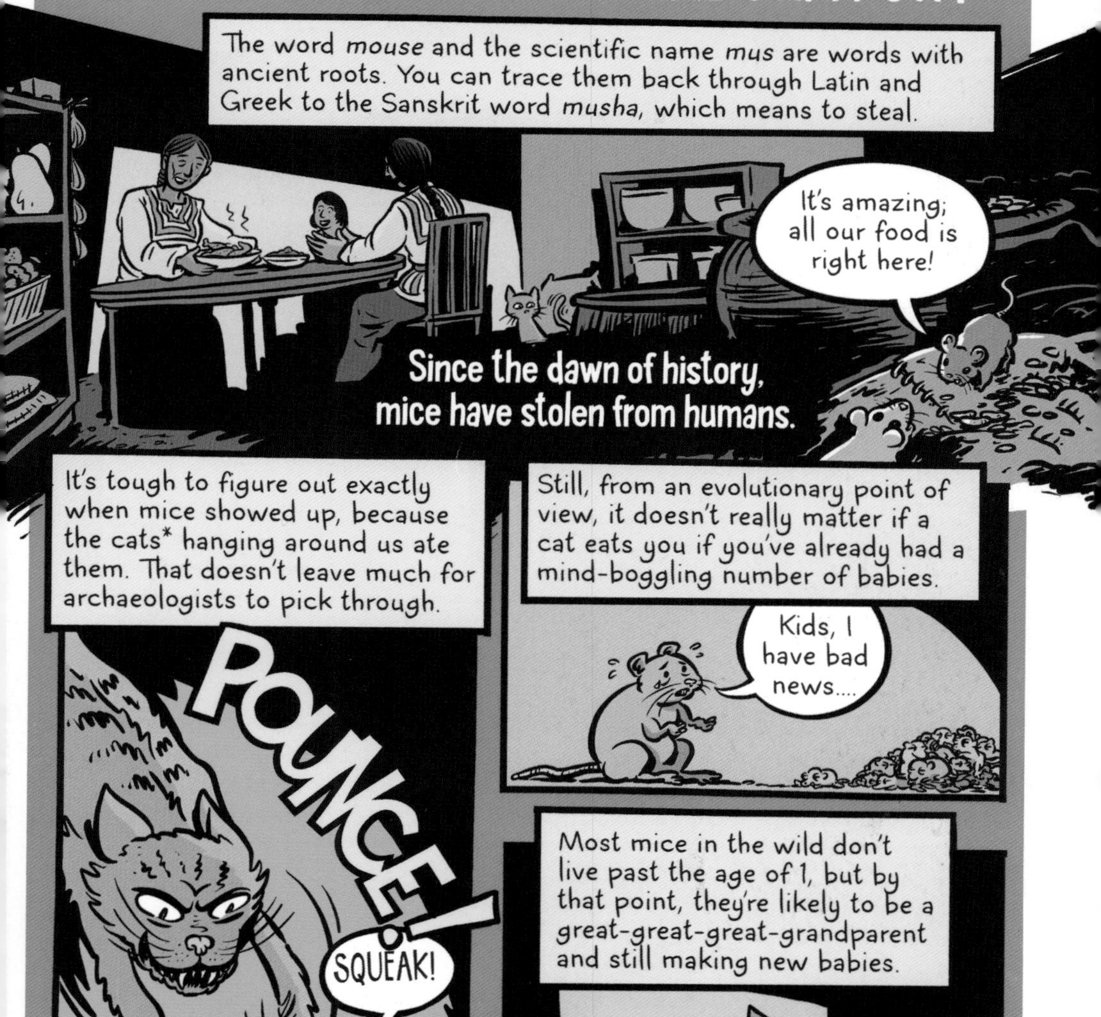

The word *mouse* and the scientific name *mus* are words with ancient roots. You can trace them back through Latin and Greek to the Sanskrit word *musha*, which means to steal.
It's amazing; all our food is right here!
Since the dawn of history, mice have stolen from humans.
It's tough to figure out exactly when mice showed up, because the cats* hanging around us ate them. That doesn't leave much for archaeologists to pick through.
POUNCE!
SQUEAK!
*See Pg. 19
Still, from an evolutionary point of view, it doesn't really matter if a cat eats you if you've already had a mind-boggling number of babies.
Kids, I have bad news....
Most mice in the wild don't live past the age of 1, but by that point, they're likely to be a great-great-great-grandparent and still making new babies.
I call it job security.

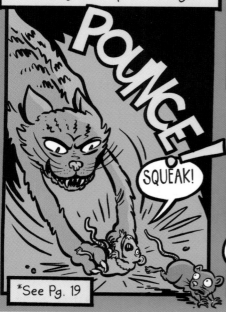

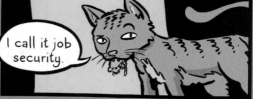

We do know where mice originally come from, though--the Indian subcontinent.

Uh, looks like the "little thieves" got into the grain pouch....

Around 11,000 years ago, people in the Middle East started farming wheat, barley, peas, lentils, and chickpeas.

Around 9,000 years ago, people in China started farming rice, soy, and millet.

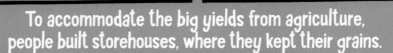

To accommodate the big yields from agriculture, people built storehouses, where they kept their grains.

These technologies spread from the west and east into the Indian subcontinent, where mice were ready and waiting to spread right back out along those same paths and into those storehouses.

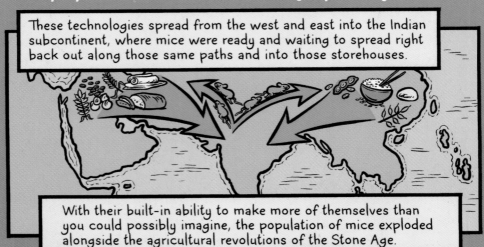

With their built-in ability to make more of themselves than you could possibly imagine, the population of mice exploded alongside the agricultural revolutions of the Stone Age.

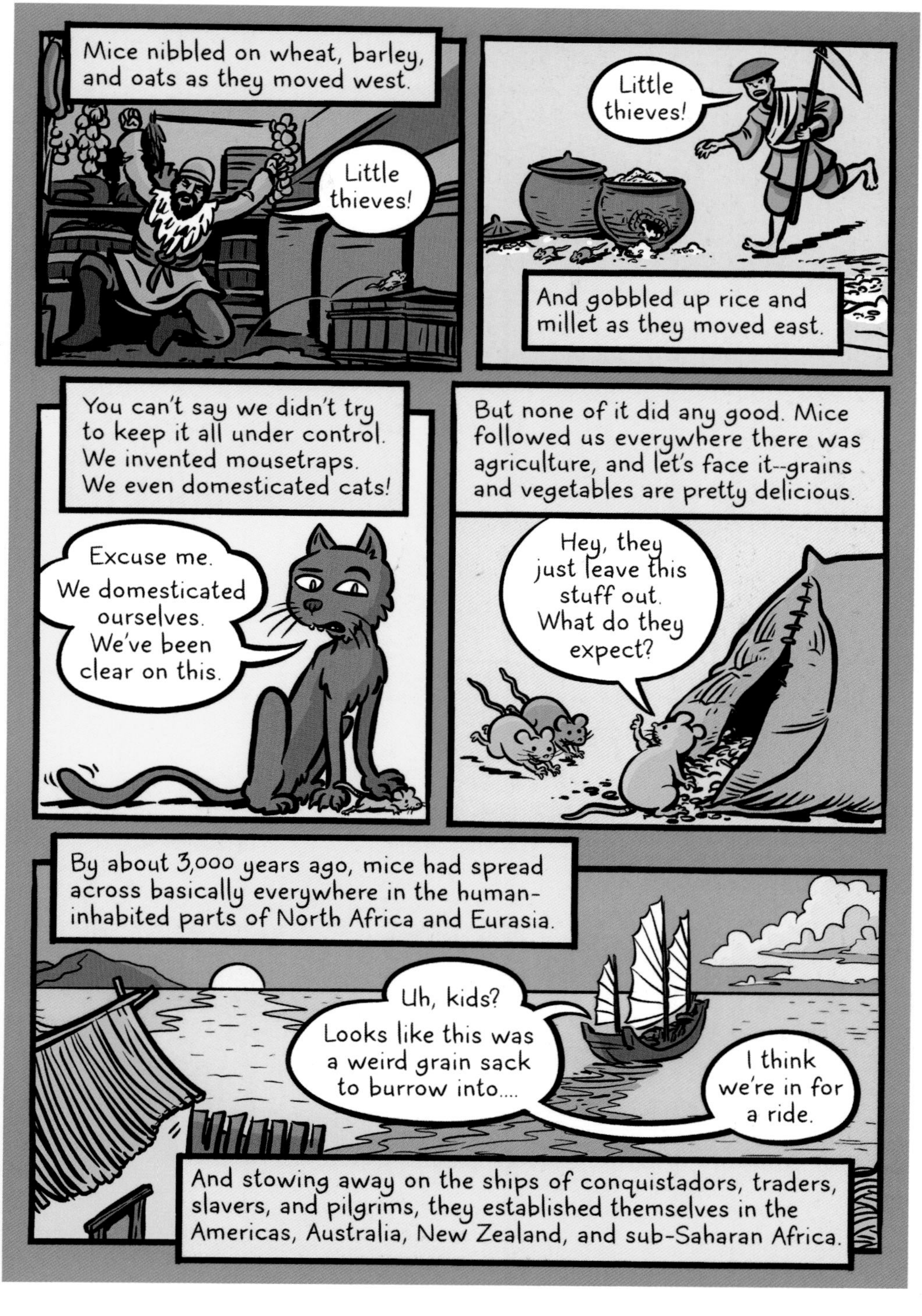

Mice nibbled on wheat, barley, and oats as they moved west.
Little thieves!
Little thieves!
And gobbled up rice and millet as they moved east.
You can't say we didn't try to keep it all under control. We invented mousetraps. We even domesticated cats!
Excuse me. We domesticated ourselves. We've been clear on this.
But none of it did any good. Mice followed us everywhere there was agriculture, and let's face it--grains and vegetables are pretty delicious.
Hey, they just leave this stuff out. What do they expect?
By about 3,000 years ago, mice had spread across basically everywhere in the human-inhabited parts of North Africa and Eurasia.
Uh, kids? Looks like this was a weird grain sack to burrow into....
I think we're in for a ride.
And stowing away on the ships of conquistadors, traders, slavers, and pilgrims, they established themselves in the Americas, Australia, New Zealand, and sub-Saharan Africa.

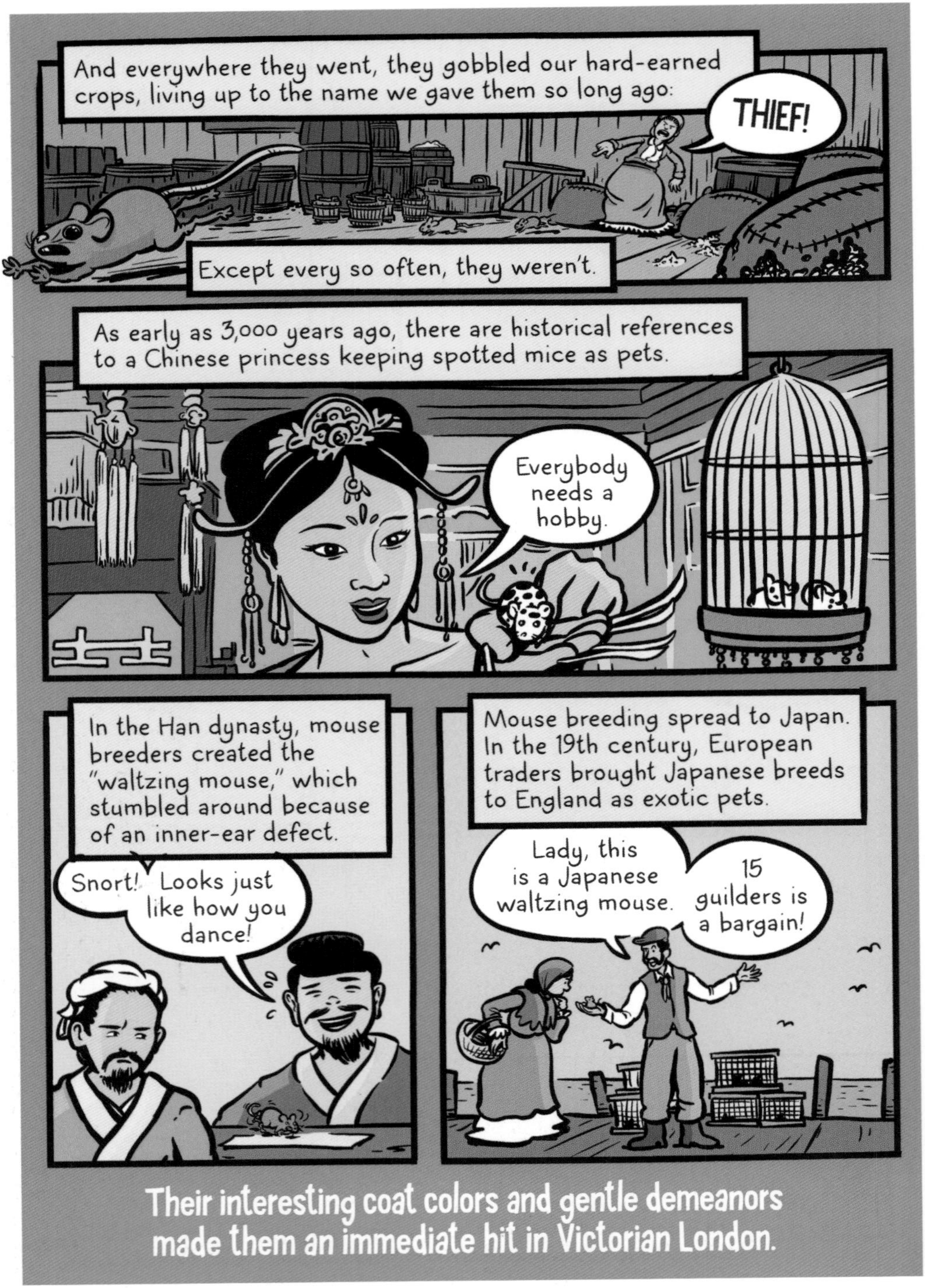
And everywhere they went, they gobbled our hard-earned crops, living up to the name we gave them so long ago:
THIEF!
Except every so often, they weren't.
As early as 3,000 years ago, there are historical references to a Chinese princess keeping spotted mice as pets.
Everybody needs a hobby.
In the Han dynasty, mouse breeders created the "waltzing mouse," which stumbled around because of an inner-ear defect.
Snort!
Looks just like how you dance!
Mouse breeding spread to Japan. In the 19th century, European traders brought Japanese breeds to England as exotic pets.
Lady, this is a Japanese waltzing mouse.
15 guilders is a bargain!
Their interesting coat colors and gentle demeanors made them an immediate hit in Victorian London.

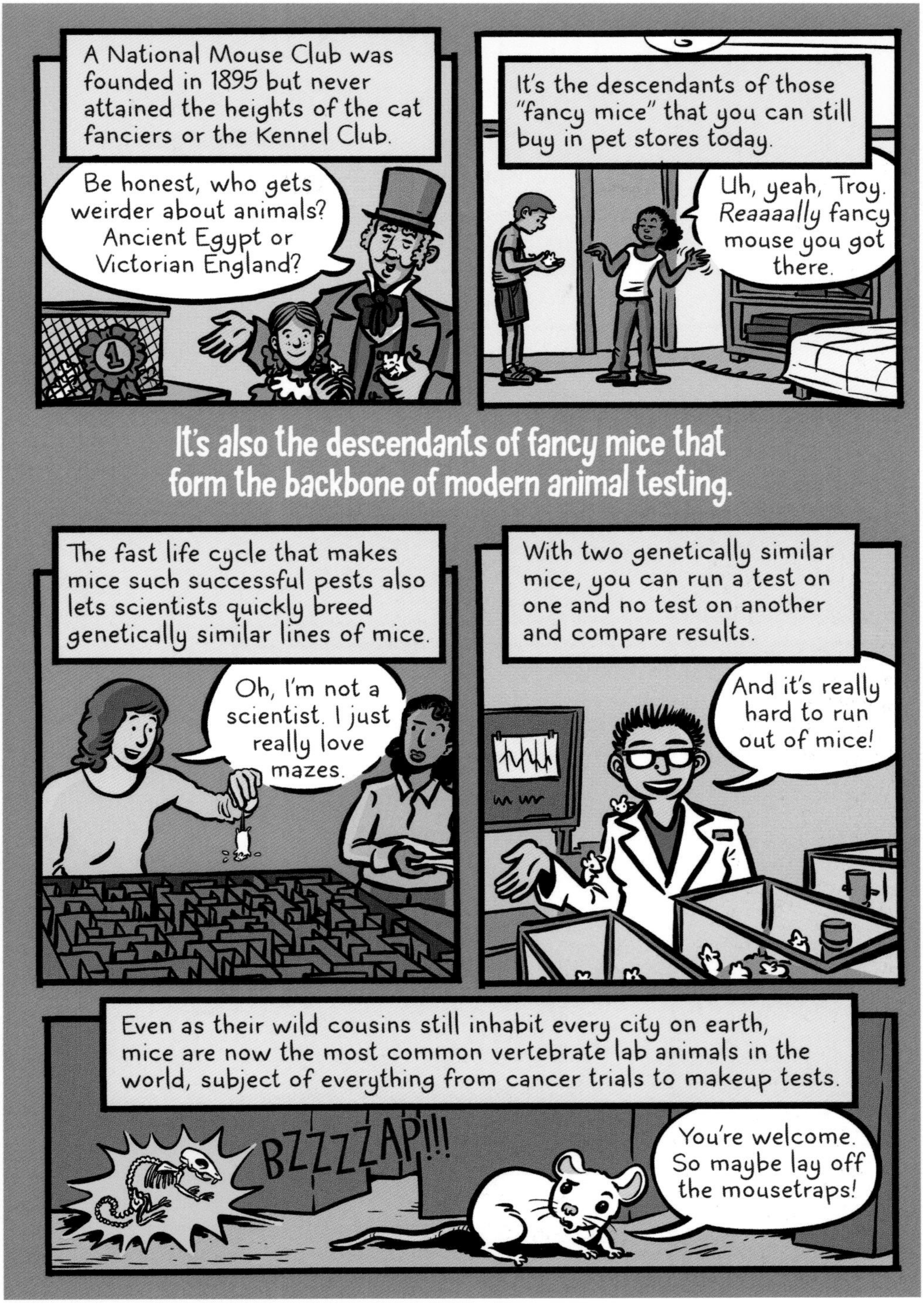
A National Mouse Club was founded in 1895 but never attained the heights of the cat fanciers or the Kennel Club.
Be honest, who gets weirder about animals? Ancient Egypt or Victorian England?
It's the descendants of those "fancy mice" that you can still buy in pet stores today.
Uh, yeah, Troy. *Reaaaally* fancy mouse you got there.
It's also the descendants of fancy mice that form the backbone of modern animal testing.
The fast life cycle that makes mice such successful pests also lets scientists quickly breed genetically similar lines of mice.
Oh, I'm not a scientist. I just really love mazes.
With two genetically similar mice, you can run a test on one and no test on another and compare results.
And it's really hard to run out of mice!
Even as their wild cousins still inhabit every city on earth, mice are now the most common vertebrate lab animals in the world, subject of everything from cancer trials to makeup tests.
BZZZZAP!!!
You're welcome. So maybe lay off the mousetraps!

PLAGUES OF MICE

Mice breed so fast that whenever there isn't a predator around to kill a lot of them, the result is a mouse plague, in which an overwhelming amount of mice overrun everything.

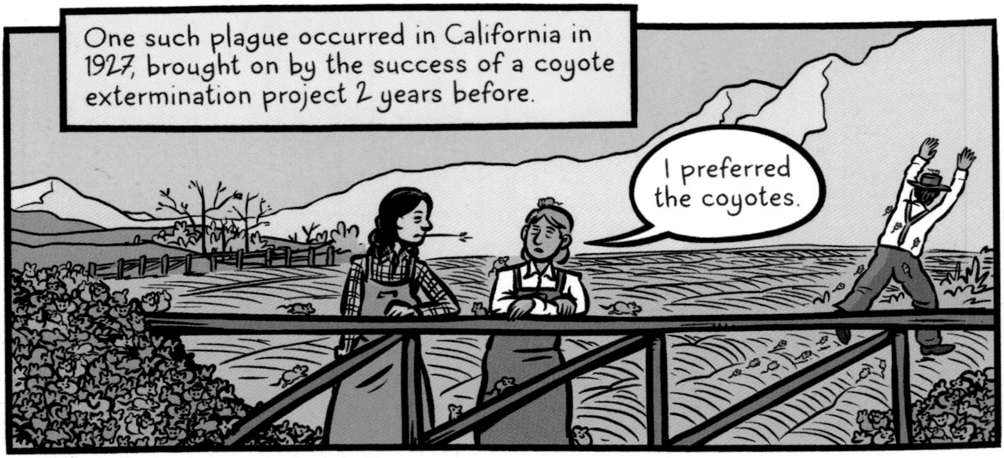
One such plague occurred in California in 1927, brought on by the success of a coyote extermination project 2 years before.
I preferred the coyotes.

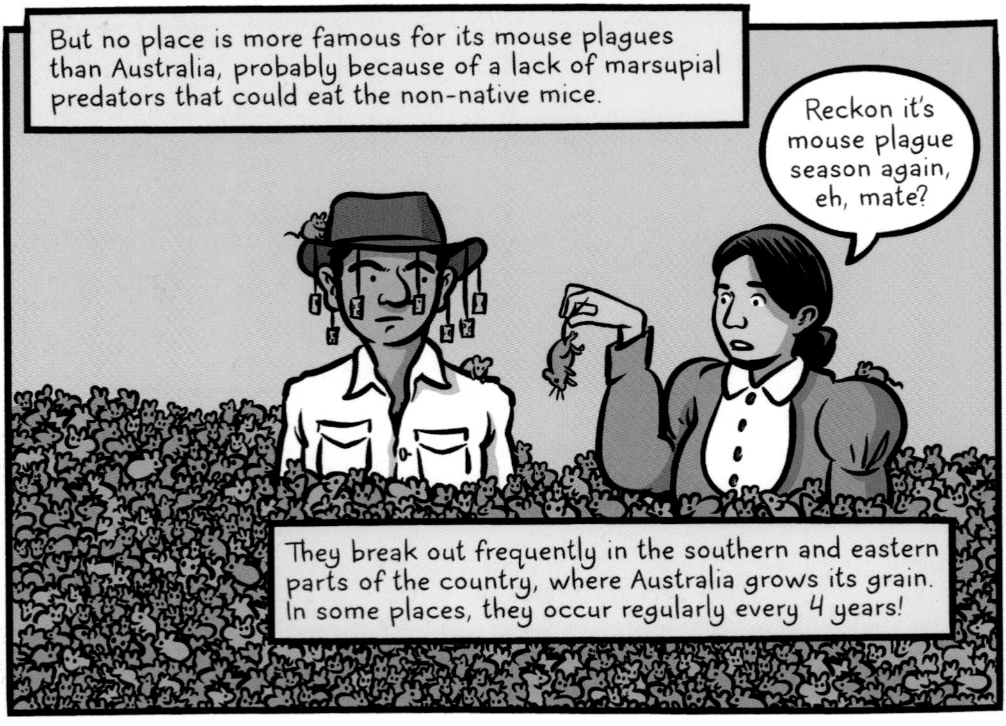
But no place is more famous for its mouse plagues than Australia, probably because of a lack of marsupial predators that could eat the non-native mice.
Reckon it's mouse plague season again, eh, mate?
They break out frequently in the southern and eastern parts of the country, where Australia grows its grain. In some places, they occur regularly every 4 years!

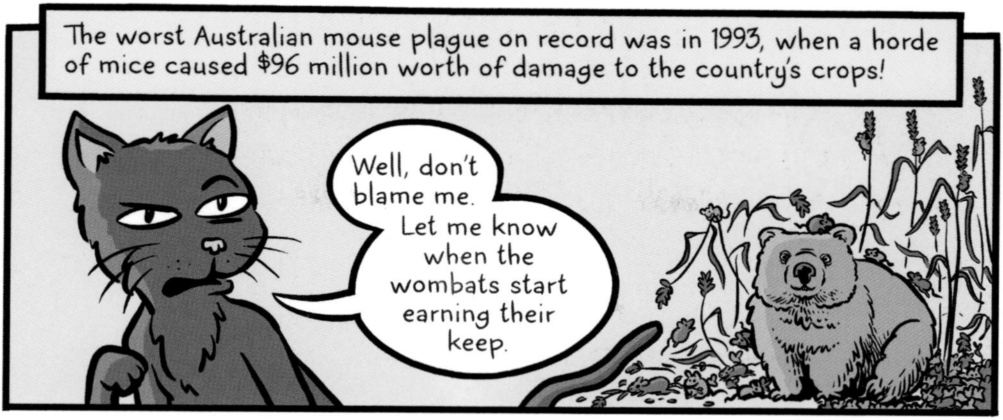
The worst Australian mouse plague on record was in 1993, when a horde of mice caused $96 million worth of damage to the country's crops!
Well, don't blame me. Let me know when the wombats start earning their keep.

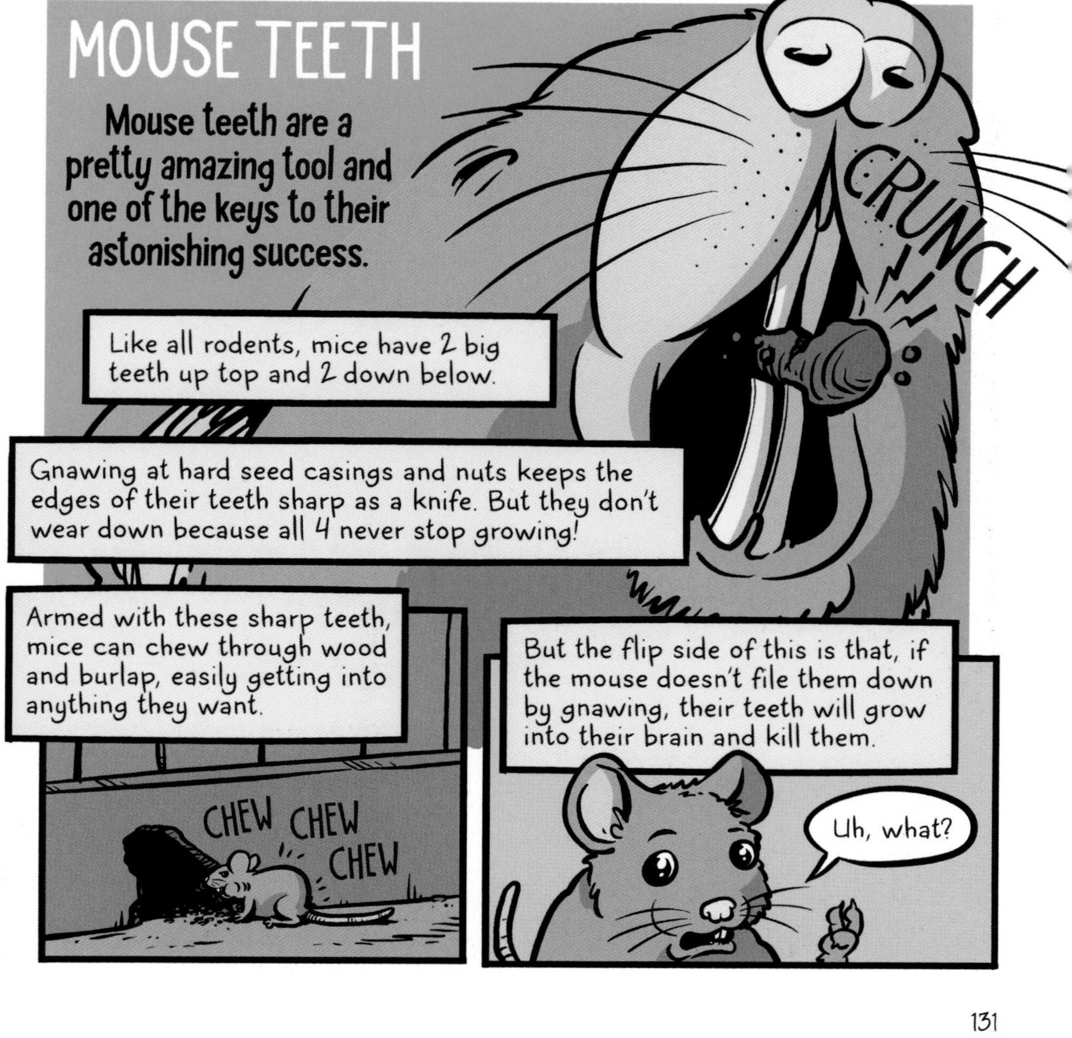
MOUSE TEETH
Mouse teeth are a pretty amazing tool and one of the keys to their astonishing success.
CRUNCH
Like all rodents, mice have 2 big teeth up top and 2 down below.
Gnawing at hard seed casings and nuts keeps the edges of their teeth sharp as a knife. But they don't wear down because all 4 never stop growing!
Armed with these sharp teeth, mice can chew through wood and burlap, easily getting into anything they want.
CHEW CHEW CHEW
But the flip side of this is that, if the mouse doesn't file them down by gnawing, their teeth will grow into their brain and kill them.
Uh, what?

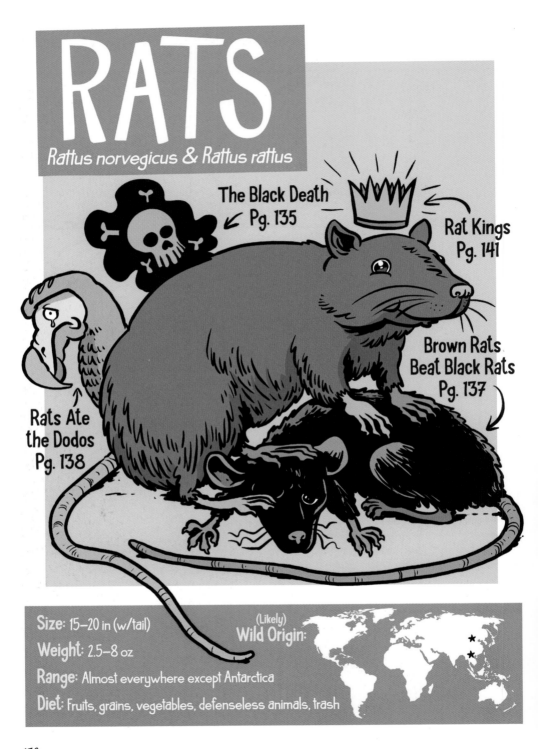

RATS

Rattus norvegicus & Rattus rattus

The Black Death
Pg. 135

Rat Kings
Pg. 141

Brown Rats
Beat Black Rats
Pg. 137

Rats Ate
the Dodos
Pg. 138

Size: 15–20 in (w/tail)

Weight: 2.5–8 oz

Range: Almost everywhere except Antarctica

Diet: Fruits, grains, vegetables, defenseless animals, trash

(Likely)
Wild Origin:

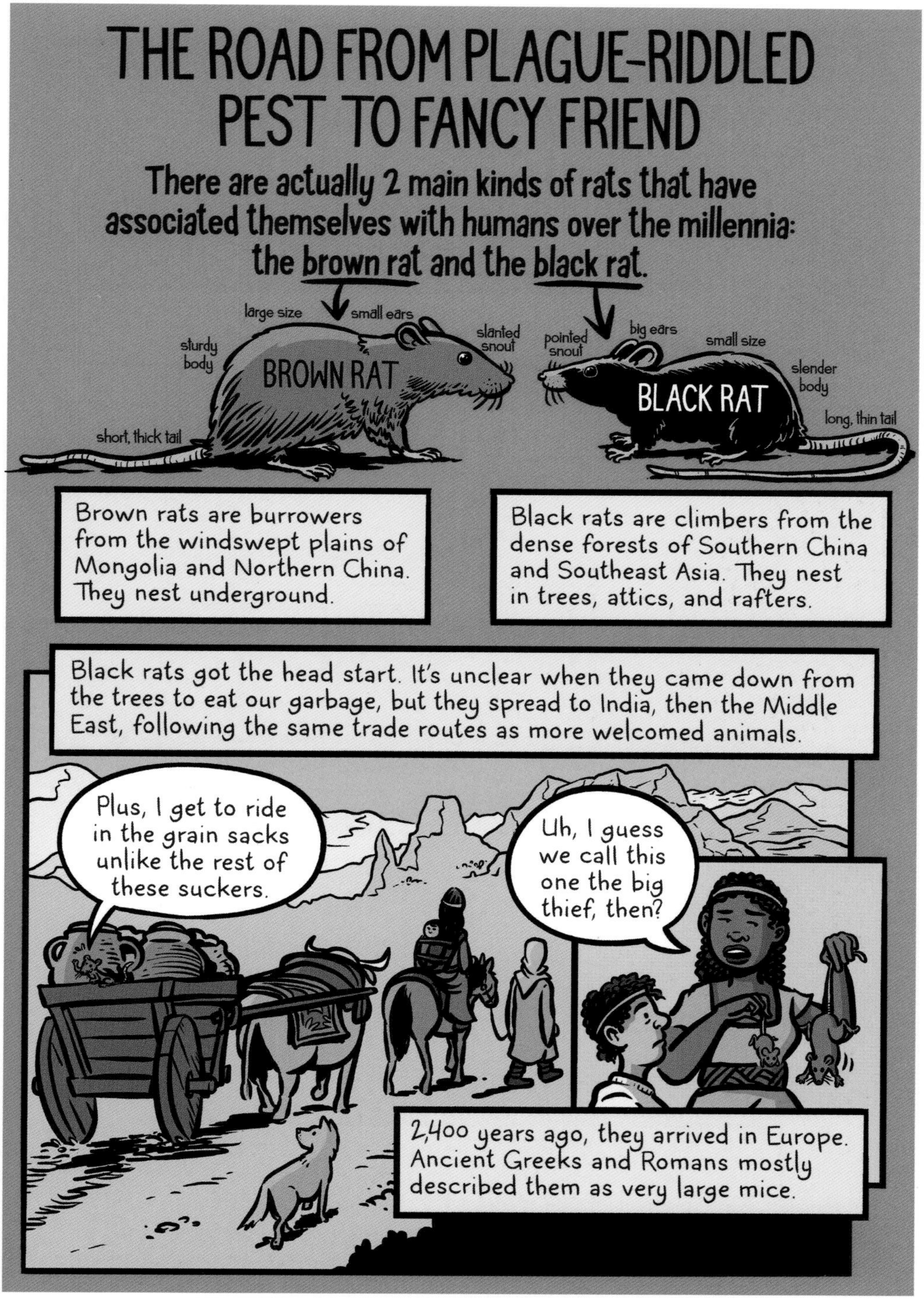
THE ROAD FROM PLAGUE-RIDDLED PEST TO FANCY FRIEND
There are actually 2 main kinds of rats that have associated themselves with humans over the millennia: the brown rat and the black rat.
large size
small ears
sturdy body
slanted snout
short, thick tail
BROWN RAT
pointed snout
big ears
small size
slender body
long, thin tail
BLACK RAT
Brown rats are burrowers from the windswept plains of Mongolia and Northern China. They nest underground.
Black rats are climbers from the dense forests of Southern China and Southeast Asia. They nest in trees, attics, and rafters.
Black rats got the head start. It's unclear when they came down from the trees to eat our garbage, but they spread to India, then the Middle East, following the same trade routes as more welcomed animals.
Plus, I get to ride in the grain sacks unlike the rest of these suckers.
Uh, I guess we call this one the big thief, then?
2,400 years ago, they arrived in Europe. Ancient Greeks and Romans mostly described them as very large mice.

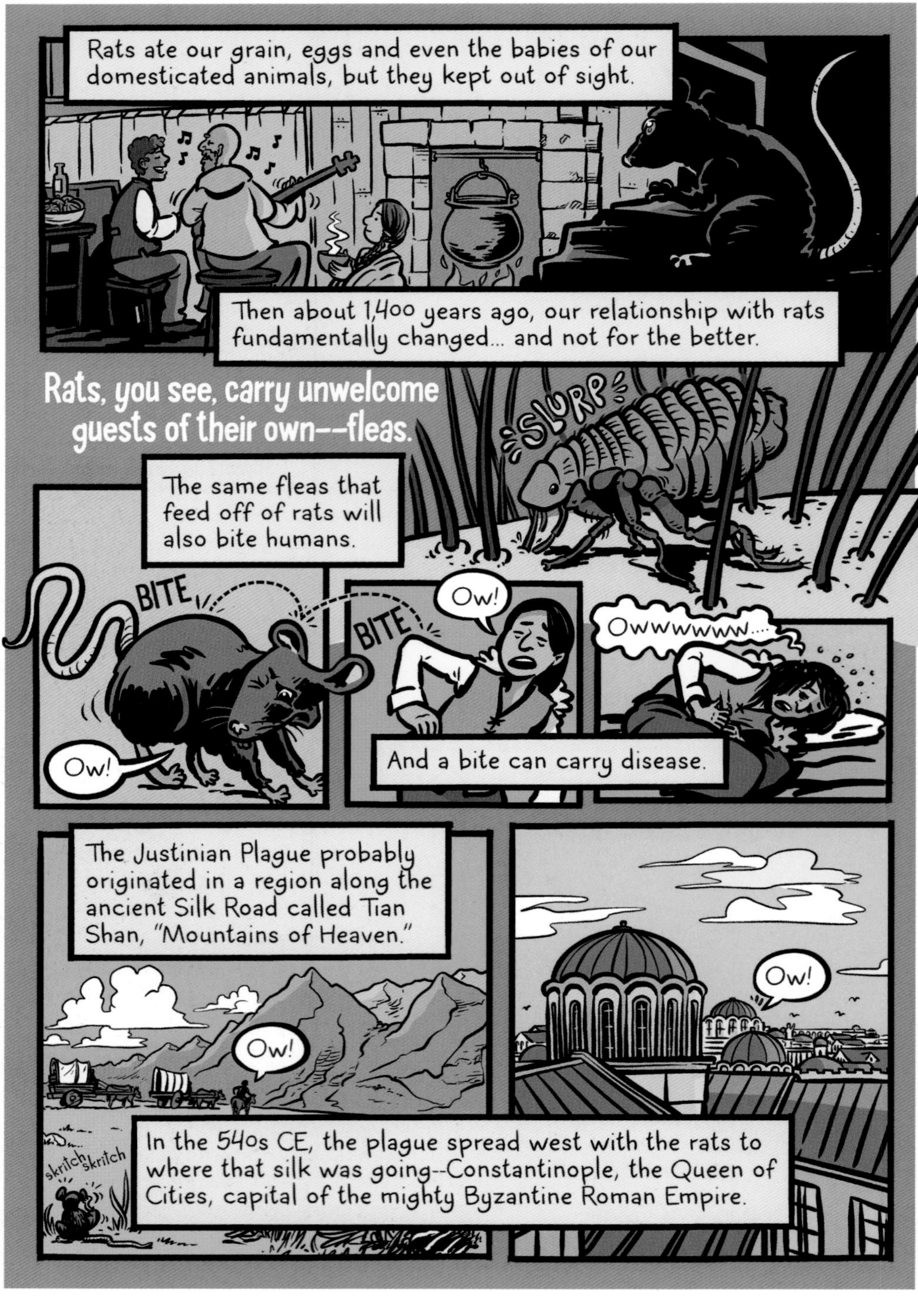

Rats ate our grain, eggs and even the babies of our domesticated animals, but they kept out of sight.
Then about 1,400 years ago, our relationship with rats fundamentally changed... and not for the better.
Rats, you see, carry unwelcome guests of their own--fleas.
SLURP
The same fleas that feed off of rats will also bite humans.
BITE
Ow!
BITE
Ow!
Owwwwww....
And a bite can carry disease.
The Justinian Plague probably originated in a region along the ancient Silk Road called Tian Shan, "Mountains of Heaven."
Ow!
skritch skritch
Ow!
In the 540s CE, the plague spread west with the rats to where that silk was going--Constantinople, the Queen of Cities, capital of the mighty Byzantine Roman Empire.

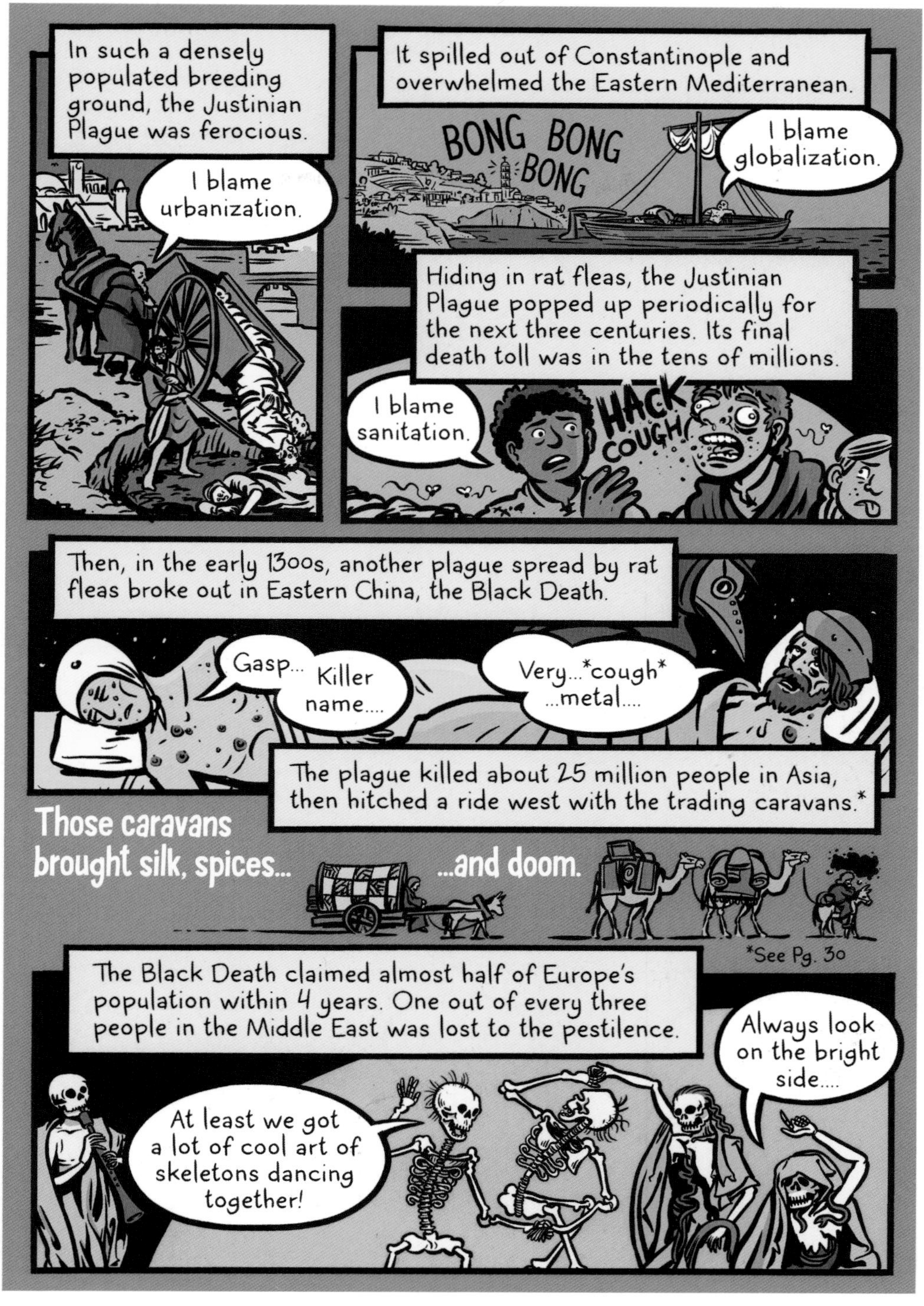
In such a densely populated breeding ground, the Justinian Plague was ferocious.
I blame urbanization.
It spilled out of Constantinople and overwhelmed the Eastern Mediterranean.
BONG BONG BONG
I blame globalization.
Hiding in rat fleas, the Justinian Plague popped up periodically for the next three centuries. Its final death toll was in the tens of millions.
I blame sanitation.
HACK COUGH
Then, in the early 1300s, another plague spread by rat fleas broke out in Eastern China, the Black Death.
Gasp... Killer name....
Very...*cough* ...metal....
The plague killed about 25 million people in Asia, then hitched a ride west with the trading caravans.*
Those caravans brought silk, spices...
...and doom.
*See Pg. 30
The Black Death claimed almost half of Europe's population within 4 years. One out of every three people in the Middle East was lost to the pestilence.
Always look on the bright side....
At least we got a lot of cool art of skeletons dancing together!

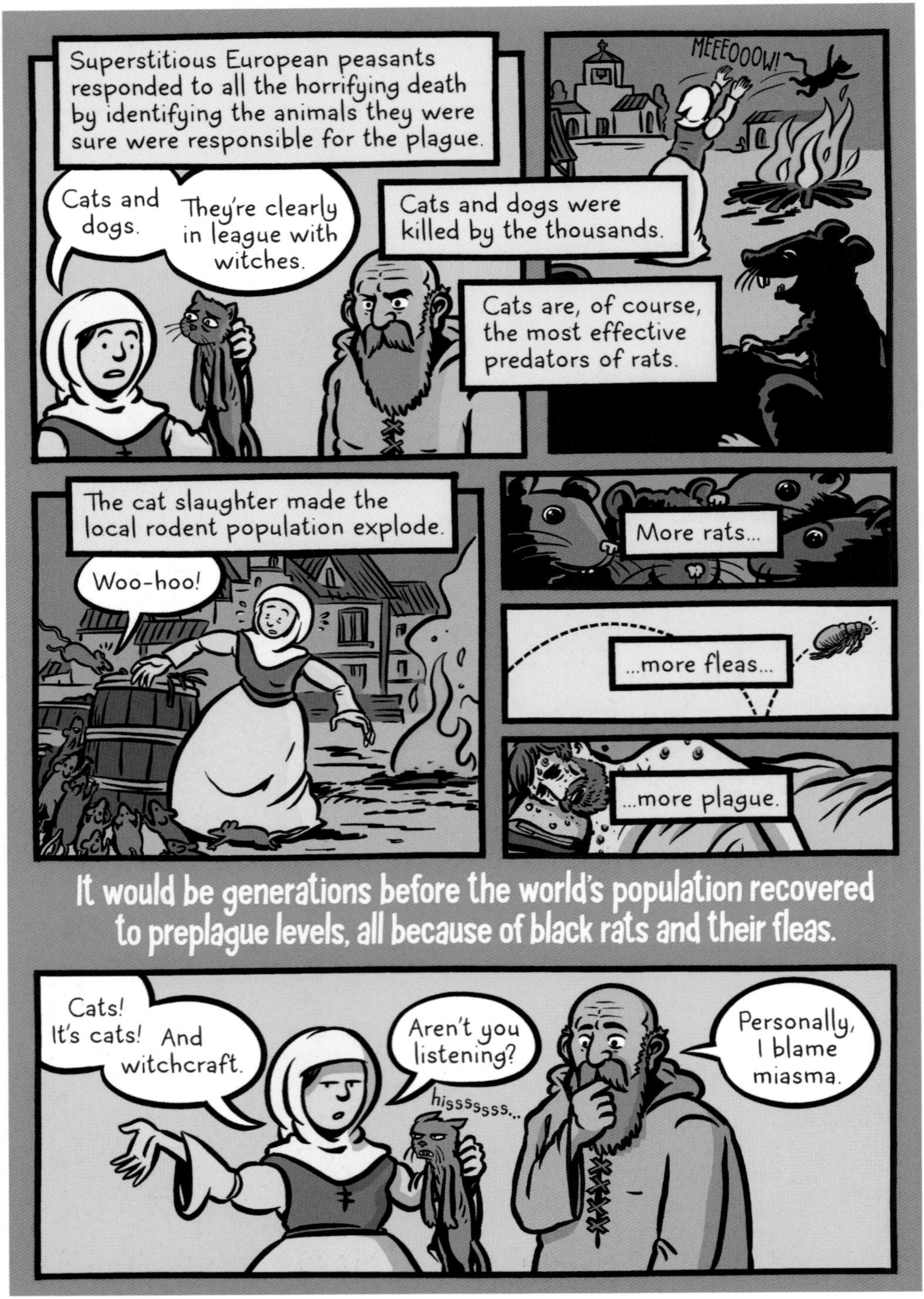
Superstitious European peasants responded to all the horrifying death by identifying the animals they were sure were responsible for the plague.
Cats and dogs.
They're clearly in league with witches.
MEEEOOOW!
Cats and dogs were killed by the thousands.
Cats are, of course, the most effective predators of rats.
The cat slaughter made the local rodent population explode.
Woo-hoo!
More rats...
...more fleas...
...more plague.
It would be generations before the world's population recovered to preplague levels, all because of black rats and their fleas.
Cats! It's cats!
And witchcraft.
Aren't you listening?
hissssssss...
Personally, I blame miasma.

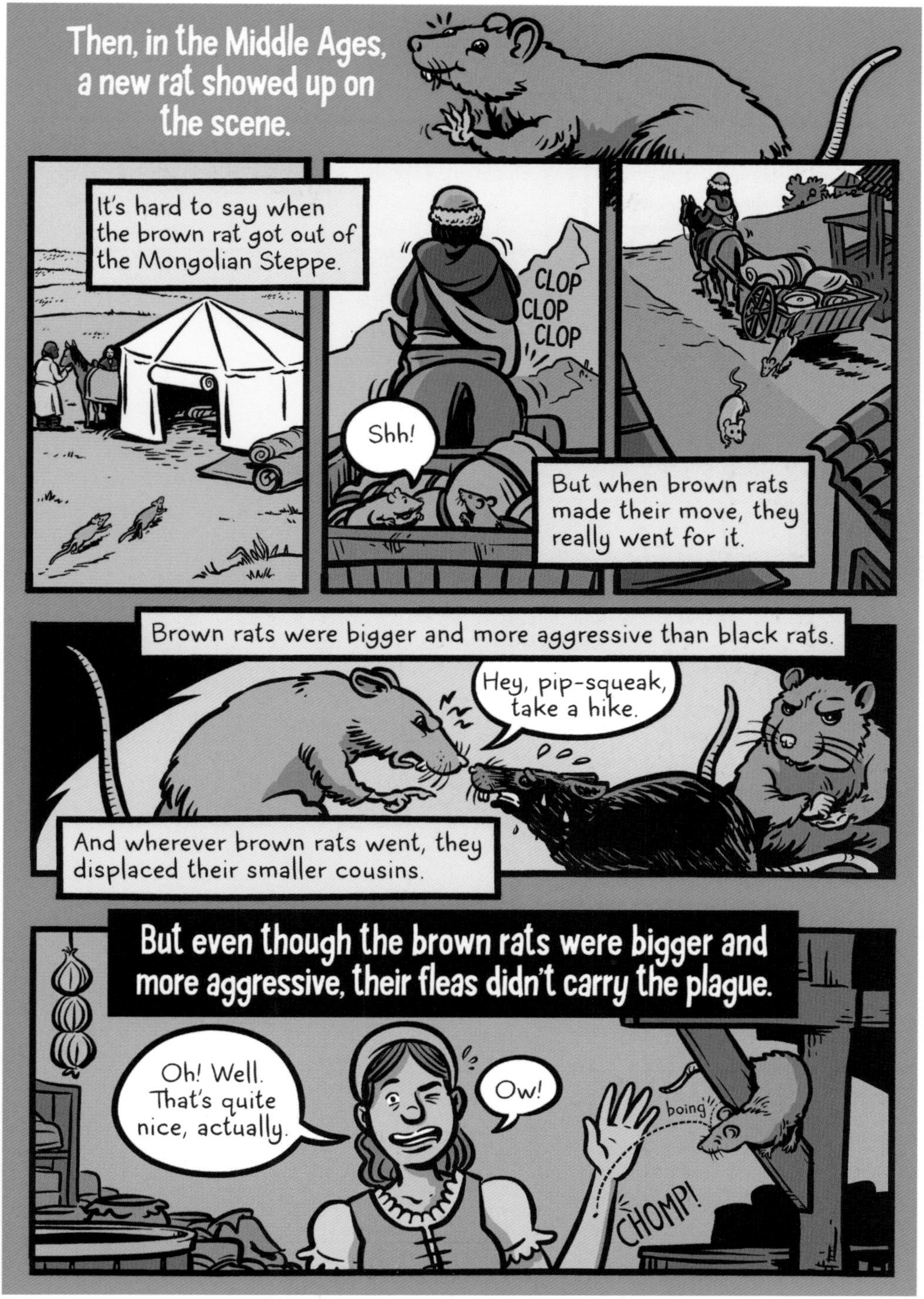
Then, in the Middle Ages, a new rat showed up on the scene.
It's hard to say when the brown rat got out of the Mongolian Steppe.
Shh!
CLOP
CLOP
CLOP
But when brown rats made their move, they really went for it.
Brown rats were bigger and more aggressive than black rats.
Hey, pip-squeak, take a hike.
And wherever brown rats went, they displaced their smaller cousins.
But even though the brown rats were bigger and more aggressive, their fleas didn't carry the plague.
Oh! Well. That's quite nice, actually.
Ow!
boing
CHOMP!

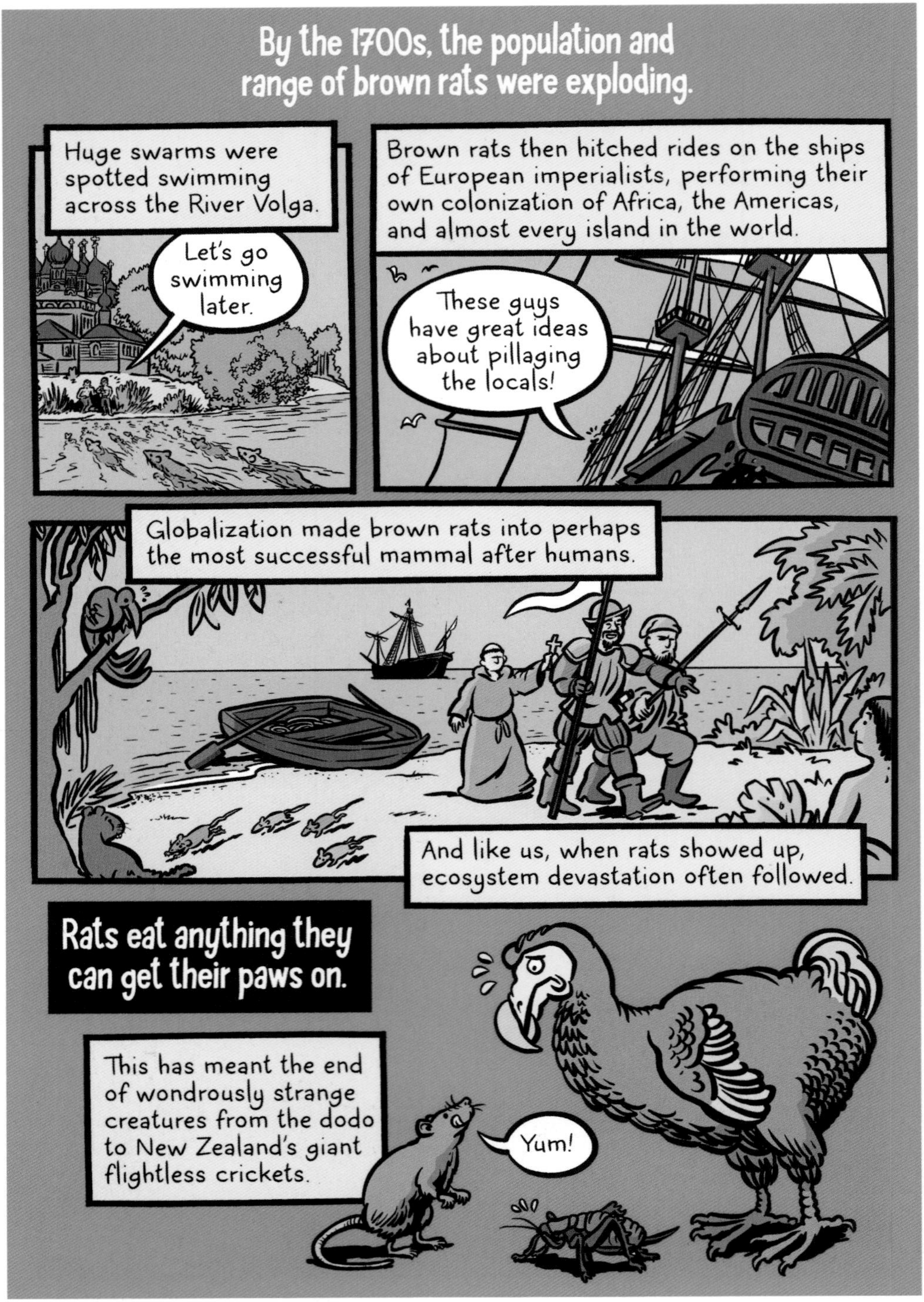
By the 1700s, the population and range of brown rats were exploding.
Huge swarms were spotted swimming across the River Volga.
Let's go swimming later.
Brown rats then hitched rides on the ships of European imperialists, performing their own colonization of Africa, the Americas, and almost every island in the world.
These guys have great ideas about pillaging the locals!
Globalization made brown rats into perhaps the most successful mammal after humans.
And like us, when rats showed up, ecosystem devastation often followed.
Rats eat anything they can get their paws on.
This has meant the end of wondrously strange creatures from the dodo to New Zealand's giant flightless crickets.
Yum!

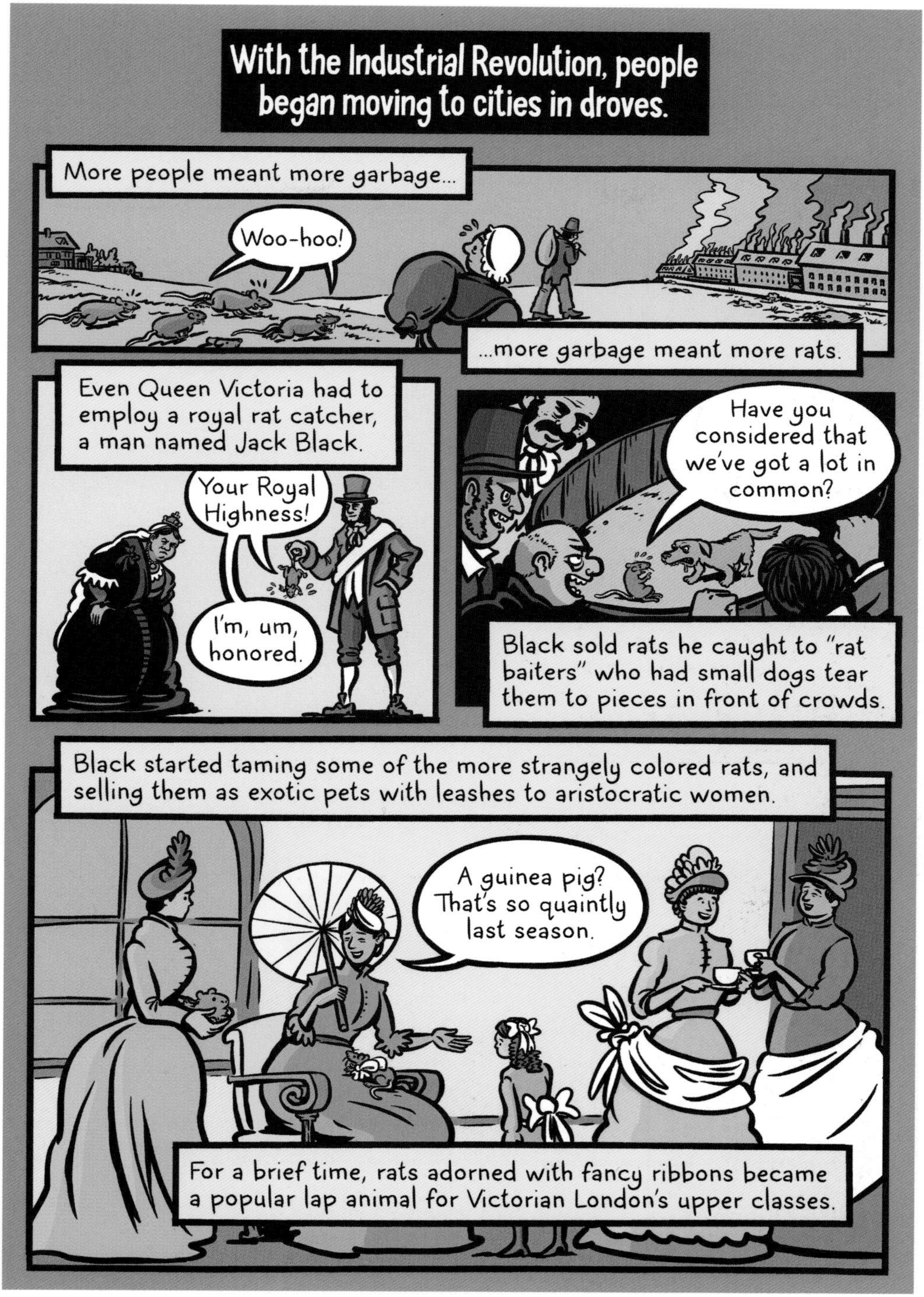
With the Industrial Revolution, people began moving to cities in droves.
More people meant more garbage...
Woo-hoo!
...more garbage meant more rats.
Even Queen Victoria had to employ a royal rat catcher, a man named Jack Black.
Your Royal Highness!
I'm, um, honored.
Have you considered that we've got a lot in common?
Black sold rats he caught to "rat baiters" who had small dogs tear them to pieces in front of crowds.
Black started taming some of the more strangely colored rats, and selling them as exotic pets with leashes to aristocratic women.
A guinea pig? That's so quaintly last season.
For a brief time, rats adorned with fancy ribbons became a popular lap animal for Victorian London's upper classes.

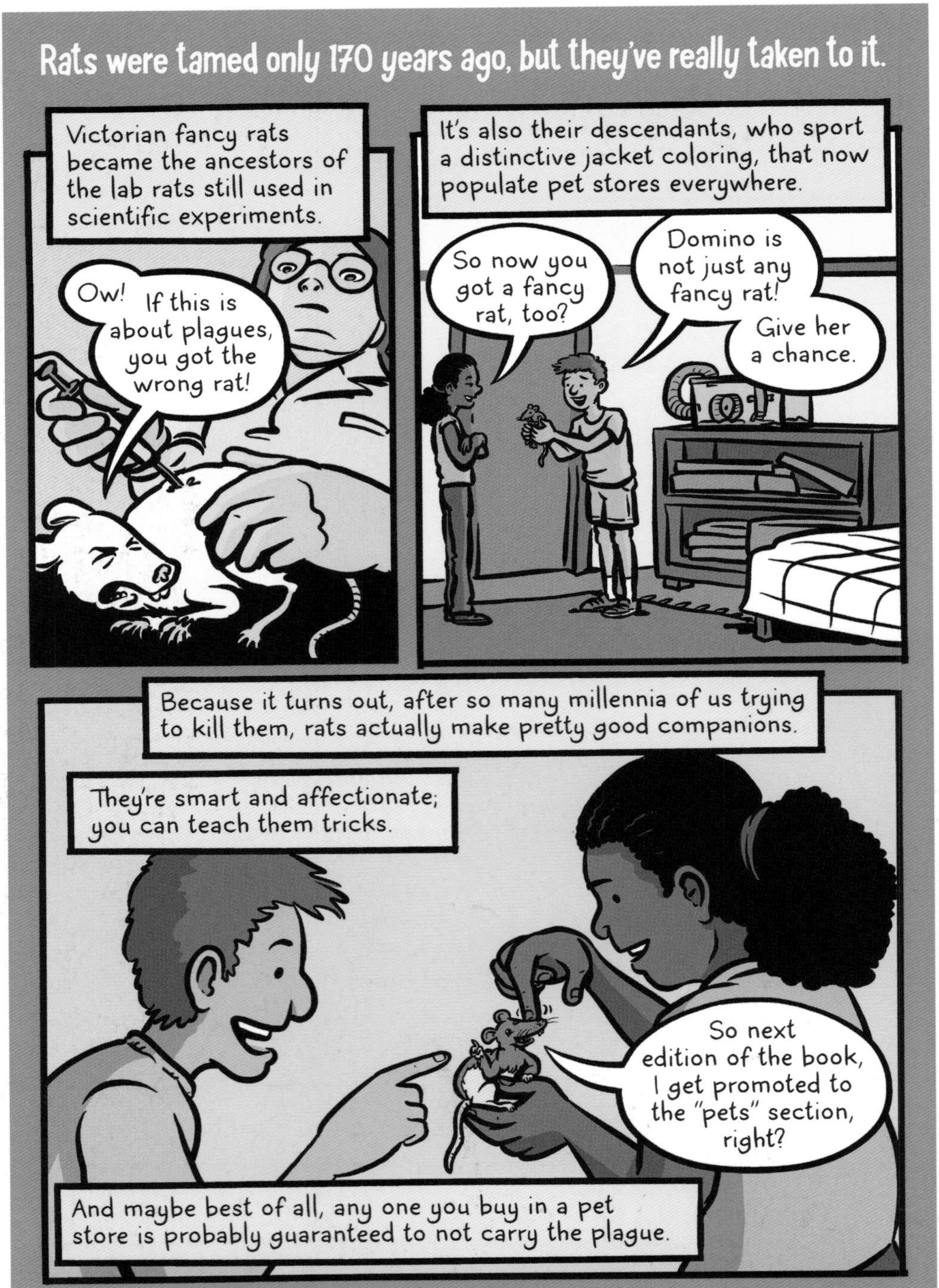

Rats were tamed only 170 years ago, but they've really taken to it.
Victorian fancy rats became the ancestors of the lab rats still used in scientific experiments.
Ow! If this is about plagues, you got the wrong rat!
It's also their descendants, who sport a distinctive jacket coloring, that now populate pet stores everywhere.
So now you got a fancy rat, too?
Domino is not just any fancy rat!
Give her a chance.
Because it turns out, after so many millennia of us trying to kill them, rats actually make pretty good companions.
They're smart and affectionate; you can teach them tricks.
So next edition of the book, I get promoted to the "pets" section, right?
And maybe best of all, any one you buy in a pet store is probably guaranteed to not carry the plague.

RAT KINGS

Medieval German peasants lived in fear of the "rat king," or *rattenkönig*, a cluster of rats tangled together by their tails.

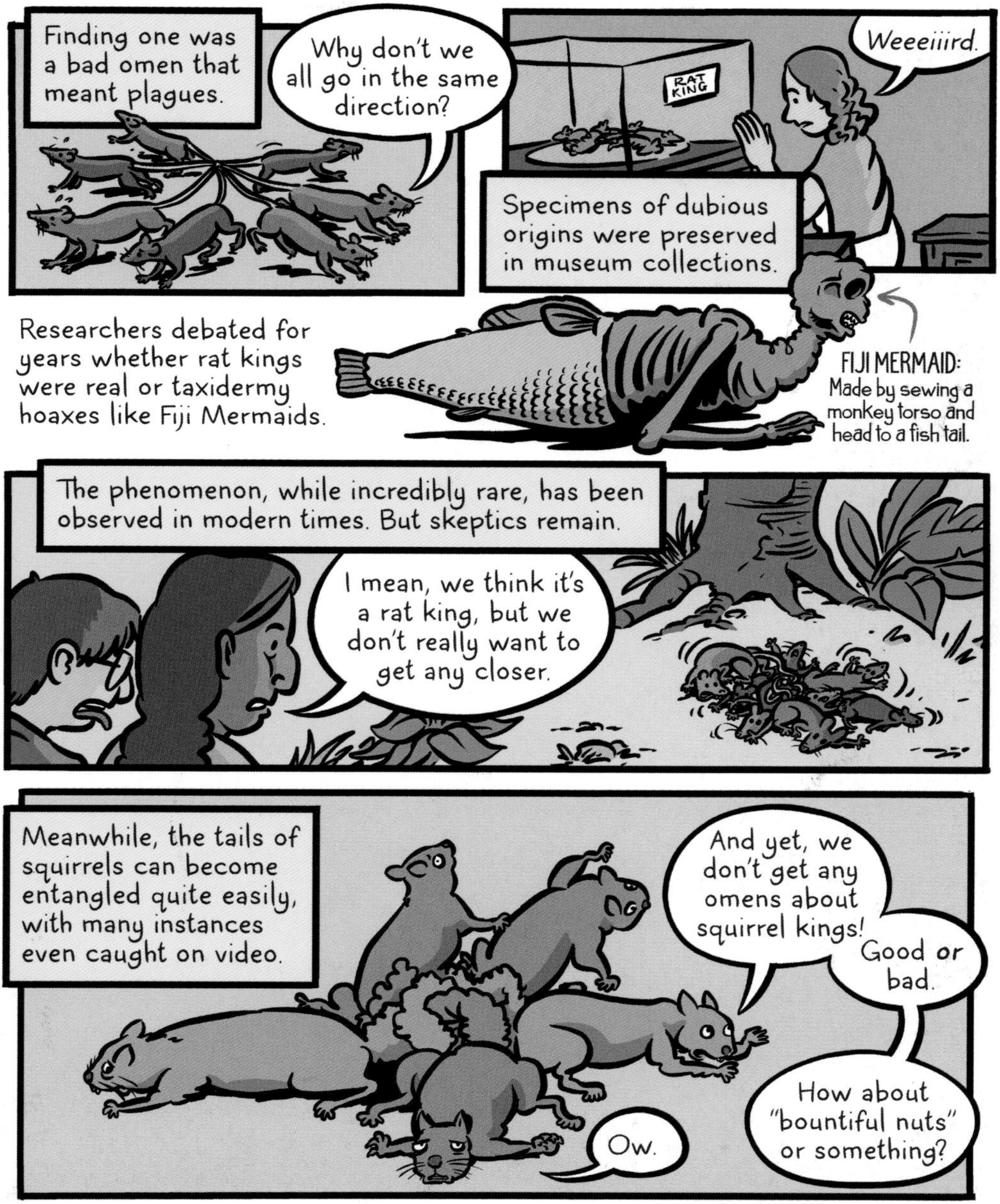

Finding one was a bad omen that meant plagues.
Why don't we all go in the same direction?
RAT KING
Weeeiiird.
Specimens of dubious origins were preserved in museum collections.
Researchers debated for years whether rat kings were real or taxidermy hoaxes like Fiji Mermaids.
FIJI MERMAID: Made by sewing a monkey torso and head to a fish tail.
The phenomenon, while incredibly rare, has been observed in modern times. But skeptics remain.
I mean, we think it's a rat king, but we don't really want to get any closer.
Meanwhile, the tails of squirrels can become entangled quite easily, with many instances even caught on video.
And yet, we don't get any omens about squirrel kings!
Good or bad.
How about "bountiful nuts" or something?
Ow.

COCKROACHES

Blattodea

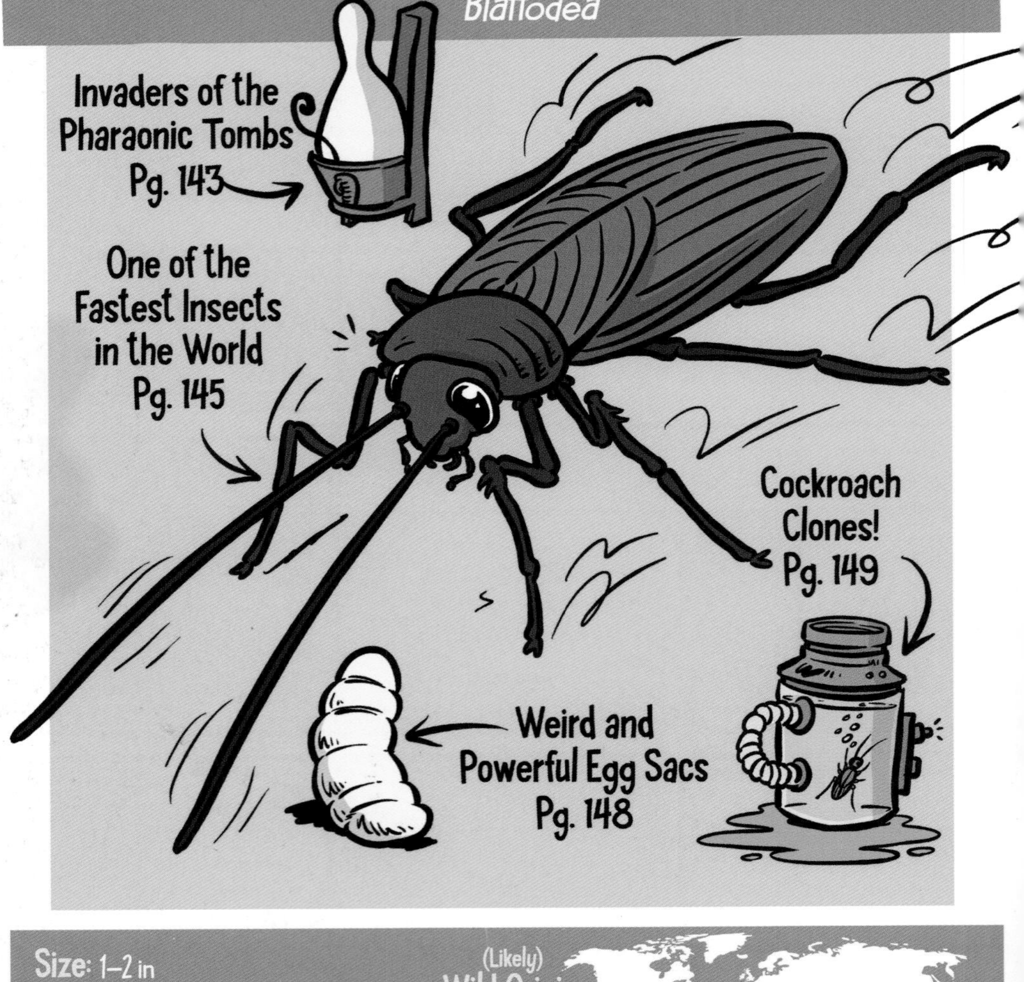
Invaders of the Pharaonic Tombs
Pg. 143
One of the Fastest Insects in the World
Pg. 145
Cockroach Clones!
Pg. 149
Weird and Powerful Egg Sacs
Pg. 148

Size: 1–2 in

Weight: 0.5–1 oz

Oddball Anatomy: Can survive a week with no head

Diet: Plant and meat food waste, rotting things, paper, hair

(Likely) **Wild Origin:**

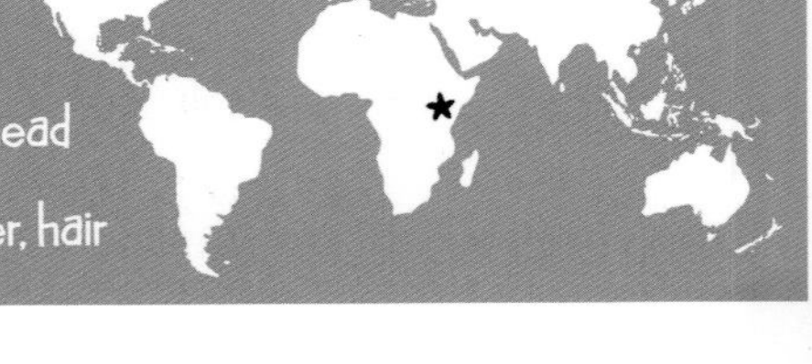

WILL COCKROACHES INHERIT THE EARTH?

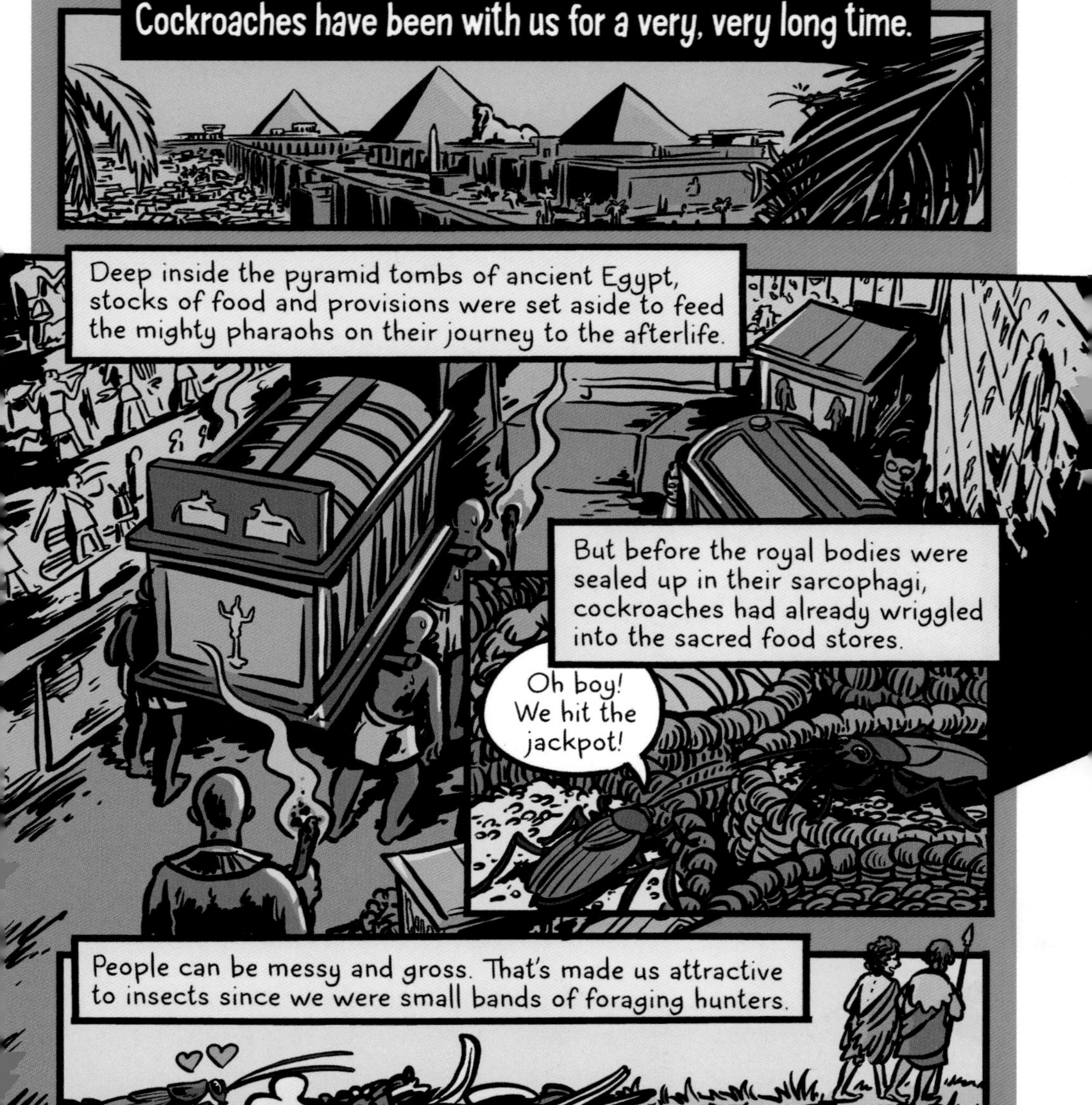
Cockroaches have been with us for a very, very long time.
Deep inside the pyramid tombs of ancient Egypt, stocks of food and provisions were set aside to feed the mighty pharaohs on their journey to the afterlife.
But before the royal bodies were sealed up in their sarcophagi, cockroaches had already wriggled into the sacred food stores.
Oh boy! We hit the jackpot!
People can be messy and gross. That's made us attractive to insects since we were small bands of foraging hunters.

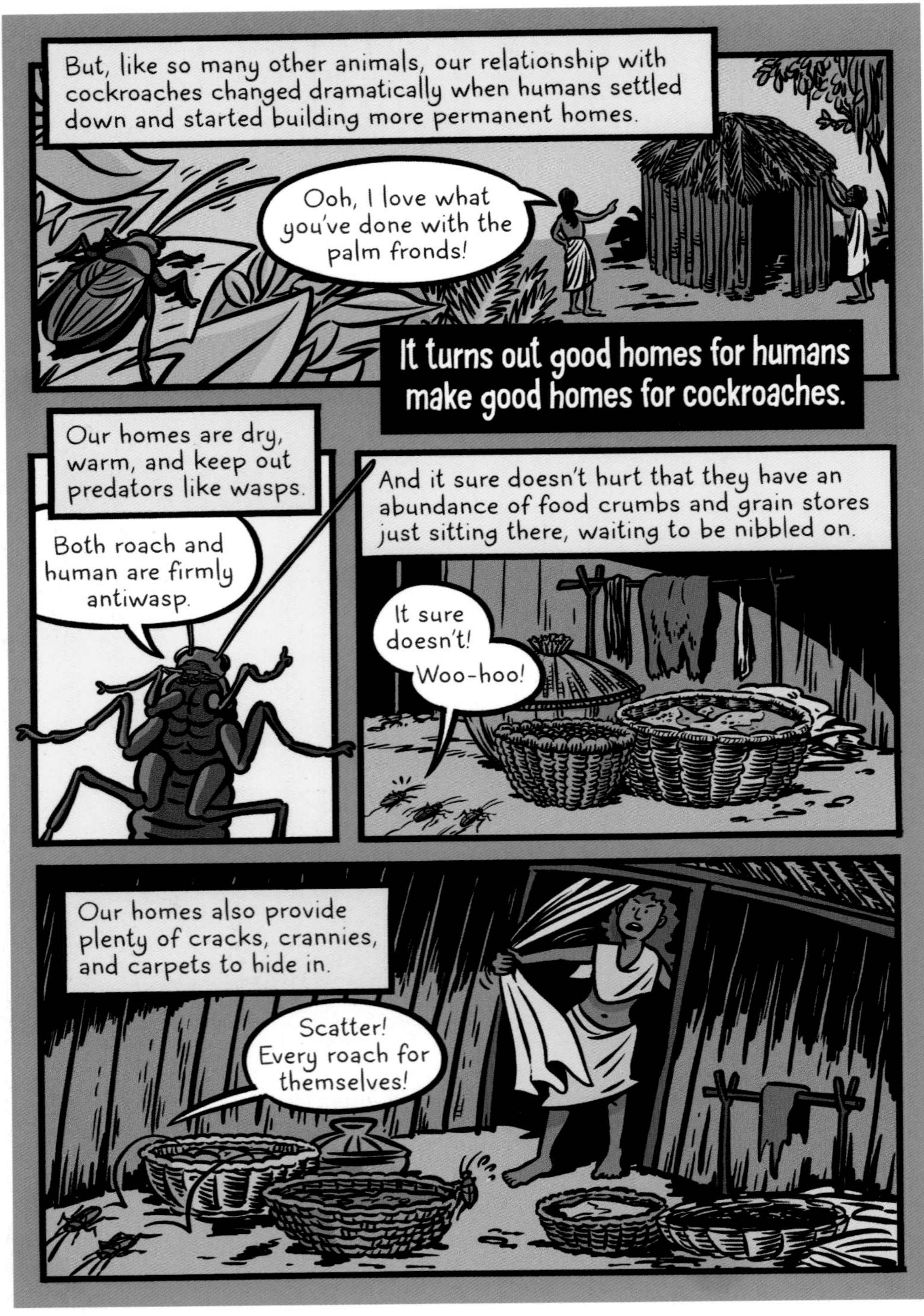

But, like so many other animals, our relationship with cockroaches changed dramatically when humans settled down and started building more permanent homes.
Ooh, I love what you've done with the palm fronds!
It turns out good homes for humans make good homes for cockroaches.
Our homes are dry, warm, and keep out predators like wasps.
Both roach and human are firmly antiwasp.
And it sure doesn't hurt that they have an abundance of food crumbs and grain stores just sitting there, waiting to be nibbled on.
It sure doesn't!
Woo-hoo!
Our homes also provide plenty of cracks, crannies, and carpets to hide in.
Scatter! Every roach for themselves!

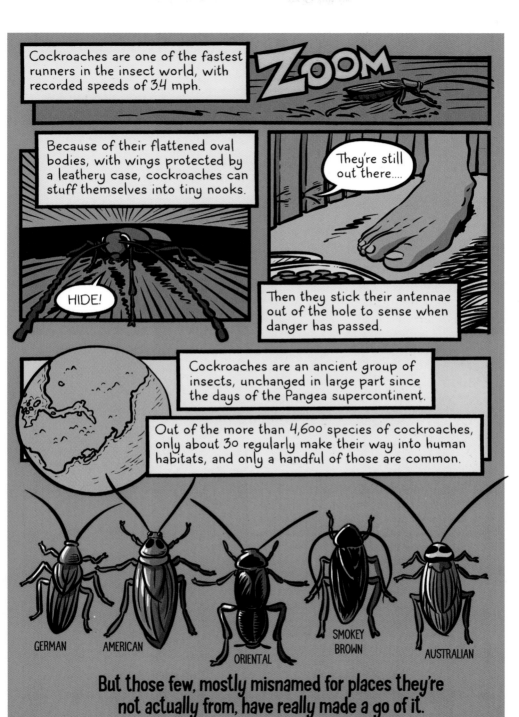

Cockroaches are one of the fastest runners in the insect world, with recorded speeds of 3.4 mph.

ZOOM

Because of their flattened oval bodies, with wings protected by a leathery case, cockroaches can stuff themselves into tiny nooks.

They're still out there....

HIDE!

Then they stick their antennae out of the hole to sense when danger has passed.

Cockroaches are an ancient group of insects, unchanged in large part since the days of the Pangea supercontinent.

Out of the more than 4,600 species of cockroaches, only about 30 regularly make their way into human habitats, and only a handful of those are common.

GERMAN   AMERICAN   ORIENTAL   SMOKEY BROWN   AUSTRALIAN

But those few, mostly misnamed for places they're not actually from, have really made a go of it.

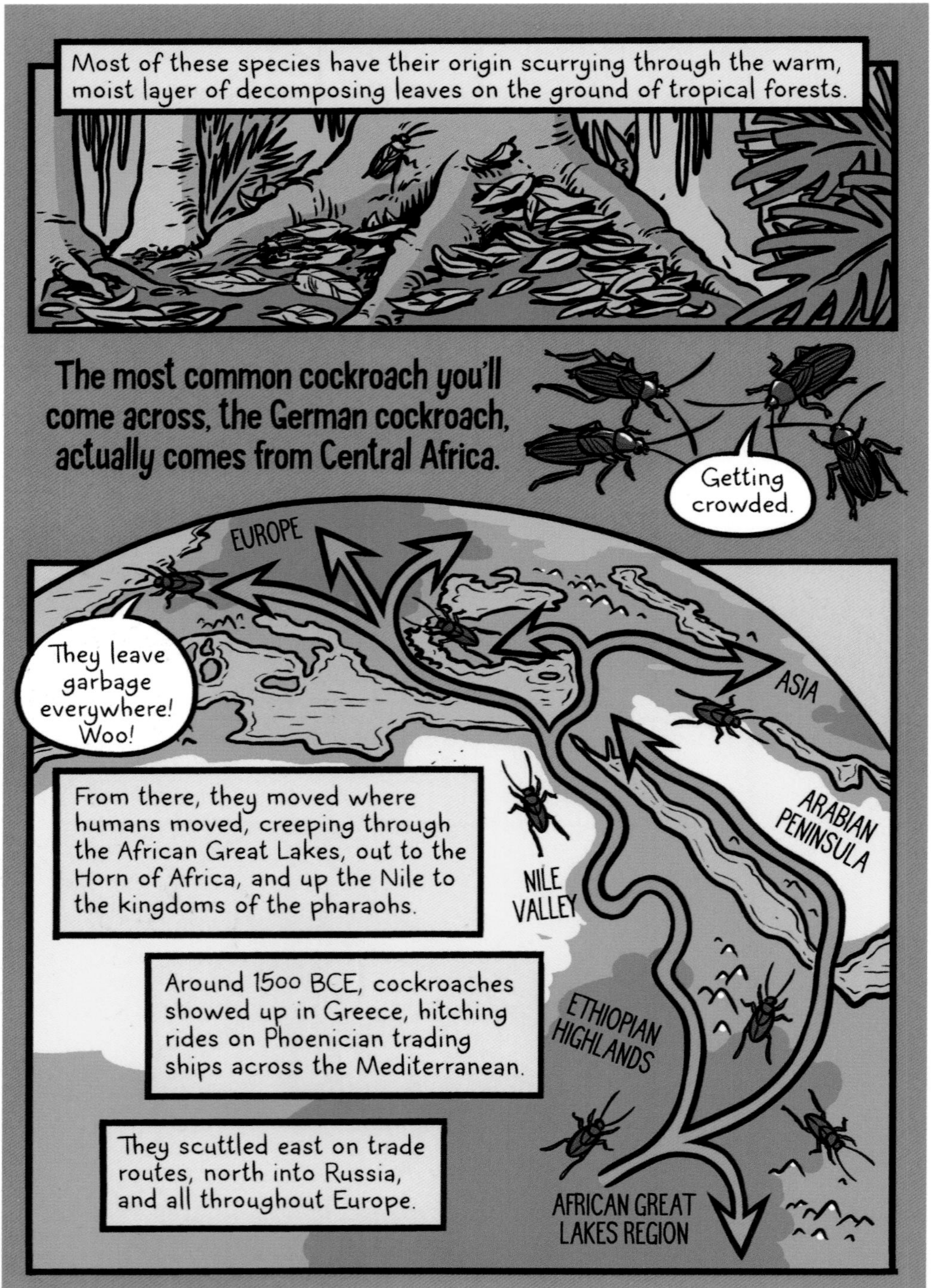
Most of these species have their origin scurrying through the warm, moist layer of decomposing leaves on the ground of tropical forests.
The most common cockroach you'll come across, the German cockroach, actually comes from Central Africa.
Getting crowded.
EUROPE
ASIA
They leave garbage everywhere! Woo!
From there, they moved where humans moved, creeping through the African Great Lakes, out to the Horn of Africa, and up the Nile to the kingdoms of the pharaohs.
NILE VALLEY
ARABIAN PENINSULA
Around 1500 BCE, cockroaches showed up in Greece, hitching rides on Phoenician trading ships across the Mediterranean.
ETHIOPIAN HIGHLANDS
They scuttled east on trade routes, north into Russia, and all throughout Europe.
AFRICAN GREAT LAKES REGION

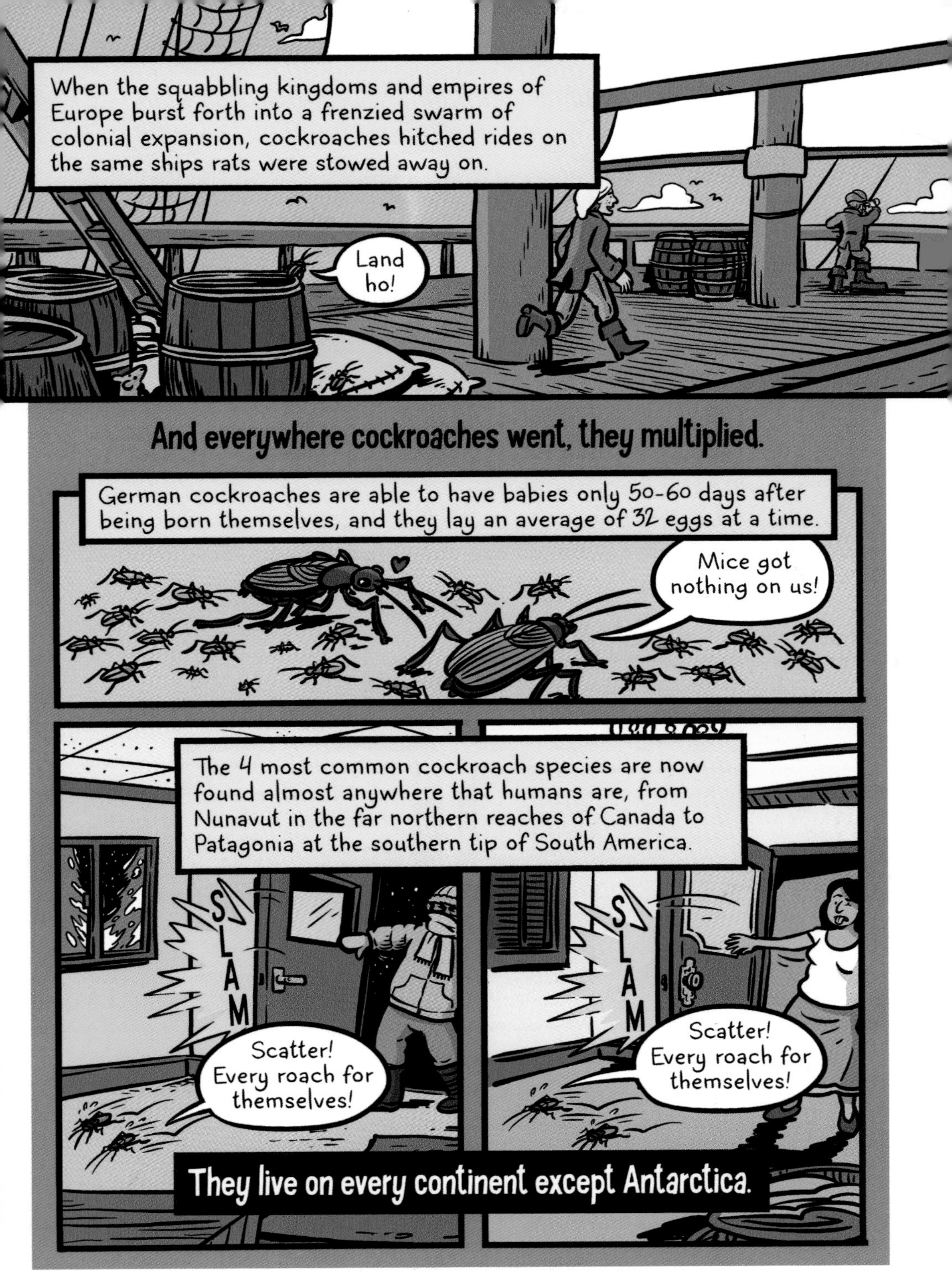
When the squabbling kingdoms and empires of Europe burst forth into a frenzied swarm of colonial expansion, cockroaches hitched rides on the same ships rats were stowed away on.
Land ho!
And everywhere cockroaches went, they multiplied.
German cockroaches are able to have babies only 50-60 days after being born themselves, and they lay an average of 32 eggs at a time.
Mice got nothing on us!
The 4 most common cockroach species are now found almost anywhere that humans are, from Nunavut in the far northern reaches of Canada to Patagonia at the southern tip of South America.
SLAM
Scatter! Every roach for themselves!
SLAM
Scatter! Every roach for themselves!
They live on every continent except Antarctica.

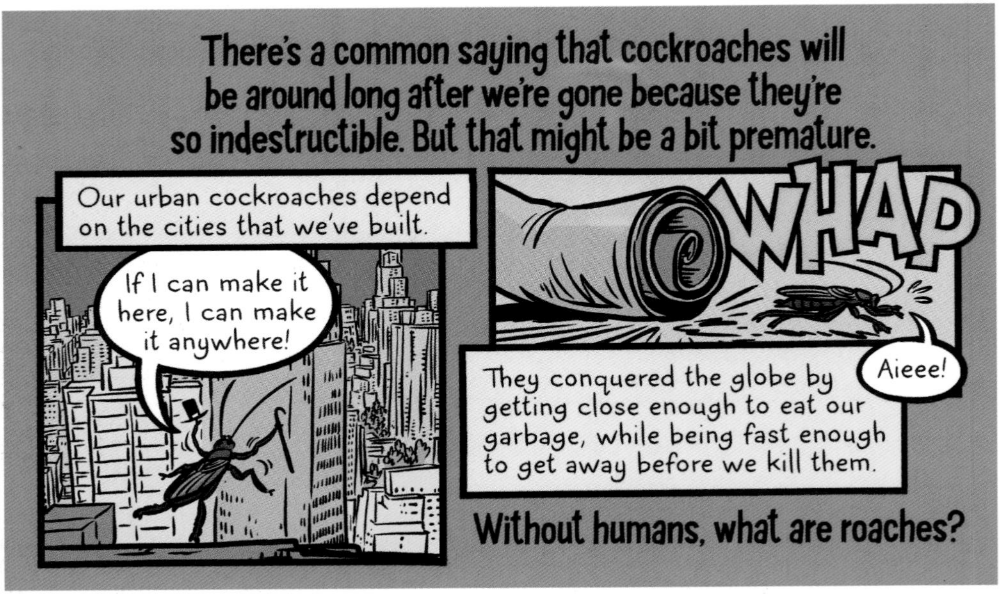
There's a common saying that cockroaches will be around long after we're gone because they're so indestructible. But that might be a bit premature.
Our urban cockroaches depend on the cities that we've built.
If I can make it here, I can make it anywhere!
WHAP
Aieee!
They conquered the globe by getting close enough to eat our garbage, while being fast enough to get away before we kill them.
Without humans, what are roaches?

OOTHECAE AND PESTICIDES

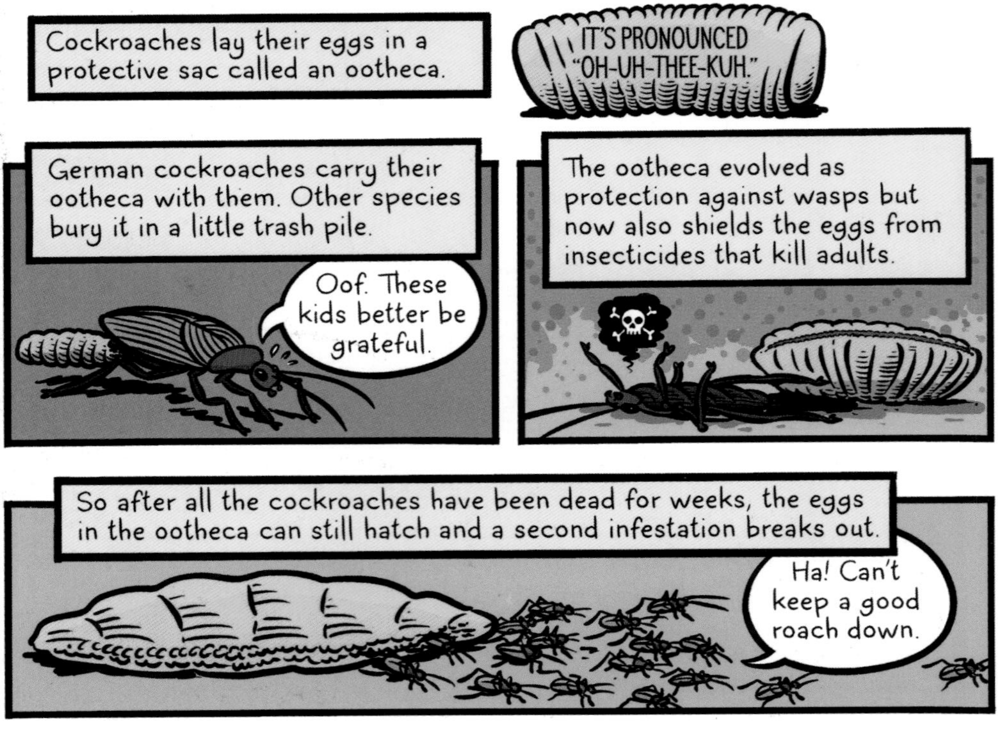
Cockroaches lay their eggs in a protective sac called an ootheca.
IT'S PRONOUNCED "OH-UH-THEE-KUH."
German cockroaches carry their ootheca with them. Other species bury it in a little trash pile.
Oof. These kids better be grateful.
The ootheca evolved as protection against wasps but now also shields the eggs from insecticides that kill adults.
So after all the cockroaches have been dead for weeks, the eggs in the ootheca can still hatch and a second infestation breaks out.
Ha! Can't keep a good roach down.

COCKROACH CLONES

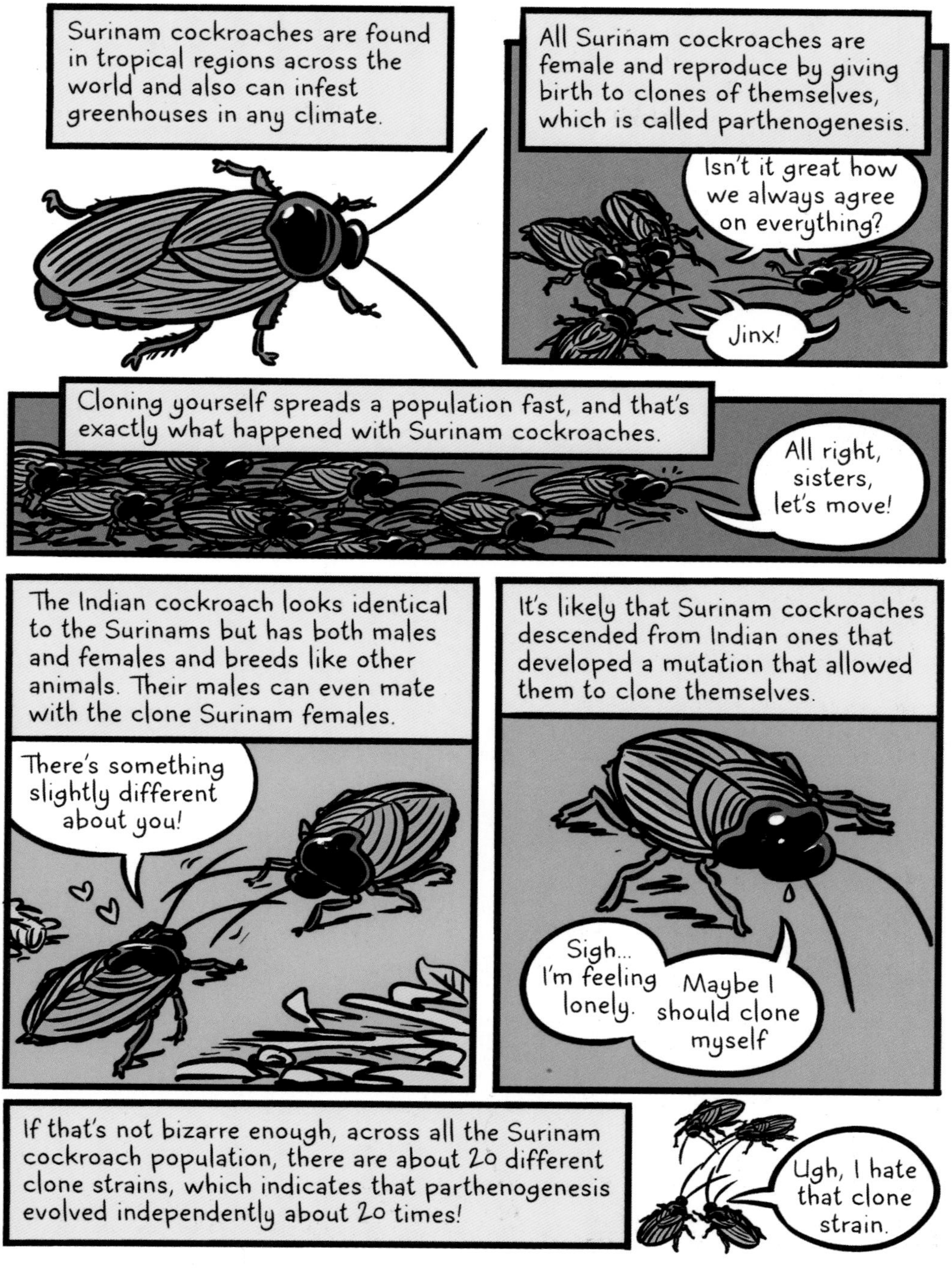
Surinam cockroaches are found in tropical regions across the world and also can infest greenhouses in any climate.
All Surinam cockroaches are female and reproduce by giving birth to clones of themselves, which is called parthenogenesis.
Isn't it great how we always agree on everything?
Jinx!
Cloning yourself spreads a population fast, and that's exactly what happened with Surinam cockroaches.
All right, sisters, let's move!
The Indian cockroach looks identical to the Surinams but has both males and females and breeds like other animals. Their males can even mate with the clone Surinam females.
There's something slightly different about you!
It's likely that Surinam cockroaches descended from Indian ones that developed a mutation that allowed them to clone themselves.
Sigh... I'm feeling lonely.
Maybe I should clone myself
If that's not bizarre enough, across all the Surinam cockroach population, there are about 20 different clone strains, which indicates that parthenogenesis evolved independently about 20 times!
Ugh, I hate that clone strain.

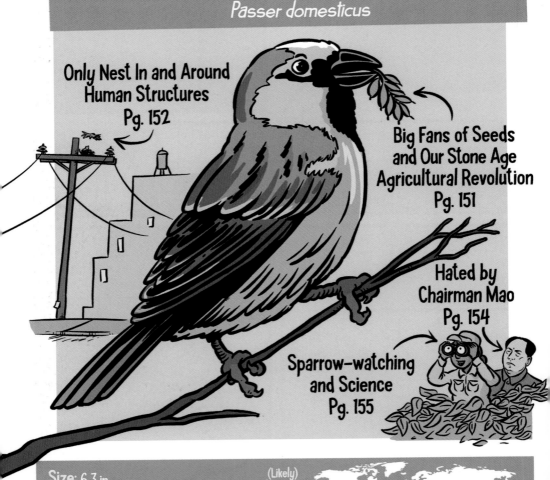

SPARROWS

Passer domesticus

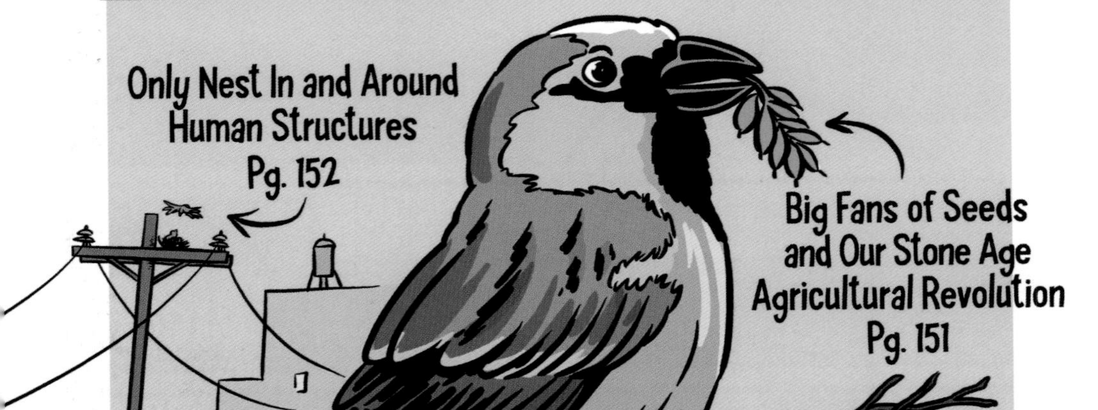
Only Nest In and Around Human Structures Pg. 152
Big Fans of Seeds and Our Stone Age Agricultural Revolution Pg. 151
Hated by Chairman Mao Pg. 154
Sparrow-watching and Science Pg. 155

Size: 6.3 in

Weight: 0.85–1.4 oz

Top Flight Speed: 28.3 mph (15 wingbeats per second)

Diet: Seeds, grains, insects, french fries

(Likely) Wild Origin:

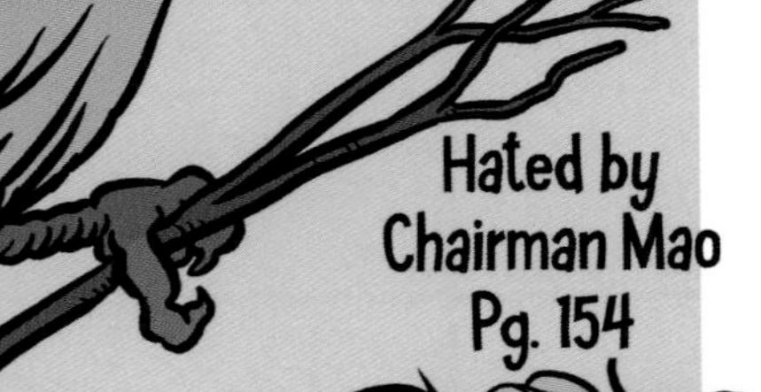

HOW THE SPARROW BECAME A NATIONAL ENEMY OF CHINA

Lots of different seed-eating birds started hanging around humans when we invented agriculture about 11,000 years ago.

Hey! What the heck?

But sparrows in particular made a go of it.

In the Middle East, the ancestors of sparrows sometimes make their homes in the lower layers of nests built by bigger birds like storks.

Could you keep it quiet up there? I'm trying to sleep.

It could be this fearlessness that gave sparrows an edge.

They're just that much braver about being around people--and stealing stuff right out from under our noses.

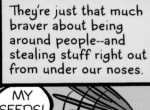

MY SEEDS!

Their population exploded alongside our agricultural revolution and they've stuck with us ever since.

Cheep?

When we started building big cities, they effortlessly made the shift from country bird to city bird.

Cheep! Cheep!

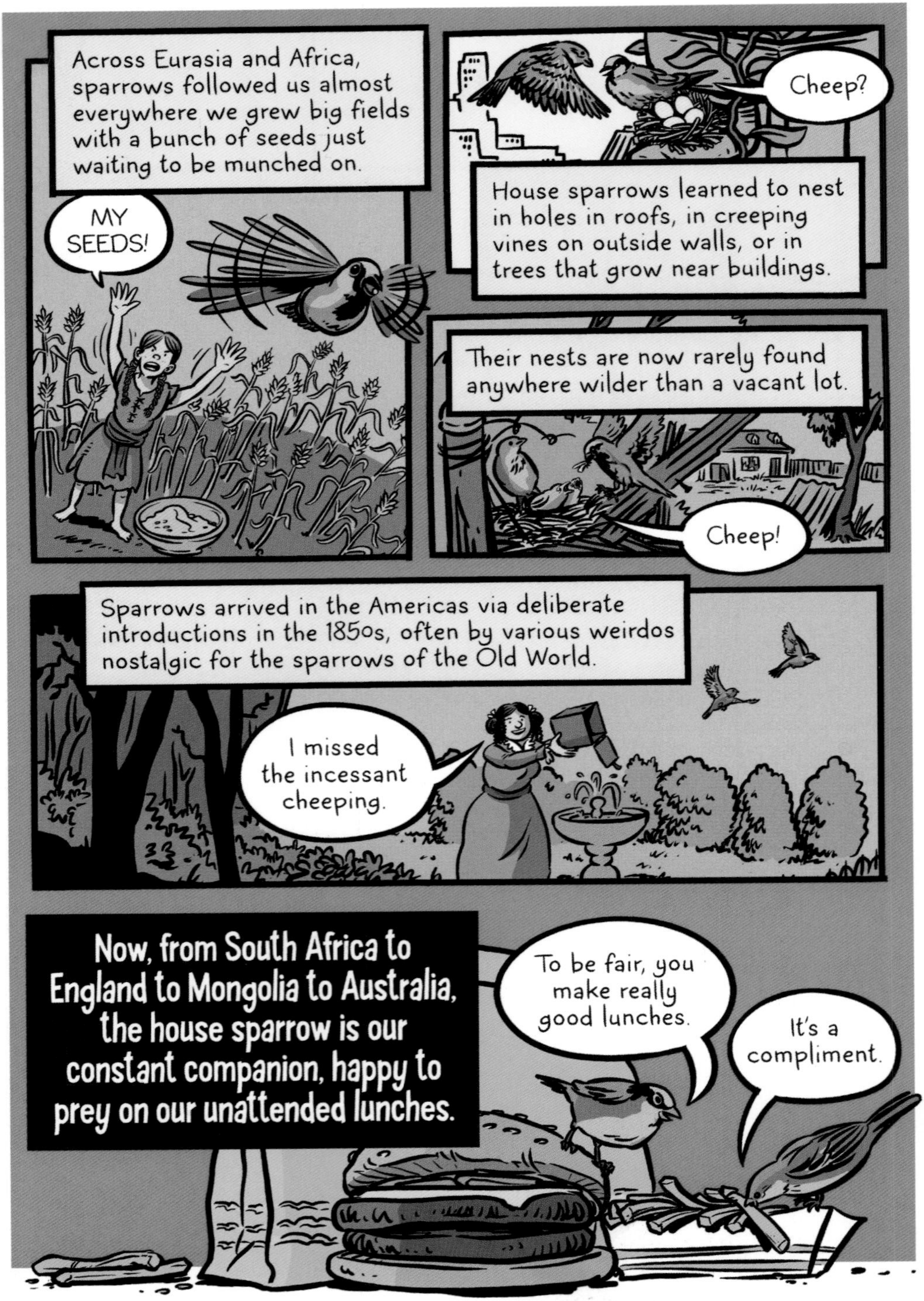
Across Eurasia and Africa, sparrows followed us almost everywhere we grew big fields with a bunch of seeds just waiting to be munched on.
MY SEEDS!
Cheep?
House sparrows learned to nest in holes in roofs, in creeping vines on outside walls, or in trees that grow near buildings.
Their nests are now rarely found anywhere wilder than a vacant lot.
Cheep!
Sparrows arrived in the Americas via deliberate introductions in the 1850s, often by various weirdos nostalgic for the sparrows of the Old World.
I missed the incessant cheeping.
Now, from South Africa to England to Mongolia to Australia, the house sparrow is our constant companion, happy to prey on our unattended lunches.
To be fair, you make really good lunches.
It's a compliment.

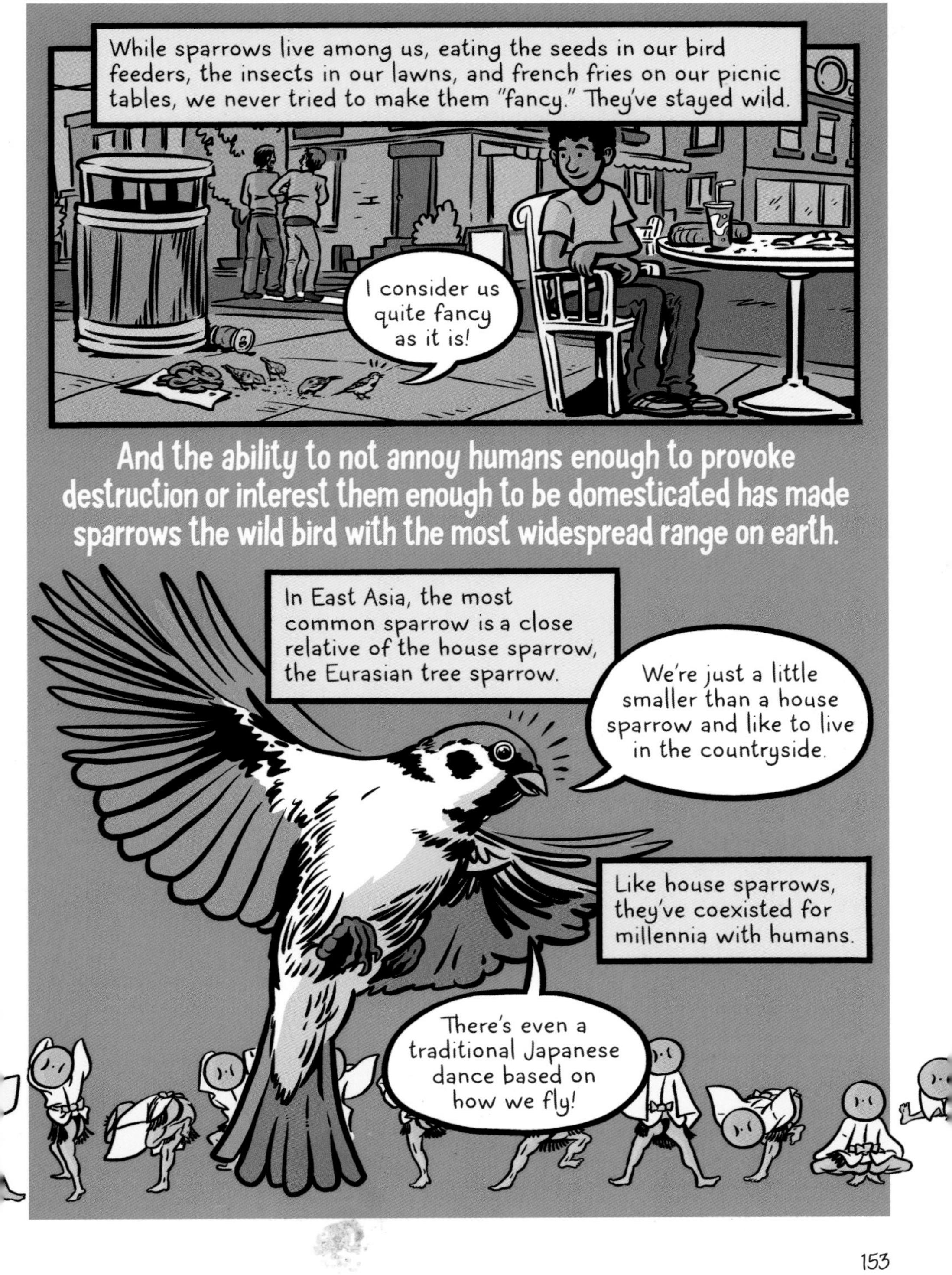
While sparrows live among us, eating the seeds in our bird feeders, the insects in our lawns, and french fries on our picnic tables, we never tried to make them "fancy." They've stayed wild.
I consider us quite fancy as it is!
And the ability to not annoy humans enough to provoke destruction or interest them enough to be domesticated has made sparrows the wild bird with the most widespread range on earth.
In East Asia, the most common sparrow is a close relative of the house sparrow, the Eurasian tree sparrow.
We're just a little smaller than a house sparrow and like to live in the countryside.
Like house sparrows, they've coexisted for millennia with humans.
There's even a traditional Japanese dance based on how we fly!

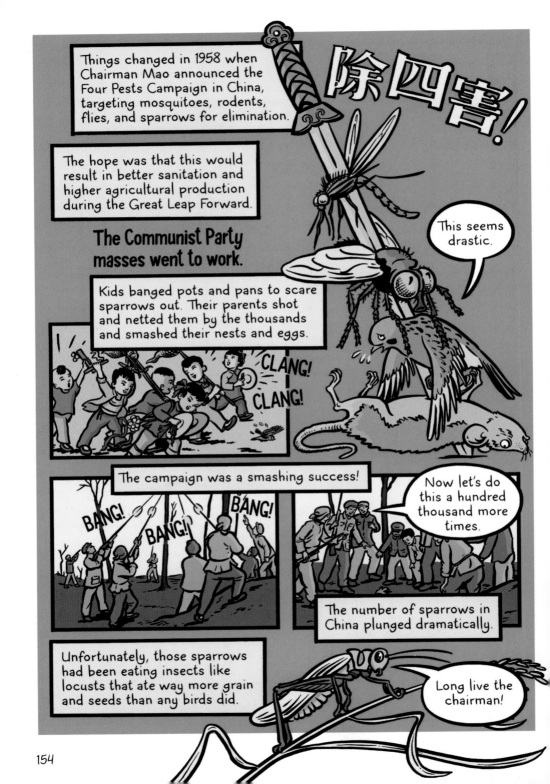

Things changed in 1958 when Chairman Mao announced the Four Pests Campaign in China, targeting mosquitoes, rodents, flies, and sparrows for elimination.

除四害!

The hope was that this would result in better sanitation and higher agricultural production during the Great Leap Forward.

The Communist Party masses went to work.

This seems drastic.

Kids banged pots and pans to scare sparrows out. Their parents shot and netted them by the thousands and smashed their nests and eggs.

CLANG!

CLANG!

The campaign was a smashing success!

BANG!

BANG!

BANG!

Now let's do this a hundred thousand more times.

The number of sparrows in China plunged dramatically.

Unfortunately, those sparrows had been eating insects like locusts that ate way more grain and seeds than any birds did.

Long live the chairman!

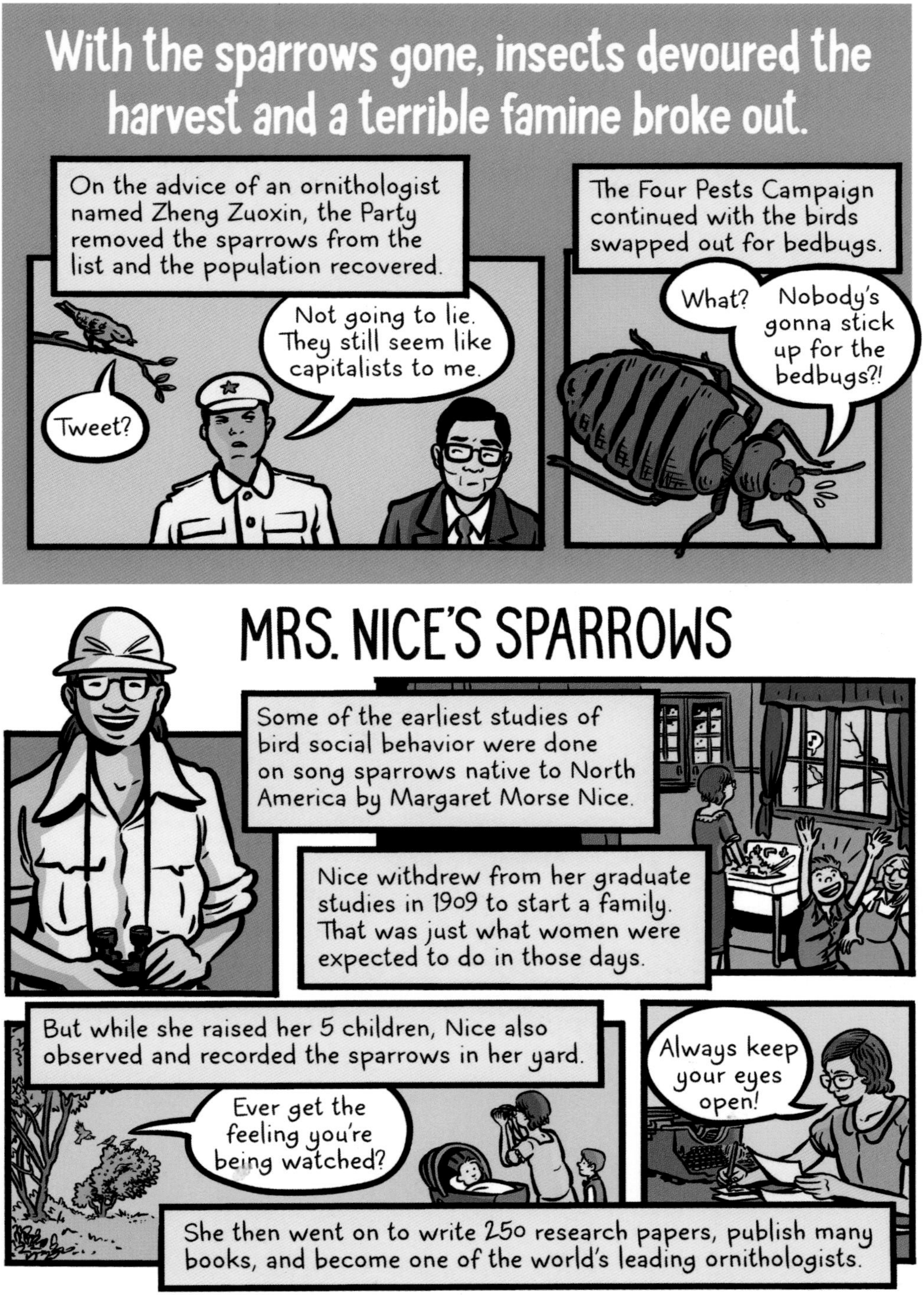

With the sparrows gone, insects devoured the harvest and a terrible famine broke out.

On the advice of an ornithologist named Zheng Zuoxin, the Party removed the sparrows from the list and the population recovered.

Not going to lie. They still seem like capitalists to me.

Tweet?

The Four Pests Campaign continued with the birds swapped out for bedbugs.

What? Nobody's gonna stick up for the bedbugs?!

MRS. NICE'S SPARROWS

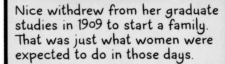

Some of the earliest studies of bird social behavior were done on song sparrows native to North America by Margaret Morse Nice.

Nice withdrew from her graduate studies in 1909 to start a family. That was just what women were expected to do in those days.

But while she raised her 5 children, Nice also observed and recorded the sparrows in her yard.

Ever get the feeling you're being watched?

Always keep your eyes open!

She then went on to write 250 research papers, publish many books, and become one of the world's leading ornithologists.

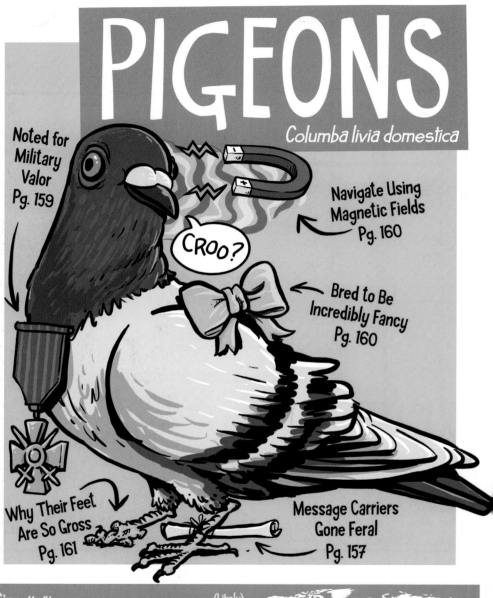

PIGEONS

Columba livia domestica

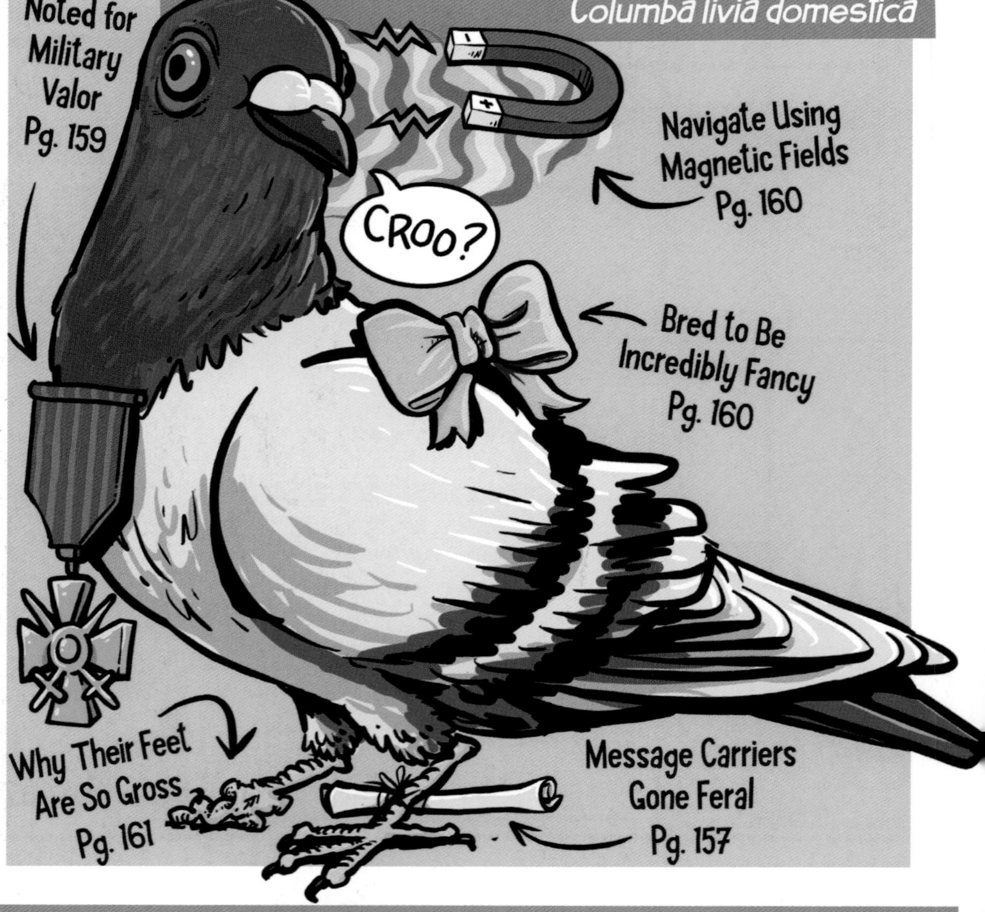
Noted for Military Valor Pg. 159
Navigate Using Magnetic Fields Pg. 160
CROO?
Bred to Be Incredibly Fancy Pg. 160
Why Their Feet Are So Gross Pg. 161
Message Carriers Gone Feral Pg. 157

Size: 11–14 in

Weight: 9–13 oz

Famous Fancier: Charles Darwin

Diet: Bread, seeds, chips, scraps...most garbage, really

(Likely) **Wild Origin:**

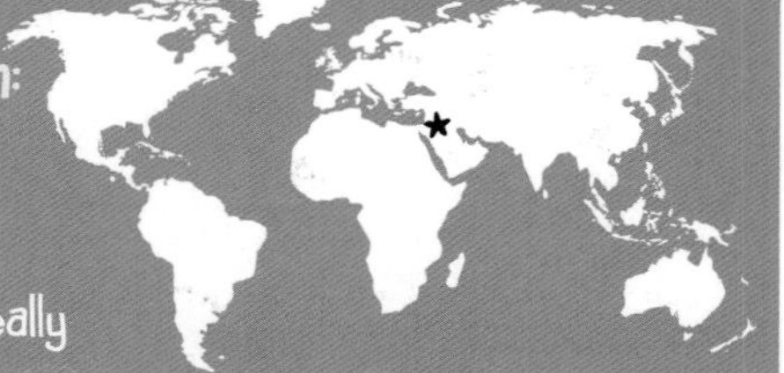

WHERE PIGEONS COME FROM

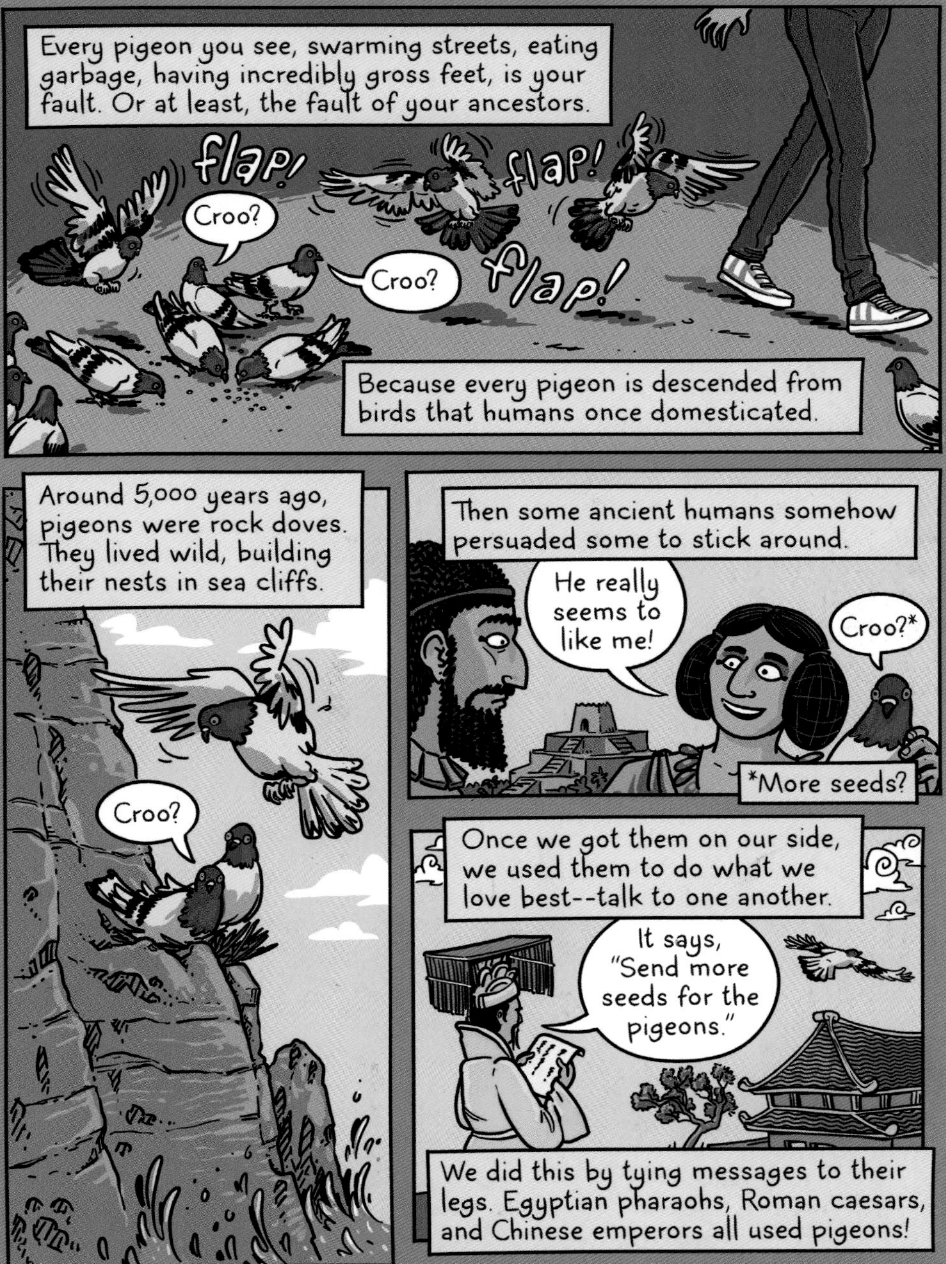

Every pigeon you see, swarming streets, eating garbage, having incredibly gross feet, is your fault. Or at least, the fault of your ancestors.
flap!
flap!
Croo?
Croo?
flap!
Because every pigeon is descended from birds that humans once domesticated.
Around 5,000 years ago, pigeons were rock doves. They lived wild, building their nests in sea cliffs.
Croo?
Then some ancient humans somehow persuaded some to stick around.
He really seems to like me!
Croo?*
*More seeds?
Once we got them on our side, we used them to do what we love best--talk to one another.
It says, "Send more seeds for the pigeons."
We did this by tying messages to their legs. Egyptian pharaohs, Roman caesars, and Chinese emperors all used pigeons!

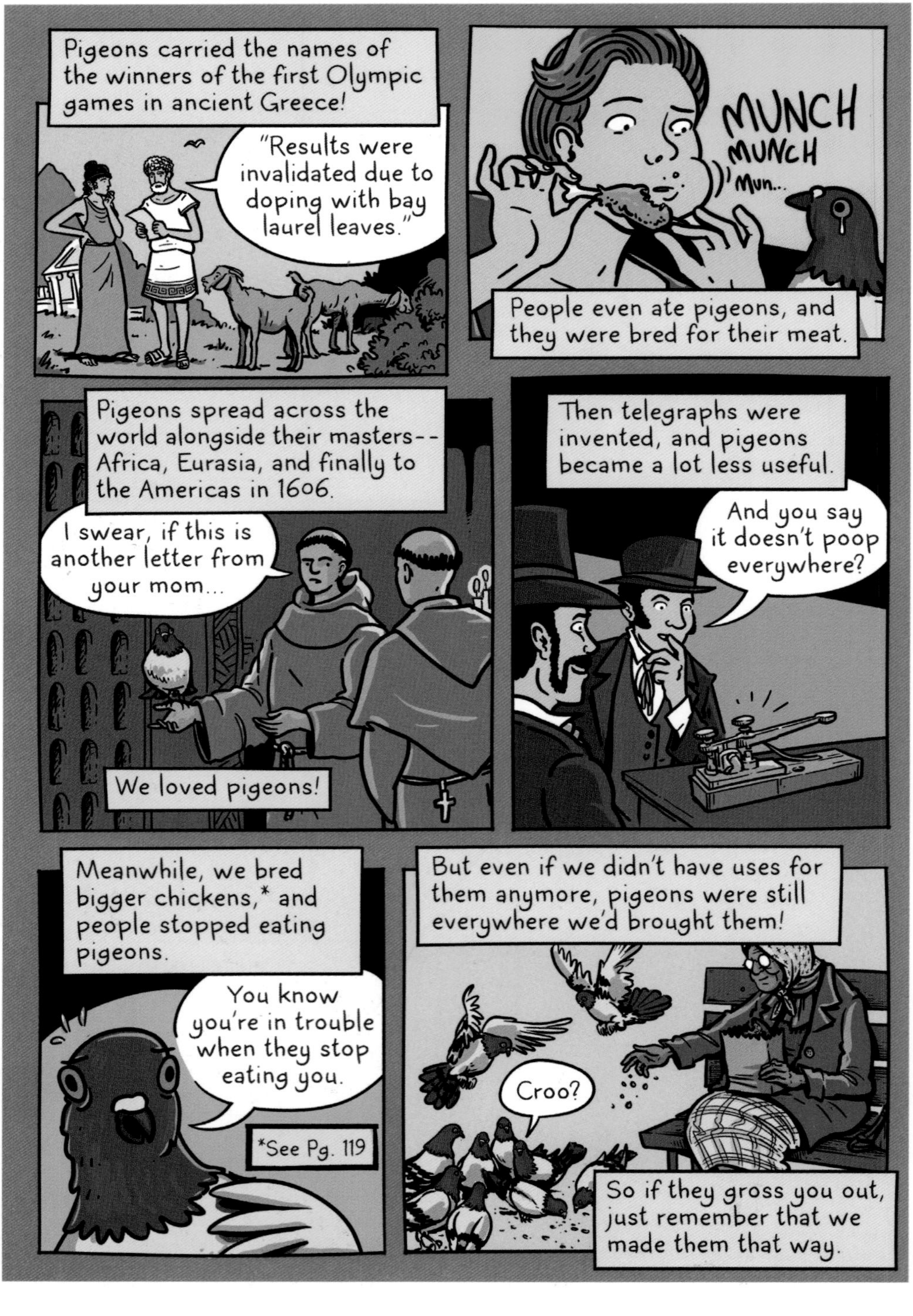
Pigeons carried the names of the winners of the first Olympic games in ancient Greece!
"Results were invalidated due to doping with bay laurel leaves."
MUNCH
MUNCH
Mun...
People even ate pigeons, and they were bred for their meat.
Pigeons spread across the world alongside their masters-- Africa, Eurasia, and finally to the Americas in 1606.
I swear, if this is another letter from your mom...
We loved pigeons!
Then telegraphs were invented, and pigeons became a lot less useful.
And you say it doesn't poop everywhere?
Meanwhile, we bred bigger chickens,* and people stopped eating pigeons.
You know you're in trouble when they stop eating you.
*See Pg. 119
But even if we didn't have uses for them anymore, pigeons were still everywhere we'd brought them!
Croo?
So if they gross you out, just remember that we made them that way.

PIGEONS AT WAR

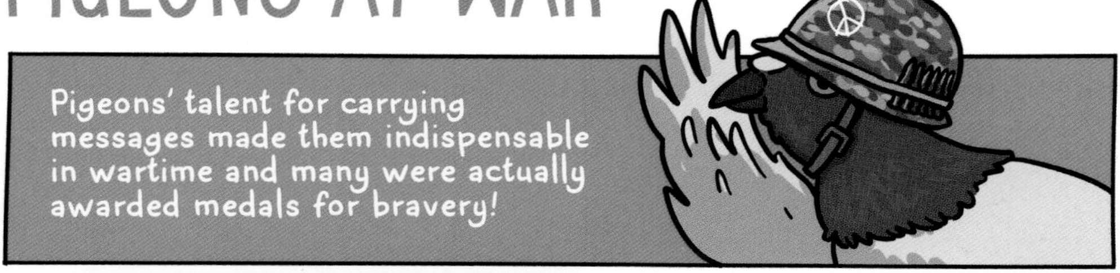
Pigeons' talent for carrying messages made them indispensable in wartime and many were actually awarded medals for bravery!

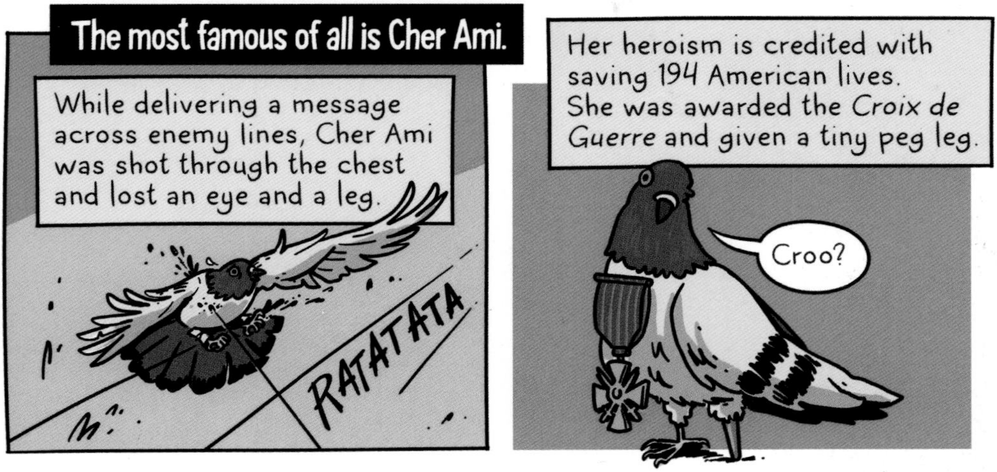
The most famous of all is Cher Ami.
While delivering a message across enemy lines, Cher Ami was shot through the chest and lost an eye and a leg.
RATATATA
Her heroism is credited with saving 194 American lives. She was awarded the *Croix de Guerre* and given a tiny peg leg.
Croo?

In fact, the Dickin Medal, awarded by the British in World War II to honor animal valor, overwhelmingly went to the birds.

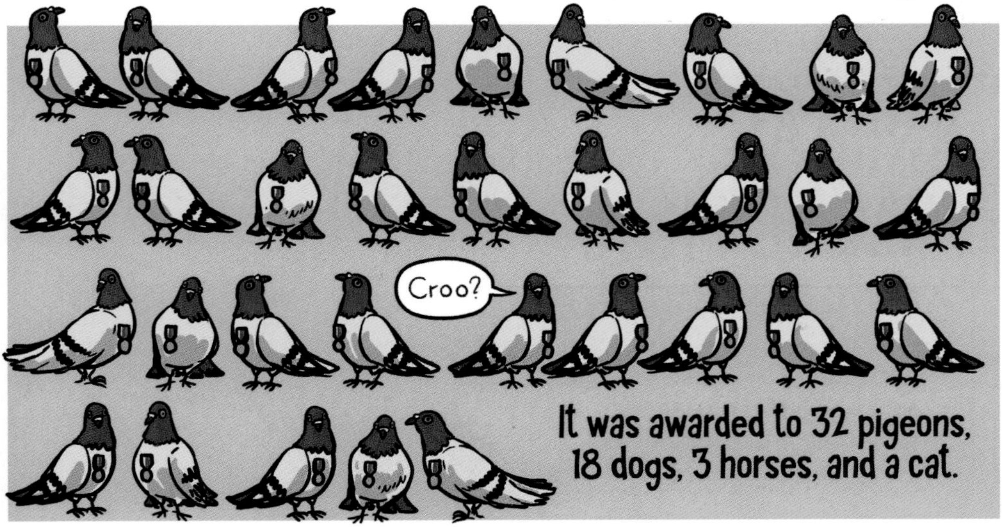
Croo?
It was awarded to 32 pigeons, 18 dogs, 3 horses, and a cat.

FANCY PIGEON BREEDING

Like dogs or cats, we also bred pigeons for weird bird beauty pageants! Here are some of the more bizarre varieties of fancy pigeons:

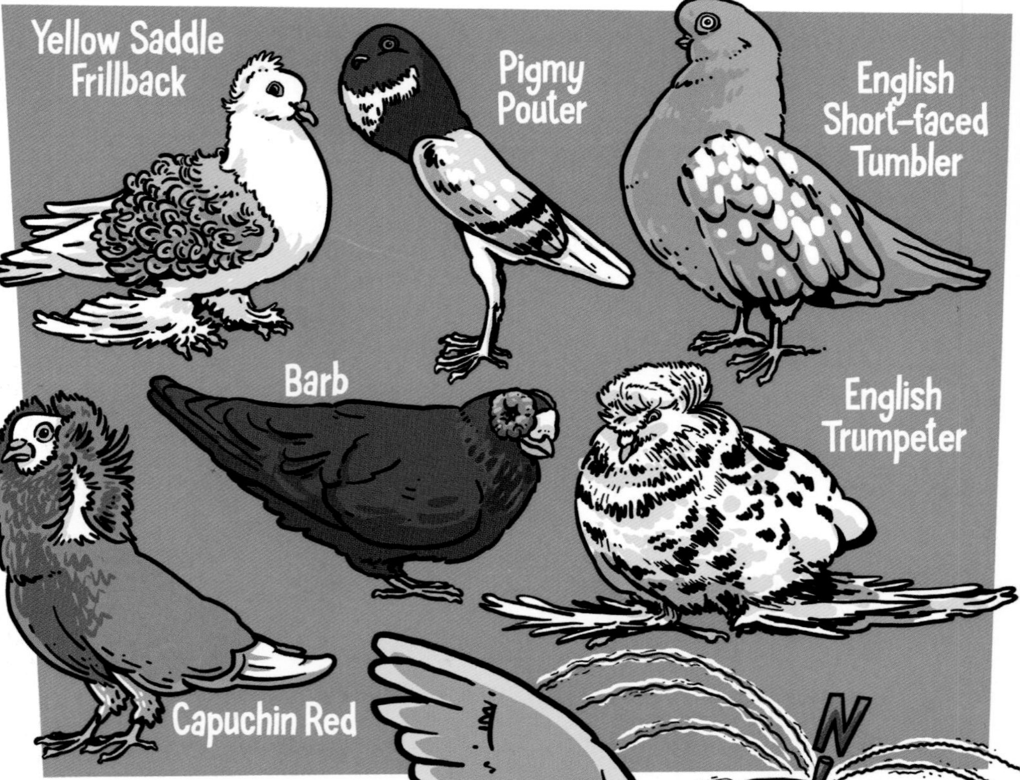

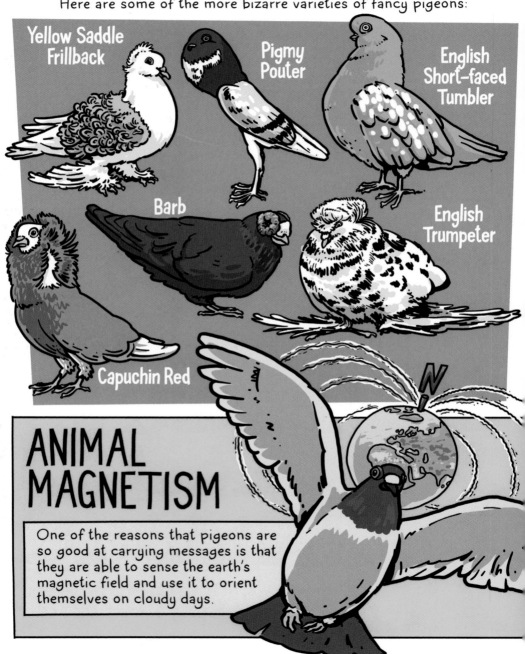

Yellow Saddle Frillback
Pigmy Pouter
English Short-faced Tumbler
Barb
English Trumpeter
Capuchin Red

ANIMAL MAGNETISM

One of the reasons that pigeons are so good at carrying messages is that they are able to sense the earth's magnetic field and use it to orient themselves on cloudy days.

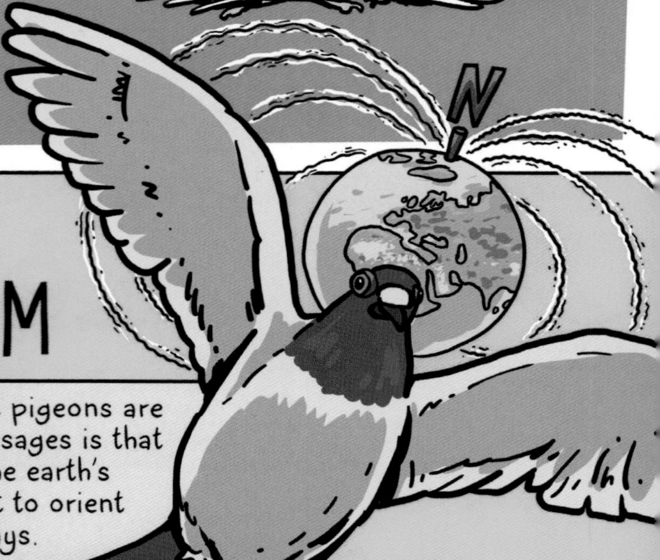

N

WHAT'S UP WITH THEIR FEET?

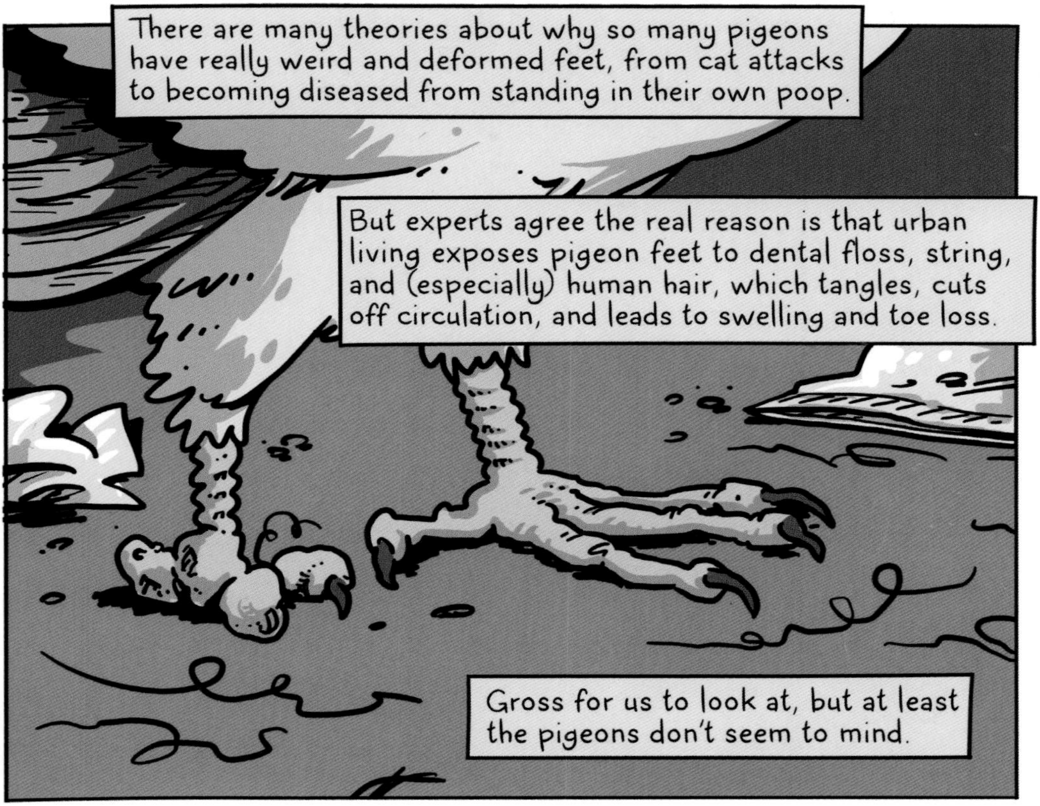

There are many theories about why so many pigeons have really weird and deformed feet, from cat attacks to becoming diseased from standing in their own poop.
But experts agree the real reason is that urban living exposes pigeon feet to dental floss, string, and (especially) human hair, which tangles, cuts off circulation, and leads to swelling and toe loss.
Gross for us to look at, but at least the pigeons don't seem to mind.

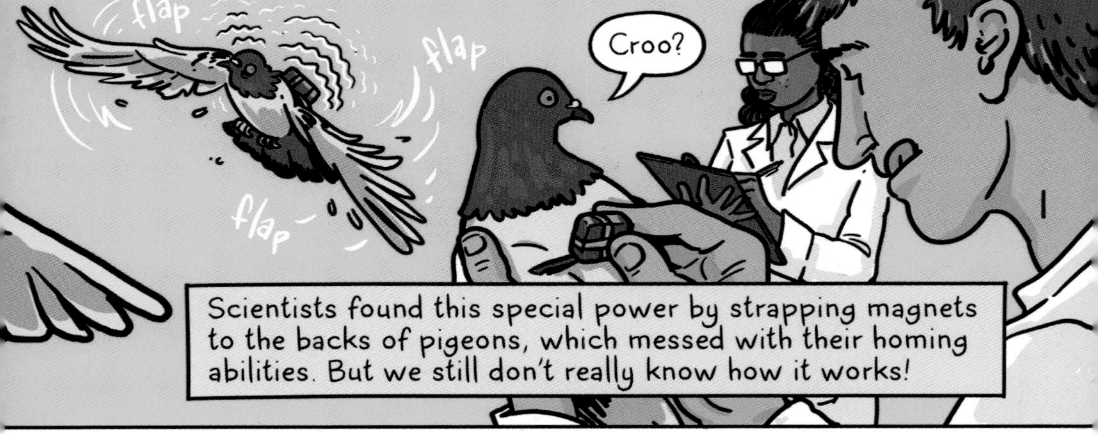

flap
flap
flap
Croo?
Scientists found this special power by strapping magnets to the backs of pigeons, which messed with their homing abilities. But we still don't really know how it works!

RACCOONS

Procyon lotor

Davy Crockett Fad
Leads to Raccoons
Taking Over Germany
Pg. 166

Animated Raccoons
and the Invasion
of Japan!
Pg. 168

Raccoon
Capital of the
World
Pg. 171

Size: 24–38 in

Weight: 14–23 lbs

Bizarre Behavior: Dunking food and objects in water

Diet: Insects, fish, frogs, worms, fruits, nuts, garbage

(Likely)
Wild Origin:

A KIDS' TV SHOW MIGHT BE WHY RACCOONS TAKE OVER THE WORLD

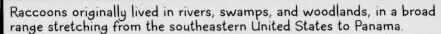

Raccoons originally lived in rivers, swamps, and woodlands, in a broad range stretching from the southeastern United States to Panama.

Just give me a river or puddle to wash my food and count me happy!

In the wild, raccoons eat fish, rodents, roots, crabs, insects, berries...honestly, whatever they can get their weird little hands on.

15,000 years ago, humans first showed up in raccoon territory.

Hey! To us, *you're* the ones with the weird hands!

When raccoons could avoid the dogs humans brought with them,* they happily added garbage to their diet.

Woo-hoo!

*See Pg. 11

But maybe because they weren't too fond of those dogs, they mostly stayed in their river valleys, swamps, and woodlands.

WOOF! WOOF! WOOF!

Aiiee!

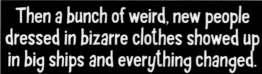

Then a bunch of weird, new people dressed in bizarre clothes showed up in big ships and everything changed.

Looks like India to me!

At first, this was bad news for raccoons. Hunting for meat, the Spaniards exterminated raccoons in Hispaniola, Cuba, and Jamaica.

Tastes like chicken.

No, it doesn't.

But aside from being occasionally hunted for fur, mainland raccoons continued on as they always had.

That hat is in really bad taste.

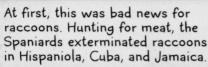

As people built farms, raccoons began to hang around them.

WOOF! WOOF! WOOF!

Aiiee!

Then in the 1700s, newcomers started building denser cities. Raccoons got interested.

The trash-to-dogs ratio is really starting to tip.

Like mice, cockroaches, dogs, and pigeons, raccoons love our trash.

Guilty as charged!

And humans' cities make a heck of a lot of trash.

Woo-hoo!

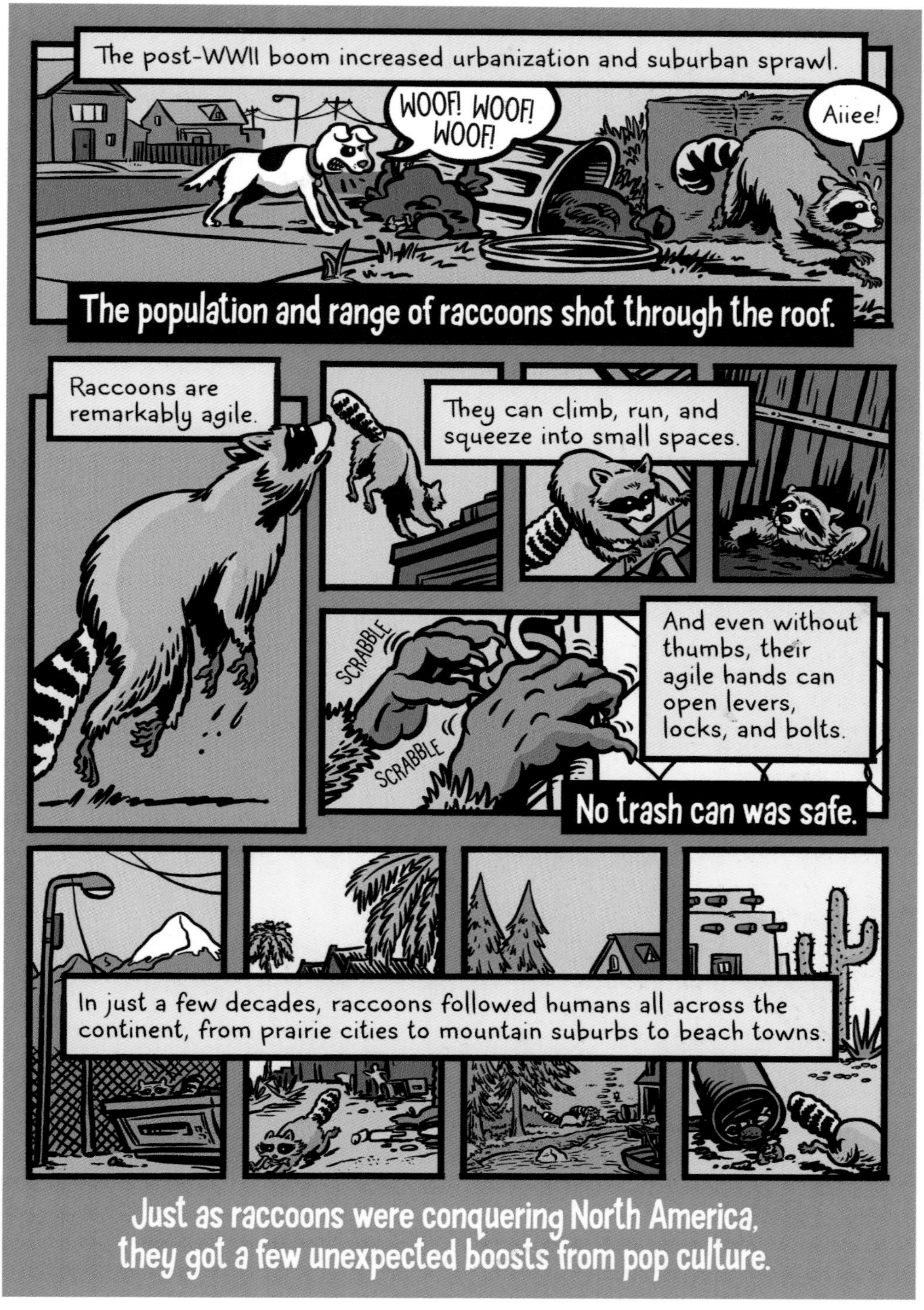
The post-WWII boom increased urbanization and suburban sprawl.
WOOF! WOOF! WOOF!
Aiiee!
The population and range of raccoons shot through the roof.
Raccoons are remarkably agile.
They can climb, run, and squeeze into small spaces.
SCRABBLE
SCRABBLE
And even without thumbs, their agile hands can open levers, locks, and bolts.
No trash can was safe.
In just a few decades, raccoons followed humans all across the continent, from prairie cities to mountain suburbs to beach towns.
Just as raccoons were conquering North America, they got a few unexpected boosts from pop culture.

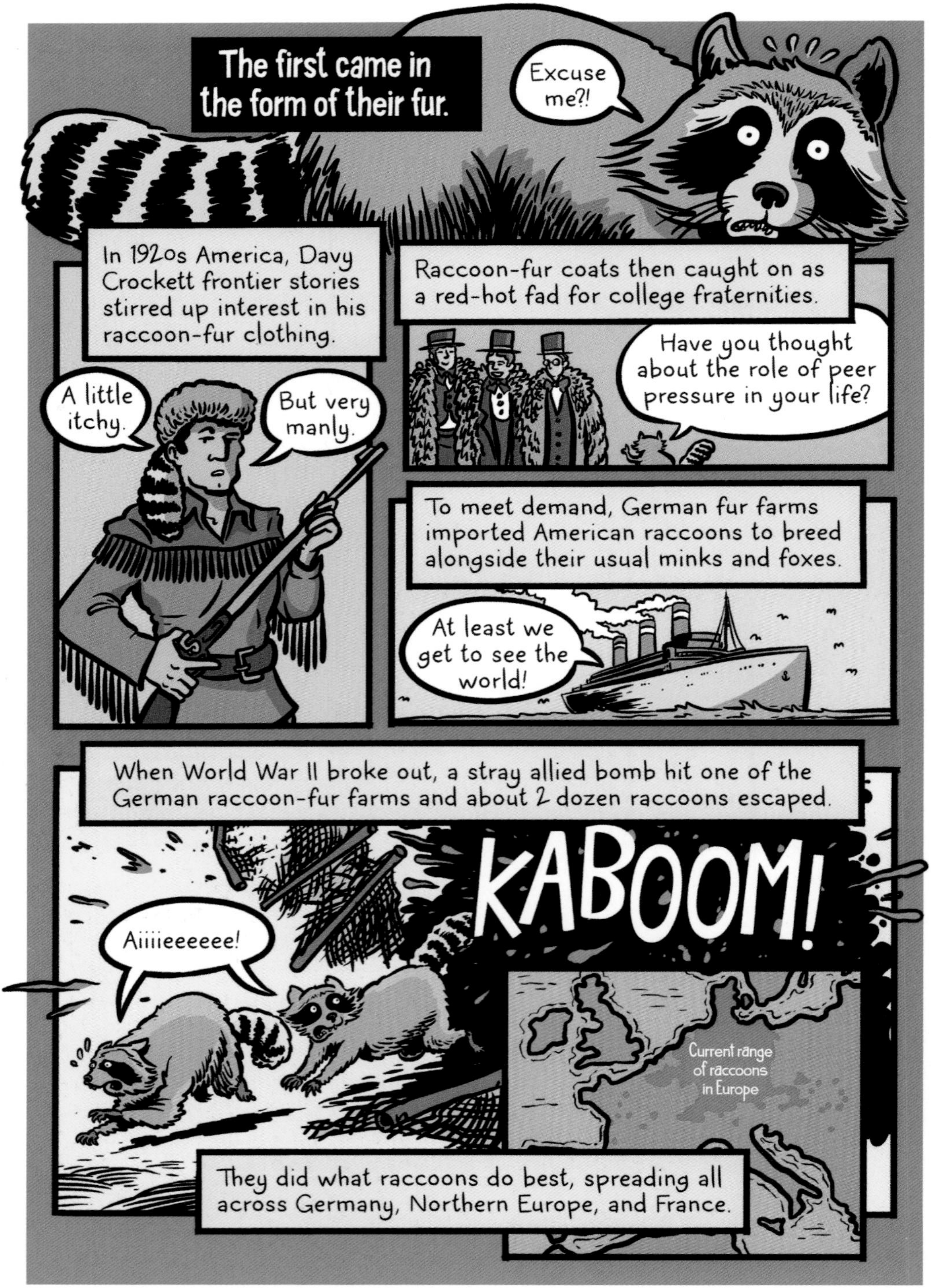

The first came in the form of their fur.
Excuse me?!
In 1920s America, Davy Crockett frontier stories stirred up interest in his raccoon-fur clothing.
A little itchy.
But very manly.
Raccoon-fur coats then caught on as a red-hot fad for college fraternities.
Have you thought about the role of peer pressure in your life?
To meet demand, German fur farms imported American raccoons to breed alongside their usual minks and foxes.
At least we get to see the world!
When World War II broke out, a stray allied bomb hit one of the German raccoon-fur farms and about 2 dozen raccoons escaped.
KABOOM!
Aiiiieeeeee!
Current range of raccoons in Europe
They did what raccoons do best, spreading all across Germany, Northern Europe, and France.

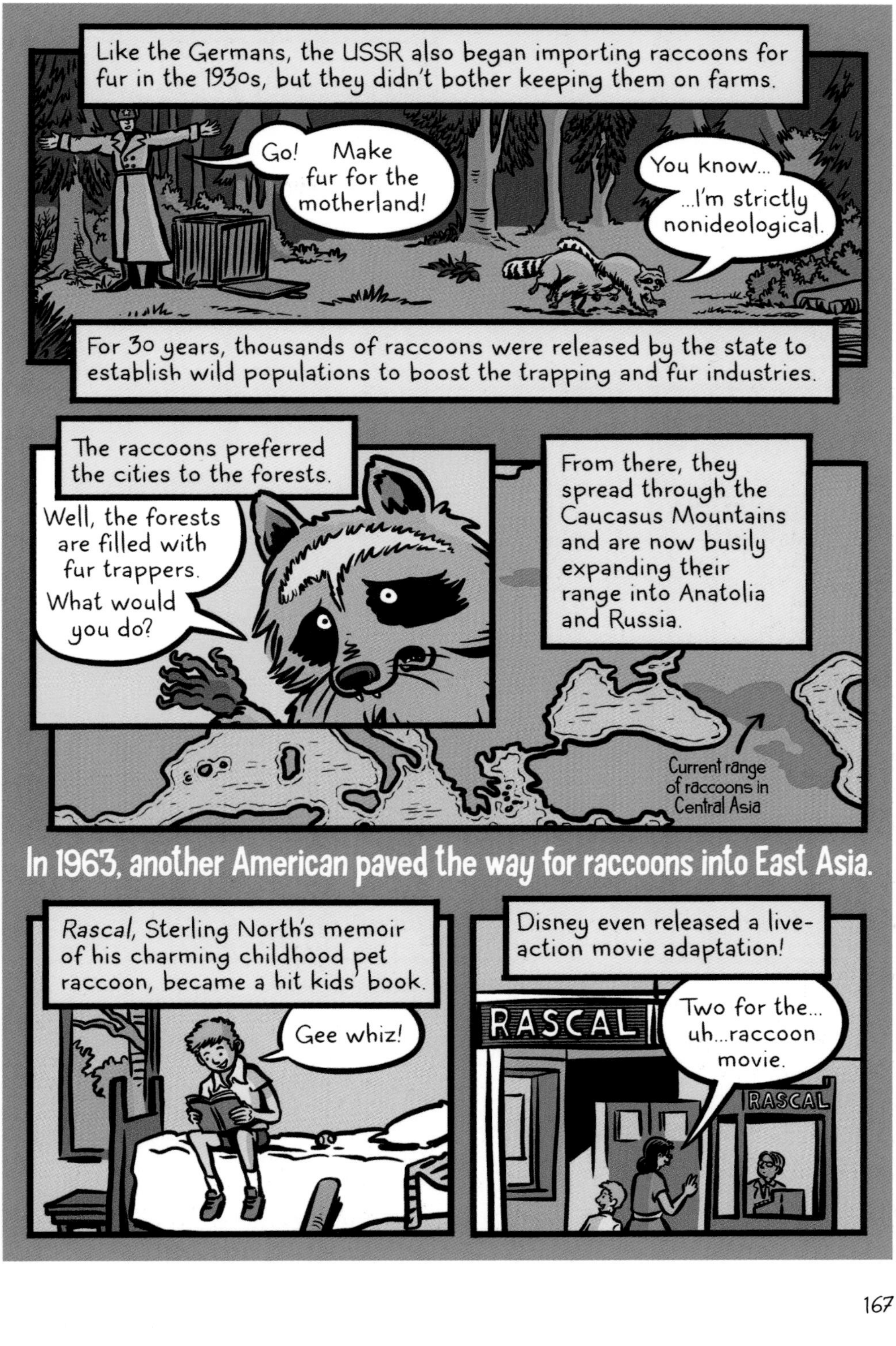
Like the Germans, the USSR also began importing raccoons for fur in the 1930s, but they didn't bother keeping them on farms.
Go! Make fur for the motherland!
You know... ...I'm strictly nonideological.
For 30 years, thousands of raccoons were released by the state to establish wild populations to boost the trapping and fur industries.
The raccoons preferred the cities to the forests.
Well, the forests are filled with fur trappers. What would you do?
From there, they spread through the Caucasus Mountains and are now busily expanding their range into Anatolia and Russia.
Current range of raccoons in Central Asia
In 1963, another American paved the way for raccoons into East Asia.
Rascal, Sterling North's memoir of his charming childhood pet raccoon, became a hit kids' book.
Gee whiz!
Disney even released a live-action movie adaptation!
RASCAL
Two for the... uh...raccoon movie.
RASCAL

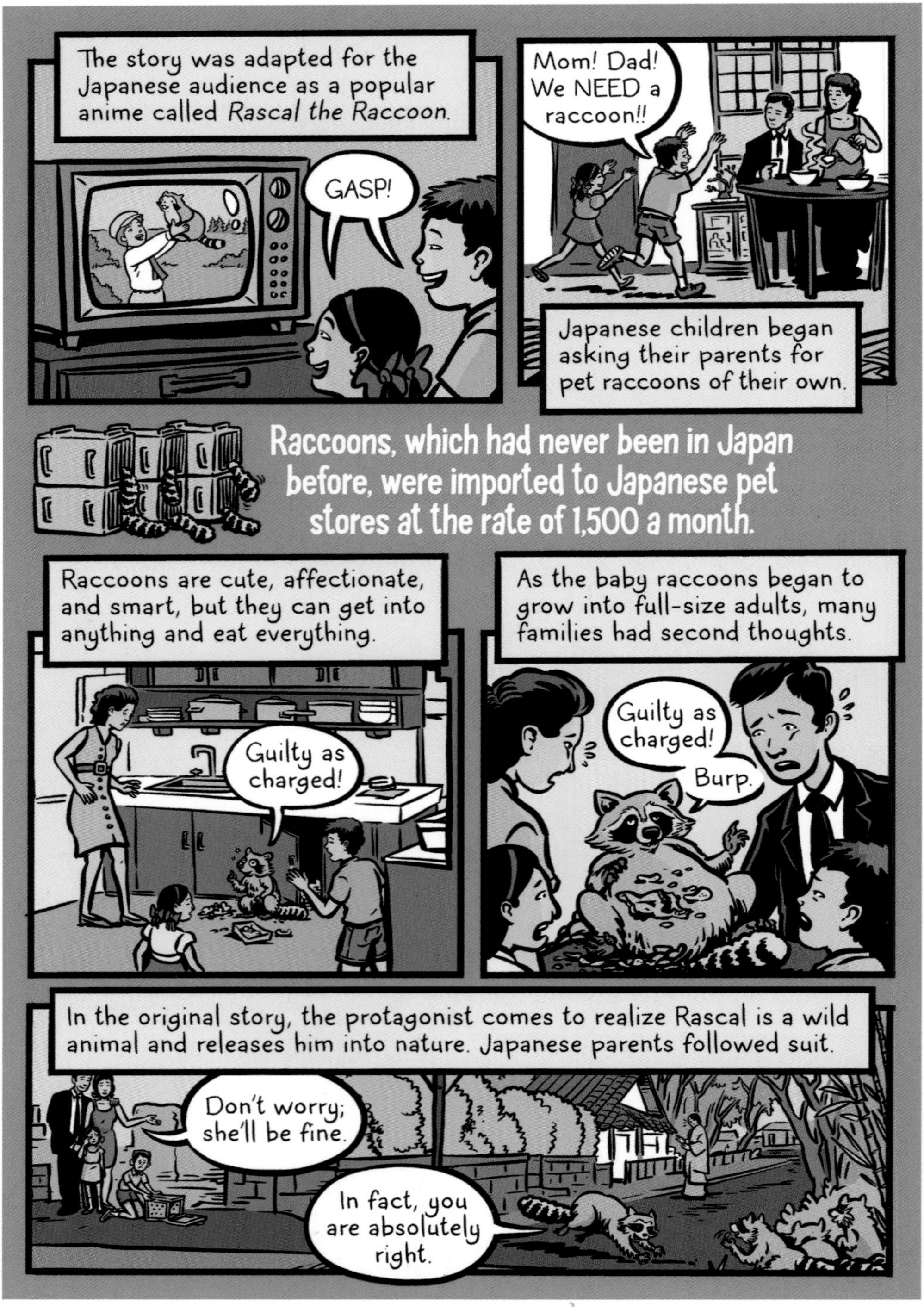
The story was adapted for the Japanese audience as a popular anime called *Rascal the Raccoon*.
GASP!
Mom! Dad! We NEED a raccoon!!
Japanese children began asking their parents for pet raccoons of their own.
Raccoons, which had never been in Japan before, were imported to Japanese pet stores at the rate of 1,500 a month.
Raccoons are cute, affectionate, and smart, but they can get into anything and eat everything.
Guilty as charged!
As the baby raccoons began to grow into full-size adults, many families had second thoughts.
Guilty as charged!
Burp.
In the original story, the protagonist comes to realize Rascal is a wild animal and releases him into nature. Japanese parents followed suit.
Don't worry; she'll be fine.
In fact, you are absolutely right.

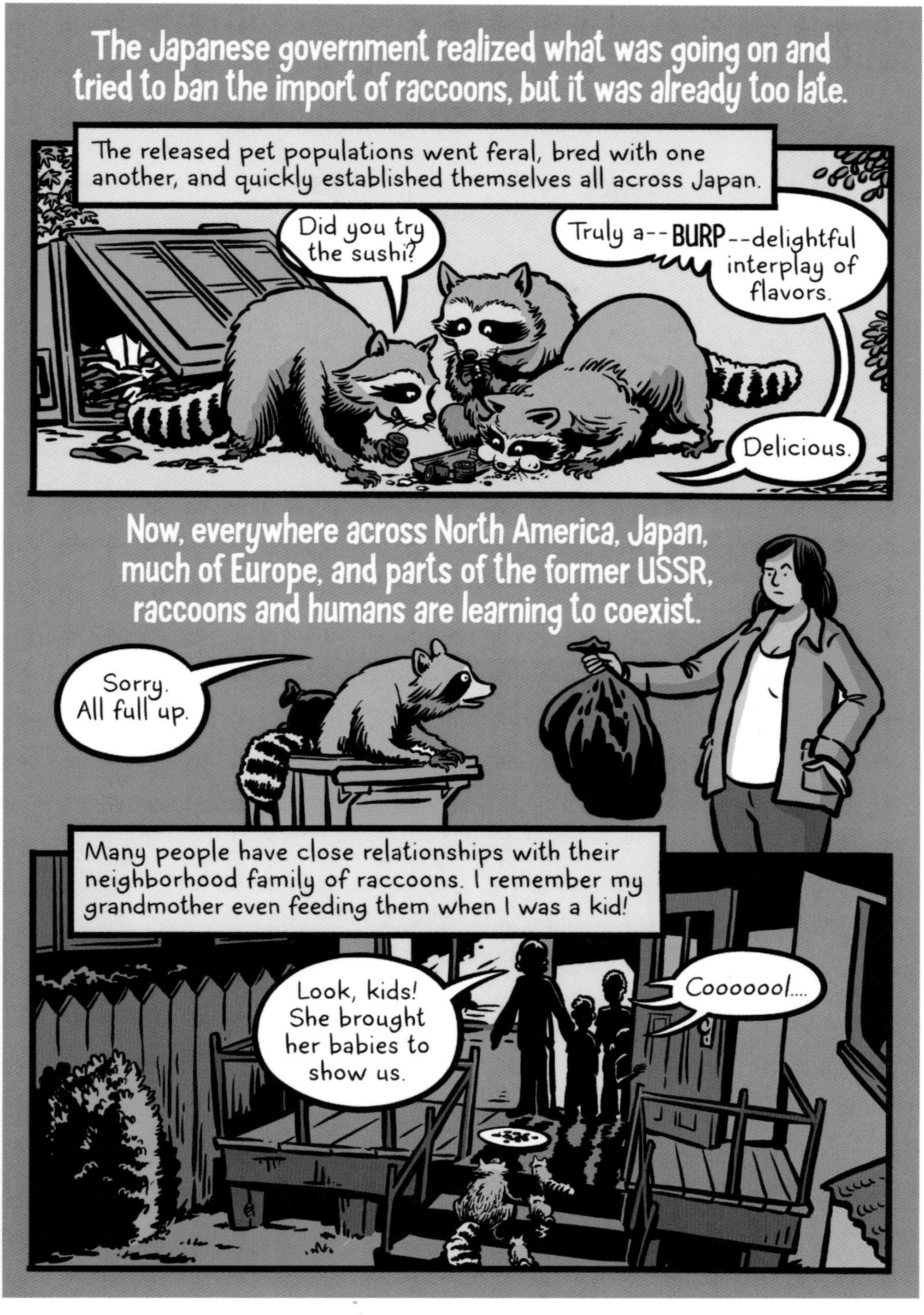
The Japanese government realized what was going on and tried to ban the import of raccoons, but it was already too late.
The released pet populations went feral, bred with one another, and quickly established themselves all across Japan.
Did you try the sushi?
Truly a--BURP--delightful interplay of flavors.
Delicious.
Now, everywhere across North America, Japan, much of Europe, and parts of the former USSR, raccoons and humans are learning to coexist.
Sorry. All full up.
Many people have close relationships with their neighborhood family of raccoons. I remember my grandmother even feeding them when I was a kid!
Look, kids! She brought her babies to show us.
Coooooool....

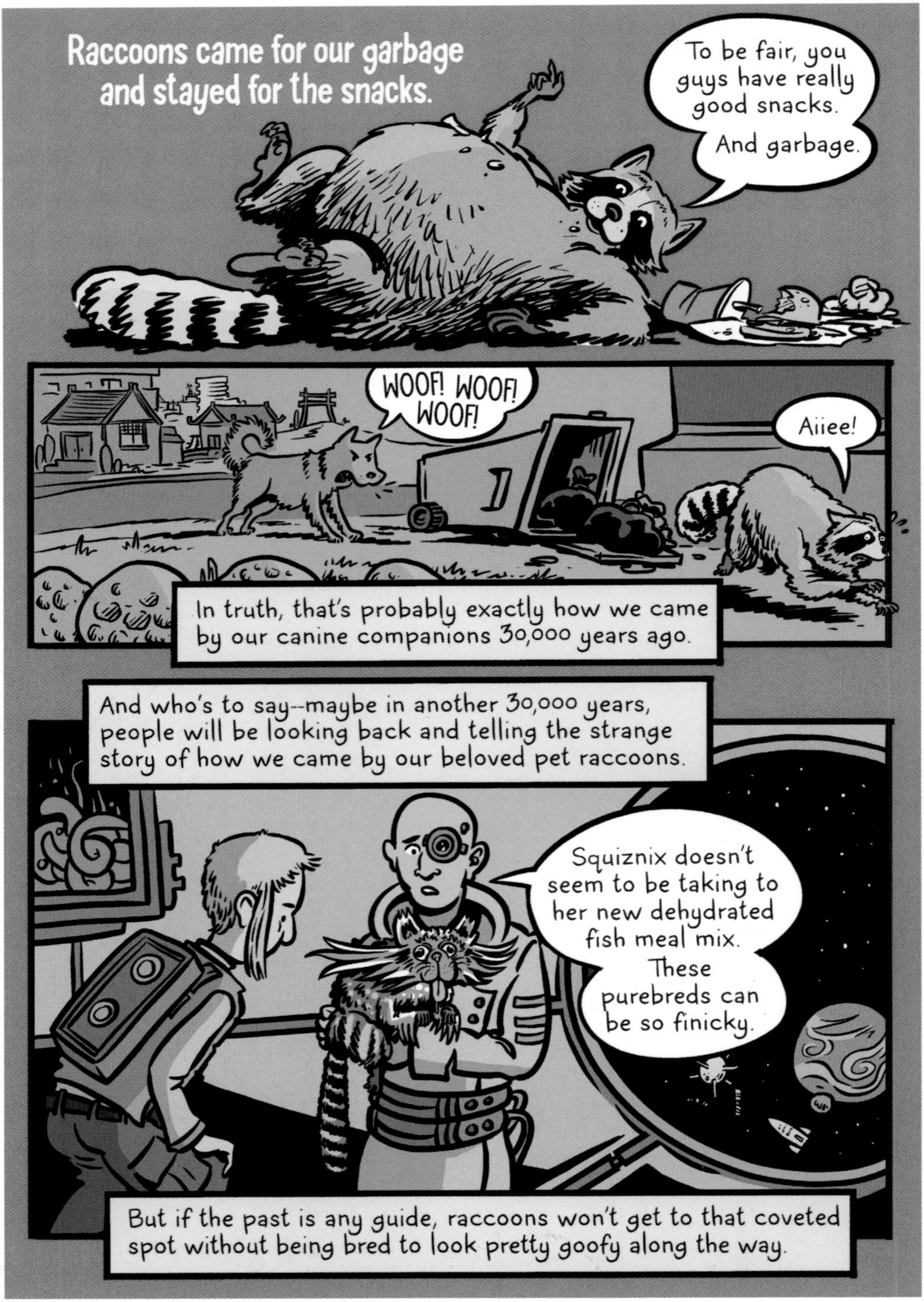
Raccoons came for our garbage and stayed for the snacks.
To be fair, you guys have really good snacks.
And garbage.
WOOF! WOOF! WOOF!
Aiiee!
In truth, that's probably exactly how we came by our canine companions 30,000 years ago.
And who's to say--maybe in another 30,000 years, people will be looking back and telling the strange story of how we came by our beloved pet raccoons.
Squiznix doesn't seem to be taking to her new dehydrated fish meal mix.
These purebreds can be so finicky.
But if the past is any guide, raccoons won't get to that coveted spot without being bred to look pretty goofy along the way.

RACCOON CAPITAL OF THE WORLD

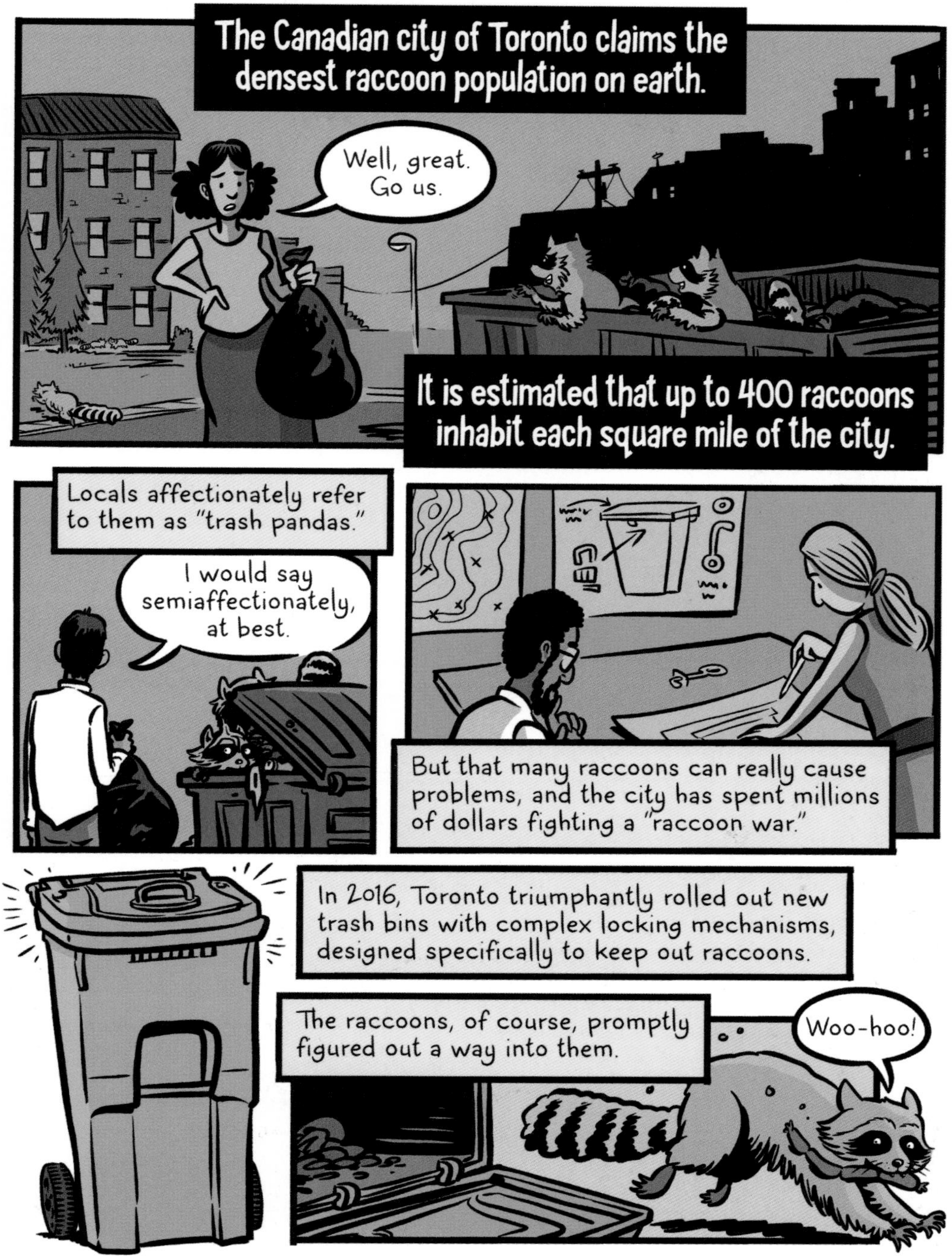

The Canadian city of Toronto claims the densest raccoon population on earth.
Well, great. Go us.
It is estimated that up to 400 raccoons inhabit each square mile of the city.
Locals affectionately refer to them as "trash pandas."
I would say semiaffectionately, at best.
But that many raccoons can really cause problems, and the city has spent millions of dollars fighting a "raccoon war."
In 2016, Toronto triumphantly rolled out new trash bins with complex locking mechanisms, designed specifically to keep out raccoons.
The raccoons, of course, promptly figured out a way into them.
Woo-hoo!

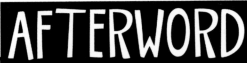

AFTERWORD

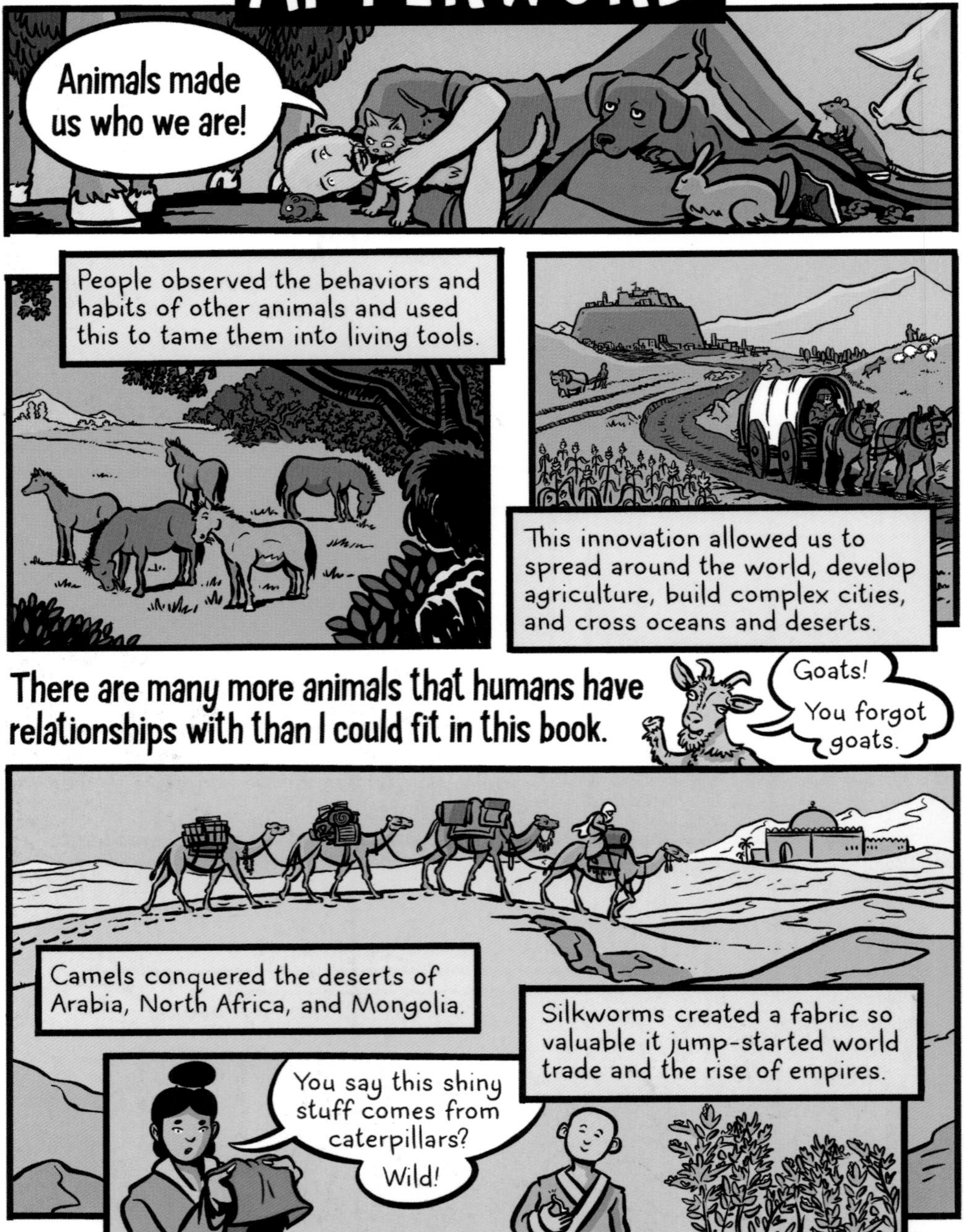

Animals made us who we are!
People observed the behaviors and habits of other animals and used this to tame them into living tools.
This innovation allowed us to spread around the world, develop agriculture, build complex cities, and cross oceans and deserts.
There are many more animals that humans have relationships with than I could fit in this book.
Goats!
You forgot goats.
Camels conquered the deserts of Arabia, North Africa, and Mongolia.
Silkworms created a fabric so valuable it jump-started world trade and the rise of empires.
You say this shiny stuff comes from caterpillars?
Wild!

Our ability to domesticate animals was not just a neat trick; it was the key to our success.

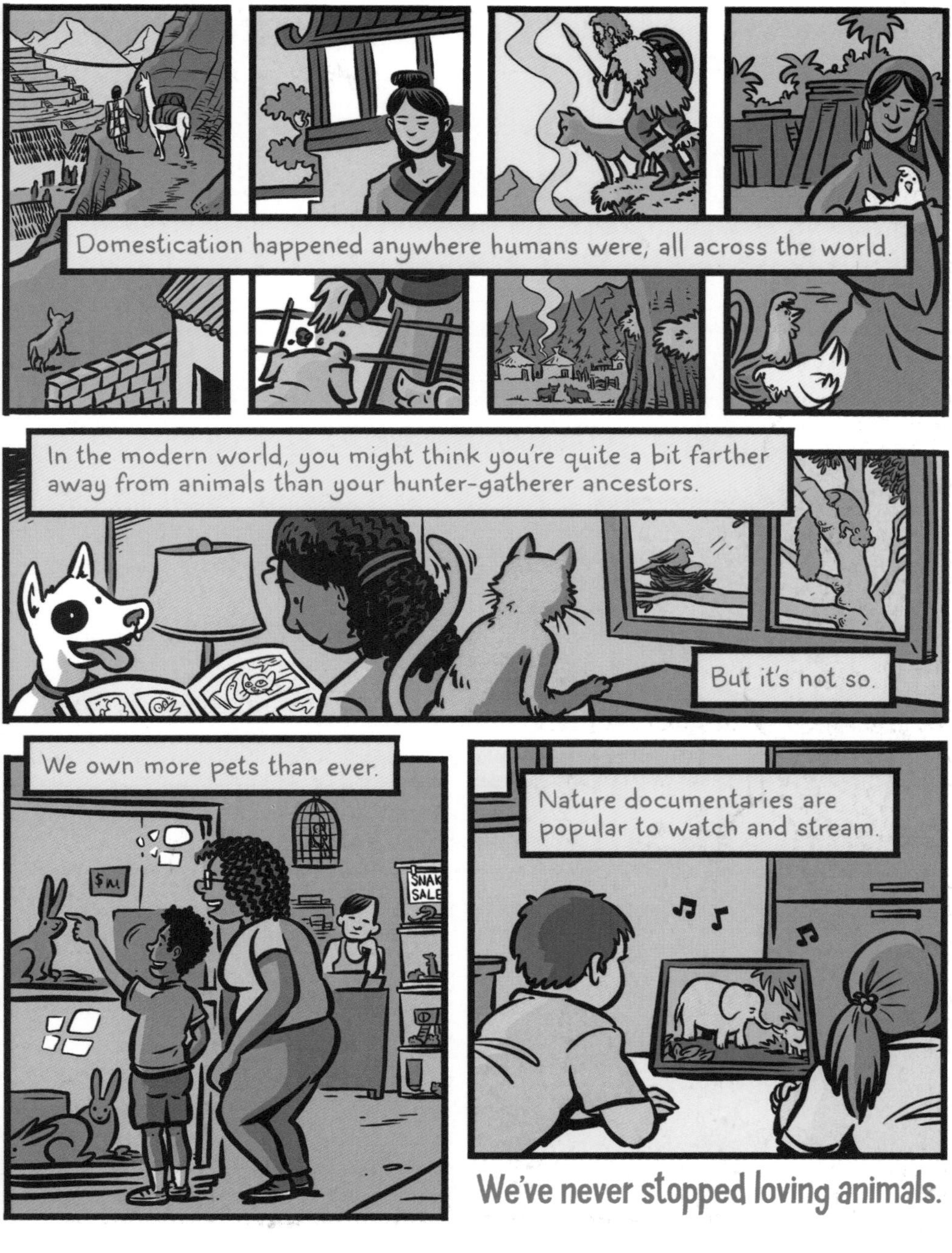
Domestication happened anywhere humans were, all across the world.
In the modern world, you might think you're quite a bit farther away from animals than your hunter-gatherer ancestors.
But it's not so.
We own more pets than ever.
Nature documentaries are popular to watch and stream.
We've never stopped loving animals.

We take care of our own. No domesticated animal has ever gone extinct, while many of their wild ancestors have vanished.

Uh, glad we went the domestication route.

Boy, this sure beats foraging through a pile of fallen leaves!

Some wild animals figured out how to make a living in our new world of cities and roads. Because of that, they're doing better than ever.

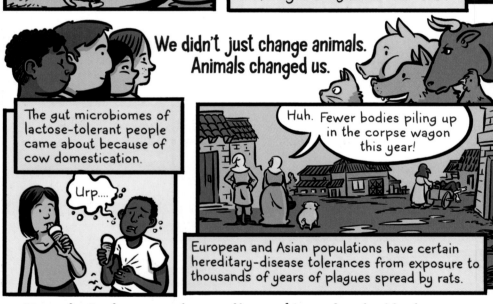

We didn't just change animals. Animals changed us.

The gut microbiomes of lactose-tolerant people came about because of cow domestication.

Urp....

Huh. Fewer bodies piling up in the corpse wagon this year!

European and Asian populations have certain hereditary-disease tolerances from exposure to thousands of years of plagues spread by rats.

No part of nature anywhere in the world is untouched by humans...

...but while our domesticated animals and those who thrive in cities are doing fine, others who come into contact with us and our entourage aren't.

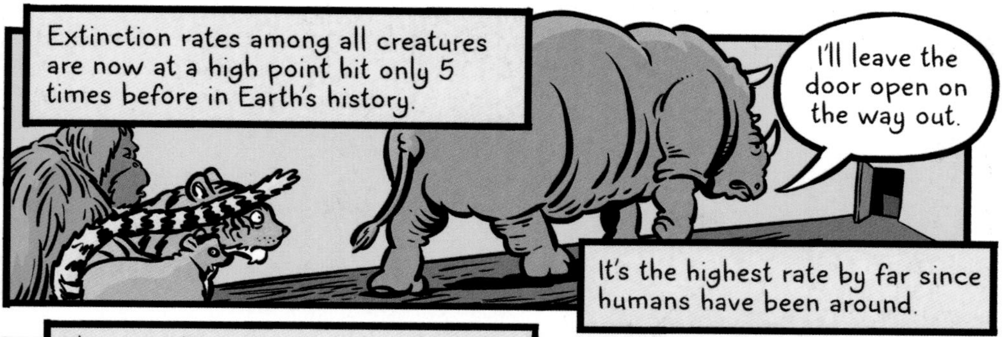
Extinction rates among all creatures are now at a high point hit only 5 times before in Earth's history.
I'll leave the door open on the way out.
It's the highest rate by far since humans have been around.

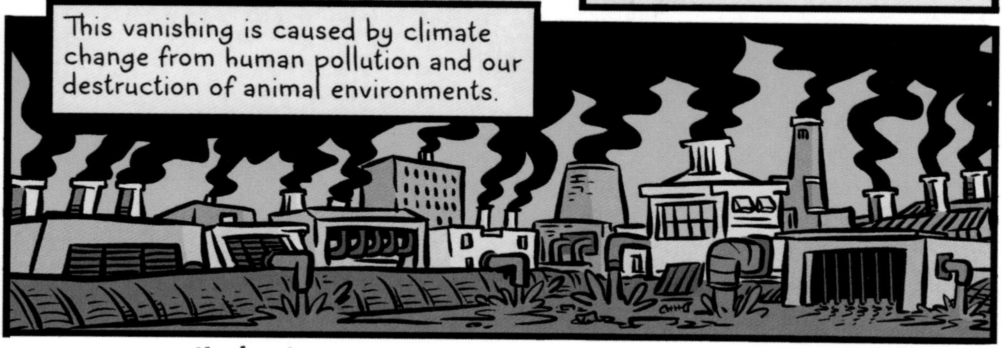
This vanishing is caused by climate change from human pollution and our destruction of animal environments.

It also happens because of the very commonplace creatures you've been reading about here.

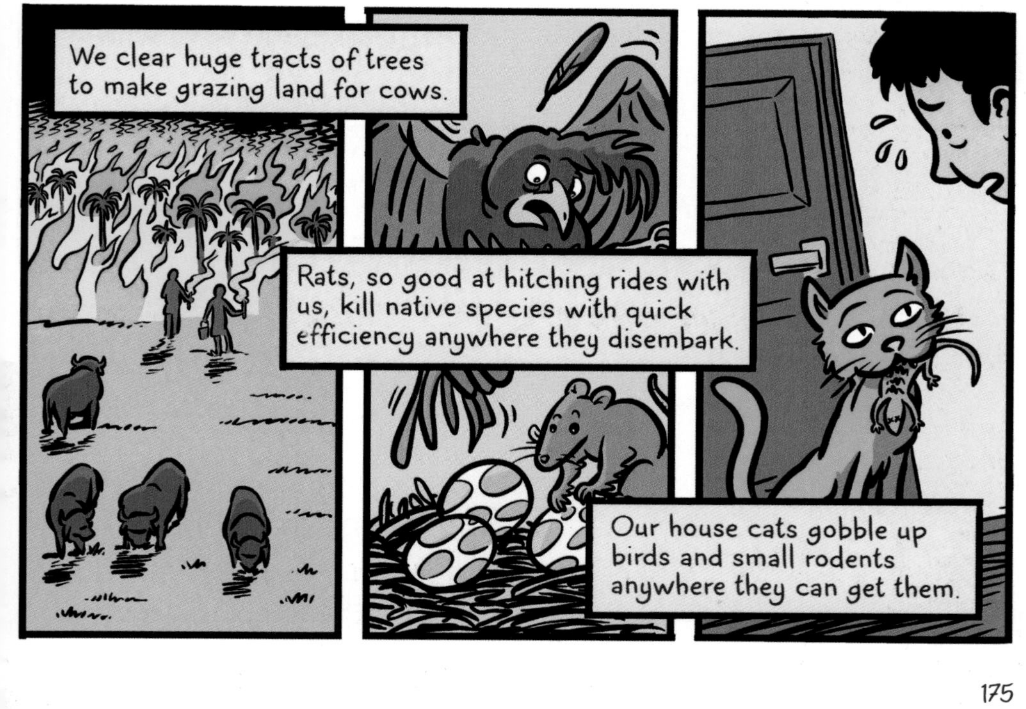
We clear huge tracts of trees to make grazing land for cows.
Rats, so good at hitching rides with us, kill native species with quick efficiency anywhere they disembark.
Our house cats gobble up birds and small rodents anywhere they can get them.

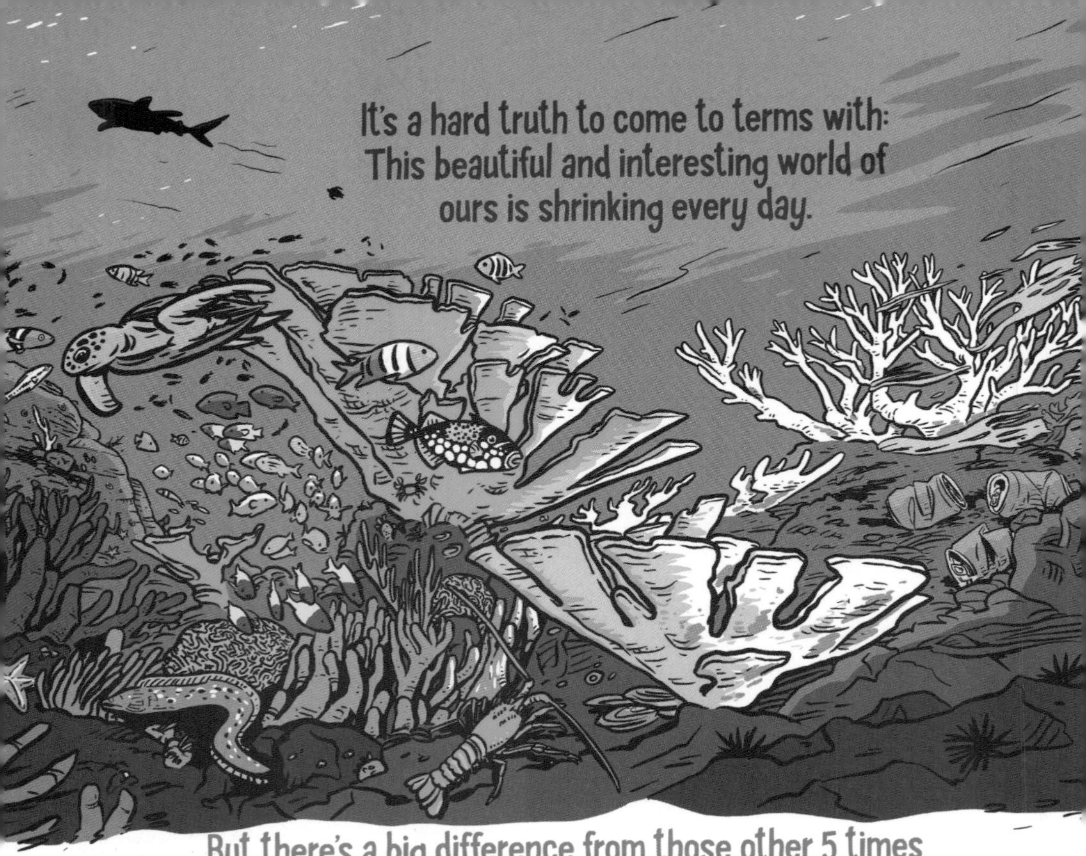
It's a hard truth to come to terms with: This beautiful and interesting world of ours is shrinking every day.

But there's a big difference from those other 5 times Earth underwent catastrophic die-offs.

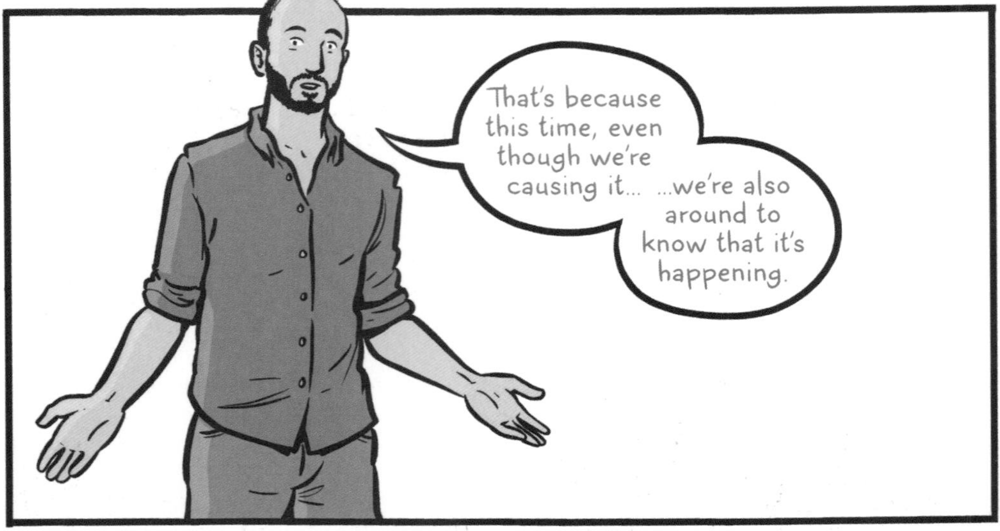
That's because this time, even though we're causing it... ...we're also around to know that it's happening.

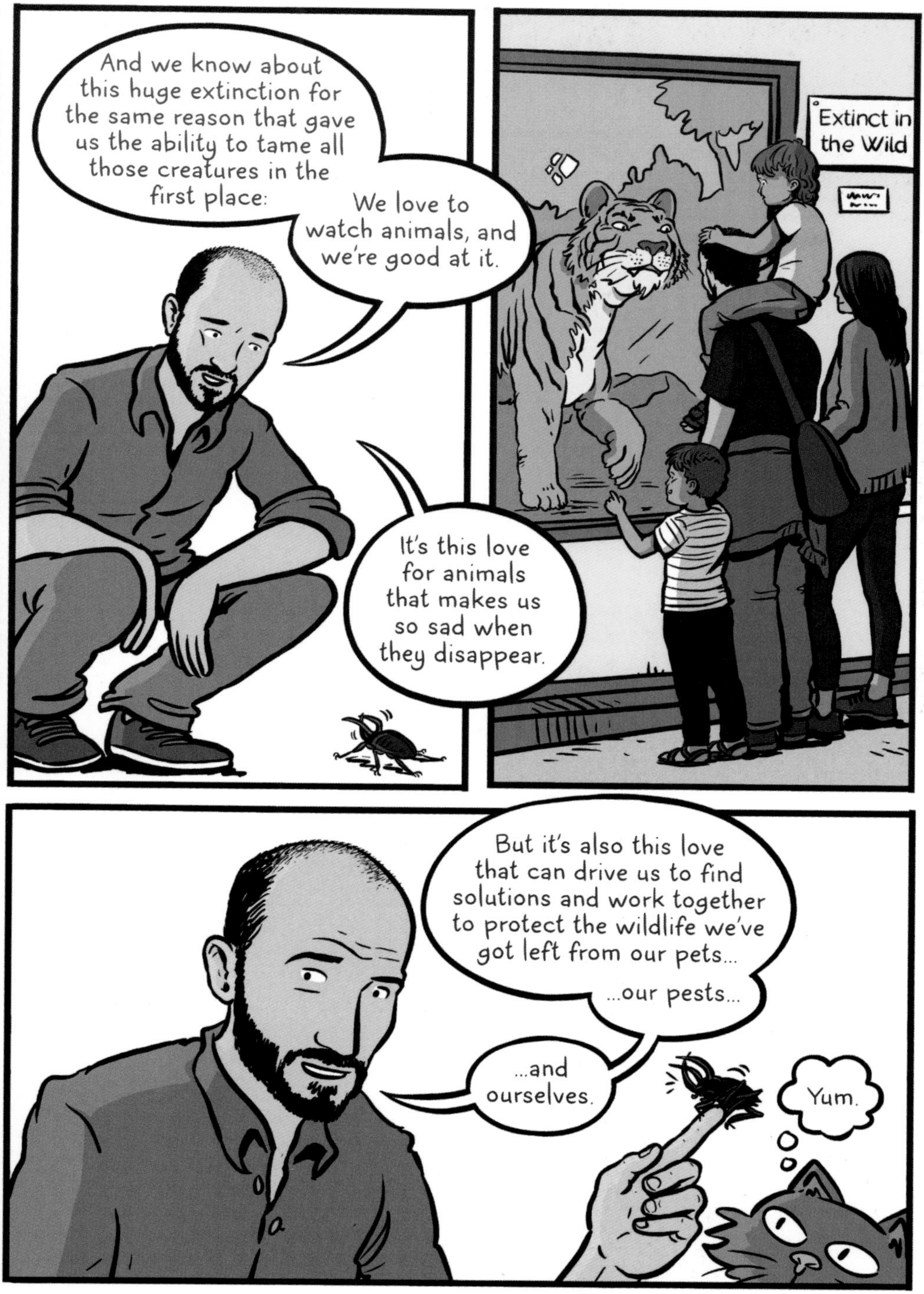
And we know about this huge extinction for the same reason that gave us the ability to tame all those creatures in the first place:
We love to watch animals, and we're good at it.
Extinct in the Wild
It's this love for animals that makes us so sad when they disappear.
But it's also this love that can drive us to find solutions and work together to protect the wildlife we've got left from our pets...
...our pests...
...and ourselves.
Yum.

APPENDIX
TIMELINE OF DOMESTICATION

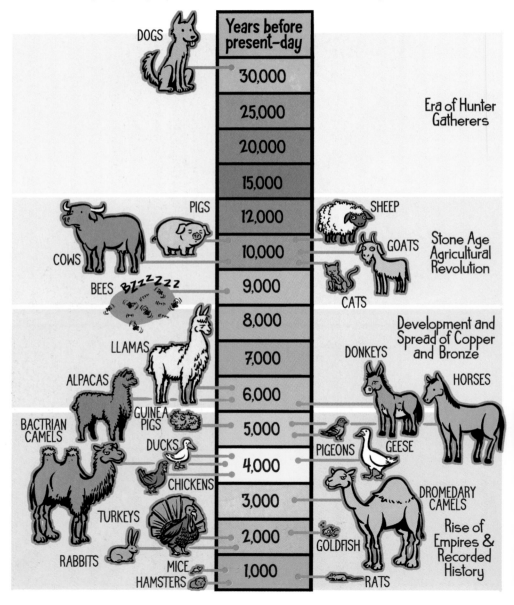

DOGS

Years before present-day	
30,000	
25,000	Era of Hunter Gatherers
20,000	
15,000	
12,000	PIGS · SHEEP
10,000	COWS · GOATS · CATS · Stone Age Agricultural Revolution
9,000	BEES BZZZZzzz
8,000	Development and Spread of Copper and Bronze
7,000	LLAMAS · DONKEYS
6,000	ALPACAS · HORSES
5,000	GUINEA PIGS · PIGEONS · GEESE
4,000	BACTRIAN CAMELS · DUCKS
3,000	CHICKENS · DROMEDARY CAMELS
2,000	TURKEYS · GOLDFISH · Rise of Empires & Recorded History
1,000	RABBITS · MICE · HAMSTERS · RATS

MAP OF (LIKELY) WILD ORIGINS

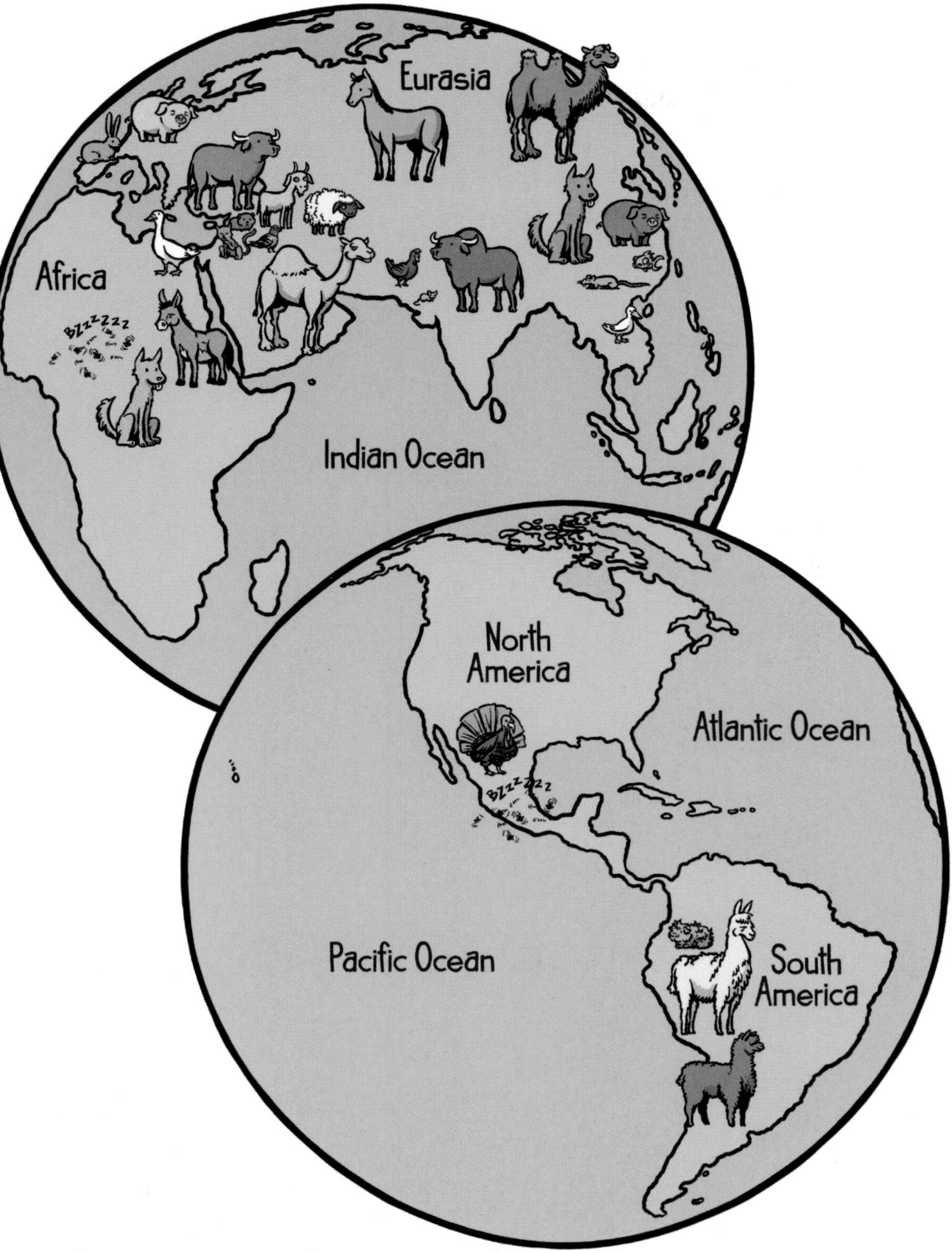

Eurasia
Africa
BZZZZZZ
Indian Ocean
North America
Atlantic Ocean
BZZZZZ
Pacific Ocean
South America

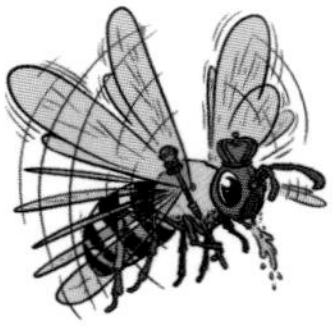

INDEX

ACKNOWLEDGMENTS

My biggest thanks to the many researchers and writers whose work I referenced to put this book together. They include Brian Fagan, Claire Preston, Richard Jones, Pat Shipman, Linda Kalof, and Richard C. Francis. The entirety of this book was drawn during the 2020 global coronavirus pandemic, and I could not have created it without the community on the compound that helped me--and my children--through that longest year. Thank you, Lisa, Steve, Teemu, and Bodie. Thank you to Kathy for listening to my never-ending research anecdotes and, in doing so, helping me figure out how they fit together. Thanks to my dad and mom for reading early drafts and lending me their years of biology and animal-behavior expertise. Thanks to my editor, Andrea Colvin, who sought me out and got excited about this idea, and thanks to my agent, Farley Chase, for making it happen.

Andy Warner

is the *New York Times* bestselling author of *Brief Histories of Everyday Objects* and *This Land Is My Land*. He is a contributing editor at the Nib and teaches cartooning at Stanford University and the Animation Workshop in Denmark. His comics have been published by *Slate*, *Fusion*, American Public Media, popsci.com, KQED, IDEO, The Center for Constitutional Rights, UNHCR, UNRWA, UNICEF, and Buzzfeed. He was a recipient of the 2018 Berkeley Civic Arts Grant and was the 2019 Hawai'i Volcanoes National Park Artist-in-Residence. He works in a garret room in South Berkeley and comes from the sea.